The Wonders of Physics

4th Edition

The Wonders of Physics

4th Edition

A.A. Varlamov

*Institute of Superconductors, Oxides and Other Innovative Materials
and Devices of the National Research Council (SPIN-CNR), Italy*

L.G. Aslamazov

National University of Science and Technology (MISiS), Russia

Translators
A.A. Abrikosov, jr., J. Vydryg, & D. Znamenski

World Scientific

NEW JERSEY · LONDON · SINGAPORE · BEIJING · SHANGHAI · HONG KONG · TAIPEI · CHENNAI

Published by

World Scientific Publishing Co. Pte. Ltd.

5 Toh Tuck Link, Singapore 596224

USA office: 27 Warren Street, Suite 401-402, Hackensack, NJ 07601

UK office: 57 Shelton Street, Covent Garden, London WC2H 9HE

British Library Cataloguing-in-Publication Data
A catalogue record for this book is available from the British Library.

THE WONDERS OF PHYSICS
4th Edition

ISBN 978-981-3273-16-0

For any available supplementary material, please visit
https://www.worldscientific.com/worldscibooks/10.1142/11067#t=suppl

Desk Editor: Christopher Teo

Typeset by Stallion Press
Email: enquiries@stallionpress.com

To our teachers and friends...

Preface

It is my pleasure to present the fourth, revised and extended, English edition of *The Wonders of Physics*. Thirty years have passed since its first publication in Russia. The book had a long life, and it was well accepted by readers. After the first release another five Russian editions followed, then three versions in English, one in Italian, Spanish, Chinese, and Japanese.

The world has undergone considerable changes during the past three decades. Our vital priorities have changed. A new generation of readers grew up; new achievements in science and technology the authors could not even dream about while working on the first edition in the mid-1980s became a part of our life. Side by side with the progress occurring in physics this book was undergoing constant development. It became three times longer, and new chapters appeared in it, telling readers about such fundamental discoveries as high-temperature superconductivity, nuclear magnetic resonance, a new vast area of modern research called nanophysics, and, naturally, about applications of new discoveries in everyday life.

For the past twenty years the author (AV) has been living in Italy, where the culture of food is one of the priorities among the public interests. My first publications in the Italian press about the physics of coffee and physical aspects of oenology came as a result of my efforts to understand how the laws of physics work in my newly discovered field of everyday human activity. These works, unexpectedly, found widespread interest and positive response; they were translated into several languages and discussed on TV. This is the reason why a new part about "gastronomic physics" appeared in the book. I have considerably extended it for this edition and hope our readers will enjoy it.

The reasons why Lev Aslamasov (1944–1986) and I wrote this book were to satisfy both our curiosity and wish to share our admiration for the beauty of physics in all its natural manifestations. We have devoted a lot of time to teaching physics at various levels, from gifted freshmen to mature Ph.D. students. All our experience has convinced us that, in addition to the obvious necessity of regular and rigorous study of this discipline, an "artistic" approach, in which the teacher (or the author) proves the importance of physics in everyday phenomena, is vital. I hope we have managed to communicate our perception of physics, not only in the title, but also in the body of the book. The book is based on articles published in the *Kvant* Magazine and other journals during the past 40 years.

I would like to express my profound gratitude to many friends and colleagues, without whom this edition would not be possible. First, my old friend Professor Alex Abrikosov (Jr.), whose enthusiasm, thorough scientific expertise, comments, and translation gave birth to the English version of this book. His work on the translation was considerably aided by the collaboration with my friends Dr. Dmitriy Znamensky, and Janine Vydrug. I was invaluably assisted by the advices of Professors D. Khmelnitsky, S. Parnovsky, Dr. R. Boyack who read this book at various stages of its readiness. Several chapters of this book were written based on joint publications with Professors G. Balestrino, A. Buzdin, Yu. Galperin, A. Glatz, S. Grasso, Yan Chen, Keizo Murata, A. Rigamonti, J. Villain. I want to add to this list of venerable scientists my young coauthor, student of Fudan University, Mr. Zheng Zhou. I would like to acknowledge their valuable contribution.

I would like to thank warmly my scientific editors: Professors Keizo Murata (Japanese edition), Attilio Rigamonti (Italian edition), Yu Lu (Chinese edition), Dr. L. Panyushkina and Dr. V. Tikhomirova (Russian edition), without whose professionalism and collaboration in the preparation of previous editions, the present one would not appear.

Finally I would like to thank my publishers: Dr. D. De Bona (La Goliardica Pavese, Italy), Professor Kok Khoo Phua (World Scientific Publishing Company, Singapore), Dr. A. Ovchinnikov (Dobrosvet, Russia), Mr. Hou (China Science Publishing & Media Ltd., China) for their enthusiasm and collaboration.

Andrey Varlamov,

(Rome, 2018)

From the Foreword to the Russian edition

The science of physics was at the head of scientific and technical revolutions in the twentieth century. Nowadays physics continues to determine the direction of human progress. The most important example is that of the recent discovery of high-temperature superconductivity, which may radically change the entire edifice of modern technology.

However, delving deeper into the mysteries of the macrocosm and microparticles, scientists move further away from traditional school physics, that treats transformers and projectiles, launched at an angle to the horizontal, namely, from what most people believe to be physics. The goal of popular literature is to bridge the gap, and to bring the excellence of modern physics to curious readers while demonstrating its major achievements. A difficult task that will not tolerate dabbling.

The book in your hands develops the best traditions of this kind of literature. Written by working theoretical physicists and, at the same time, the dedicated popularizers of scientific knowledge, clear and captivating in manner, it brings to the reader the latest achievements in quantum solid-state physics; but on the way it shows how the laws of physics reveal themselves even in seemingly trivial episodes and natural phenomena around us. Most importantly, it shows the world through the eyes of scientists, capable of "proving harmony with algebra".

It was a great loss that one of the authors of the book, the well-known specialist in the theory of superconductivity, Professor L. G. Aslamazov, who for a long time was the vice-editor of the popular magazine *Kvant*, did not live to see the book come out.

I hope that readers, ranging from high-school students to professional physicists, will find this book, marked by its extremely vast scope of encompassed questions, a really interesting, enjoyable, and rewarding read.

A. A. Abrikosov, 2003 Nobel Prize winner,

(Moscow, 1987)

Contents

Color Inset

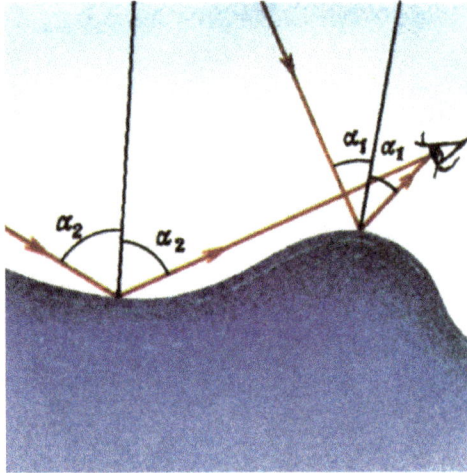

Fig. 1: Schematic image of wave.

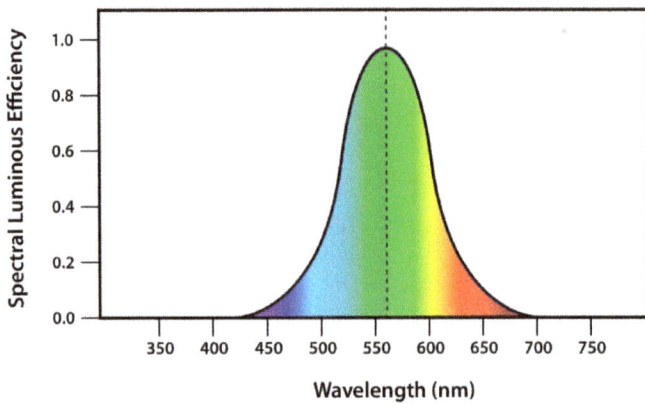

Fig. 2: Dependence of human eye perception on wavelength of light.

Fig. 3: Arkadii Rylov "In the Blue".

PART 1

Outdoor physics

From the first part of the book, the reader will learn why rivers wind and how they wash their banks out, why the sky is blue and white horses are white. We are going to tell you about the properties of the ocean, winds and the role of the Earth's rotation.

In short we shall present examples of how the laws of physics work on a global scale.

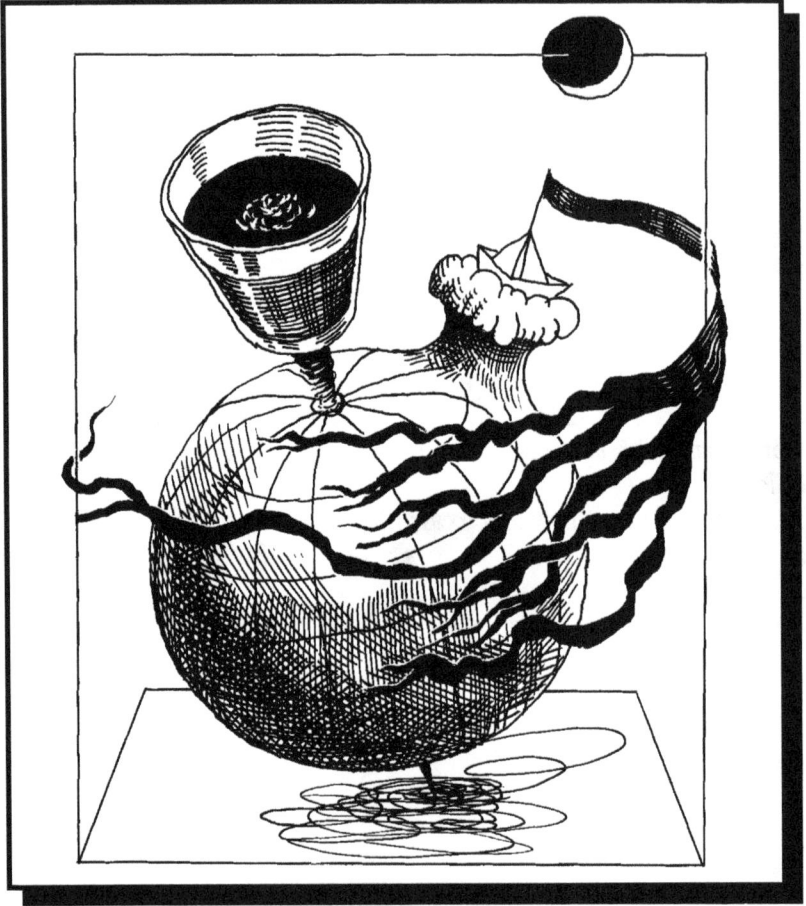

Chapter 1

Meandering down to the sea

Oh you river, twisty river, tell me what will be your
fate?
Will you twist yourself apart? Have you ever had a
spate?
Those sandbanks on your inside bank dont care if you
are late.
So flow and let flow river, your meanders are just great!

Kenton M. Stewart. *The River That Meanders.*

Have you ever seen a straight river without bends? Of course a short
section of a river may cut straight but no rivers are entirely without bends.
Even if the river flows through a plain, it usually loops around and the
bends repeat periodically. Moreover, as a rule, one bank at the bend is steep
while the other slopes gently. How could one explain these peculiarities of
river behavior? Hydrodynamics is the branch of physics that deals with
the motion of liquids. Although it is a well-developed science, rivers are
complicated natural objects and even hydrodynamics cannot explain every
feature of their behavior. Nevertheless, it can answer many questions.

You may be surprised to learn that even the great Albert Einstein[a]
gave time to the problem of meanders. In the report delivered in 1926 at a
meeting of the Prussian Academy of Sciences, he compared the motion of
river water to swirling water in a glass. The analogy allowed him to explain
why rivers choose twisted paths.

---

[a]A. Einstein, (1879–1955), German-born physicist, lived in Switzerland, US citizen from
1940; creator of the theory of relativity; Nobel Prize 1921 in physics.

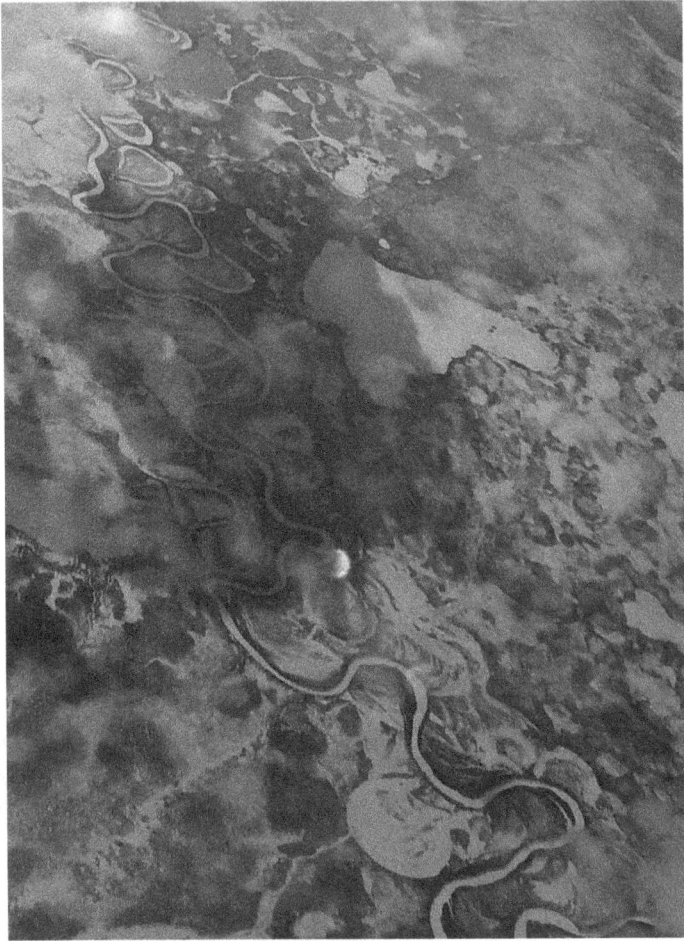

Fig. 1.1: One of the Siberian rivers visible from
the airplane porthole (courtesy of K. Murata).

Let us try to understand this too, at least qualitatively, by starting with
a glass of tea.

1.1 Tea-leaves in a glass

Make a glass of tea with loose tea-leaves (no tea-bags!), stir it well, and take the spoon out. The brew will gradually stop and the tea-leaves will gather in the center of the bottom. What made them settle there? To answer this question let us first determine what shape the surface takes as the liquid swirls in the glass.

The tea-cup experiment shows that the surface of the tea becomes curved. The reason is clear. In order to make particles of water move circularly, the net force acting on each of them must provide a centripetal acceleration. Consider a cube situated in the liquid at a distance r from the axis of rotation, Figure 1.2, a. Let the mass of tea in it be Δm. If the angular speed of rotation is ω then the centripetal acceleration of the cube is $\omega^2 r$. It comes as the result of the difference of the pressures acting on the faces of the cube (the left and right faces in Figure 1.2, a). So,

$$m\,\omega^2\,r = F_1 - F_2 = (P_1 - P_2)\,\Delta S, \tag{1.1}$$

where ΔS is the area of the face. The pressures P_1 and P_2 are determined by the distances h_1 and h_2 from the surface of the liquid:

$$P_1 = \rho\,g\,h_1 \quad \text{and} \quad P_2 = \rho\,g\,h_2, \tag{1.2}$$

where ρ is the density of the liquid and g is the free fall acceleration. As soon as the force F_1 is greater than F_2, then h_1 exceeds h_2 and the surface of the rotating liquid must be curved, as shown in Figure 1.2. The faster the rotation is, the greater is the curvature of the surface.

One can find the shape of the curved surface of the revolving liquid. It turns out to be a paraboloid — that is, a surface with a parabolic cross section.

As long as we continue stirring the tea with the spoon, we keep it swirling. But when we remove the spoon the viscous friction between layers of the liquid and the friction against the walls and bottom of the glass will convert the kinetic energy of liquid into thermal energy, and the motion will gradually come to rest.

As the rotation slows down, the surface of the liquid flattens. In the mean time vortex currents directed as shown in Figure 1.2, b appear in the liquid. The vortex currents are formed because of the nonuniform deceleration of the liquid at the bottom of the glass and at the surface.

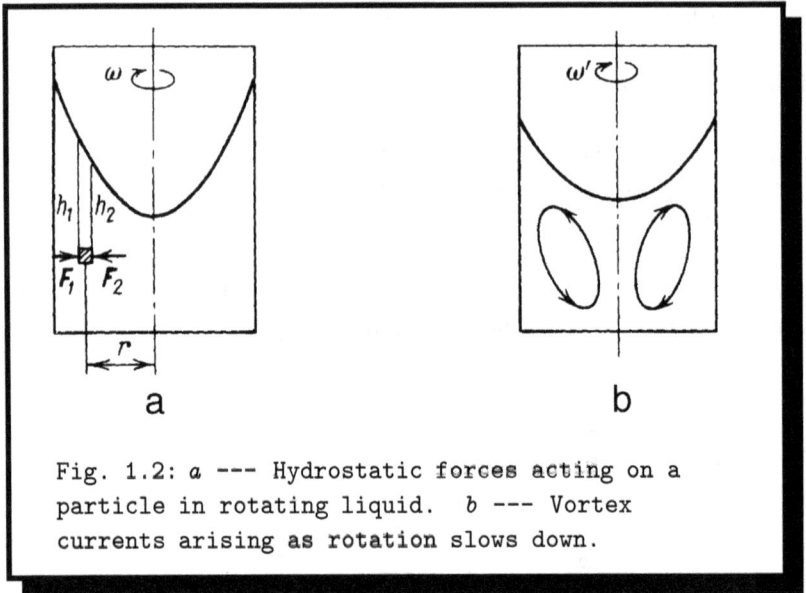

Fig. 1.2: *a* --- Hydrostatic forces acting on a
particle in rotating liquid. *b* --- Vortex
currents arising as rotation slows down.

Near the bottom, where the friction is stronger, the liquid slows down more effectively than at the surface. So, despite being at equal distances from the axis of rotation, particles of liquid acquire different speeds: the ones that are closer to the bottom become slower than those near the surface. However the net force due to the pressure differences is the same for all these particles. This force cannot cause the required centripetal acceleration of all the particles at once (as was the case with uniform rotation with the same angular speed). Near the surface, the angular speed is too large, and particles of water are thrown to the sides of the glass; near the bottom the angular speed is too low, and the resultant force makes water move to the center.

Now it is clear why tea-leaves gather in the middle of the bottom, Figure 1.3. They are drawn there by vortex currents that arise due to the nonuniform deceleration. Of course, our analysis is simplified but it accurately grasps the main points.

Fig. 1.3: The tea-cup experiment. Vortex currents
drive tea-leaves to the center of the bottom.

1.2 How river-beds change

Let us consider the motion of water in a river bend. The picture resembles
what we have observed in our glass of tea. The surface of water inclines
inside the bend in order that pressure differences produce the necessary
centripetal accelerations. (Figure 1.4 shows schematically a cross section
of bending river.) Quite similarly to the tea glass, the velocity of water
near the bottom is less than that near the surface of the river (distribution
of velocities with depth is shown by vectors in Figure 1.4). Near the
surface the net difference in hydrostatic pressures cannot make the faster
water particles follow the curve and the water is "thrown" to the outer shore
(away from the center of the bend). Near the bottom, on the other hand, the
velocity is small, so the water moves toward the inner shore (to the center
of the bend). Hence additional circulation of water appears in addition to
the main flow. Figure 1.4 shows the direction of water circulation in the
transverse plane.

The circulation of water causes soil erosion. As a result, the outer bank
is undermined and washed out while the soil gradually settles along the
inner shore, forming an ever thickening layer (remember the tea-leaves in
the glass!). The shape of the river-bed changes so that the cross section
resembles the one shown in Figure 1.5. It is also interesting to observe how
the velocity of the stream varies across the river (from bank to bank). In
straight stretches water runs most quickly in the middle of the river. At

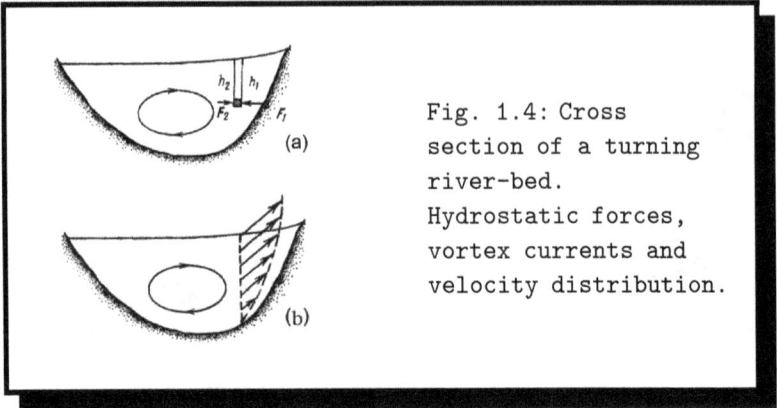

Fig. 1.4: Cross
section of a turning
river-bed.
Hydrostatic forces,
vortex currents and
velocity distribution.

bends, the line of fastest flow shifts outwards. This happens because it is
more difficult to turn fast-moving water particles than slow-moving ones.
A larger centripetal acceleration is necessary. But the greater the velocity
of the flow the greater the circulation of the water, and consequently, the
greater the soil erosion. This is why the fastest place in a river-bed is
usually the deepest one — a fact well known by river pilots.

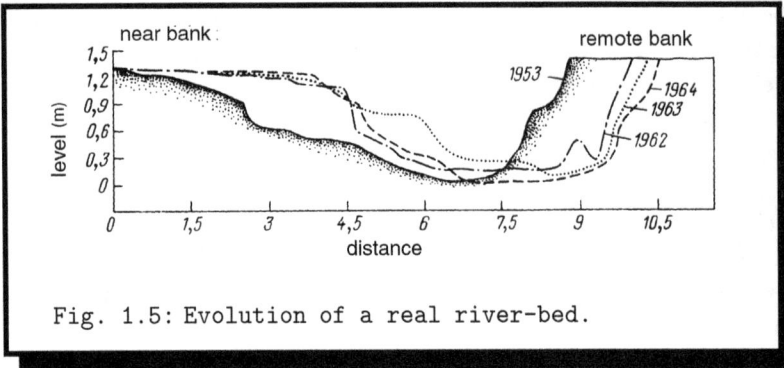

Fig. 1.5: Evolution of a real river-bed.

Soil erosion along the outer bank and sedimentation along the inner one
result in a gradual shift of the entire river-bed away from the center of the
bend thereby increasing the river meander. Figure 1.5 shows the very same
cross section of a real river-bed at several year intervals. You can clearly
notice the shift of the river-bed and the increase in its meander.

So even an occasional slight river bend — created, for example, by a landslide or by a fallen tree — will grow. Straight flow of a river across a plain is unstable.

1.3 How meanders are formed

The shape of a river-bed is largely determined by the relief of the terrain it crosses. A river passing a hilly landscape winds in order to avoid heights and follows valleys. It "looks for" a path with the maximum slope.

But how do rivers flow in open country? How does the instability of a straight river-bed with respect to bending described above influence the course of a stream? The instability must increase the length of the path and make the river wind. It is natural to think that in the ideal case (an absolutely flat, homogeneous terrain), a periodic curve must appear. What will it look like?

Geologists have put forth the idea that at their turns, paths of rivers flowing through plains should take the form of a bent ruler.

Take a steel ruler and bring its ends together. The ruler will bend like it is shown in Figure 1.6. This special form of elastic curve is called the *Euler curve* after the great mathematician Leonard Euler[b] who analyzed it theoretically. The shape of the bent ruler has a wonderful property: of all possible curves of a fixed length connecting two given points, it has the minimum average curvature. If we measure the angular deflection θ_k (Figure 1.6) at equal intervals along the curve and add up their squares then the sum $\theta_1^2 + \theta_2^2 + \ldots$ will be minimal for the Euler curve. This "economic" feature of the Euler curve was fundamental for the river-bed shape hypothesis.

To test this hypothesis geologists modeled a changing river-bed. They passed water through an artificial channel in a lightly erodible homogeneous medium composed of small, weakly-held-together particles. Soon the straight channel began to wander, and the shape of the bend was described by the Euler curve (Figure 1.7). Of course, nobody has ever seen such a perfect river-bed in nature (because of the heterogeneity of the soil, for instance). But rivers flowing through plains usually do meander and form periodic structures. In Figure 1.7 you can see a real river-bed and the Euler curve (the dashed line) that approximates its shape best of all.

---

[b]L. Euler, (1707–1783), Swiss-born mathematician and physicist; member of Berlin, Paris, St. Petersburg academies and of the London Royal Society; worked a long time in Russia. Buried in St. Petersburg.

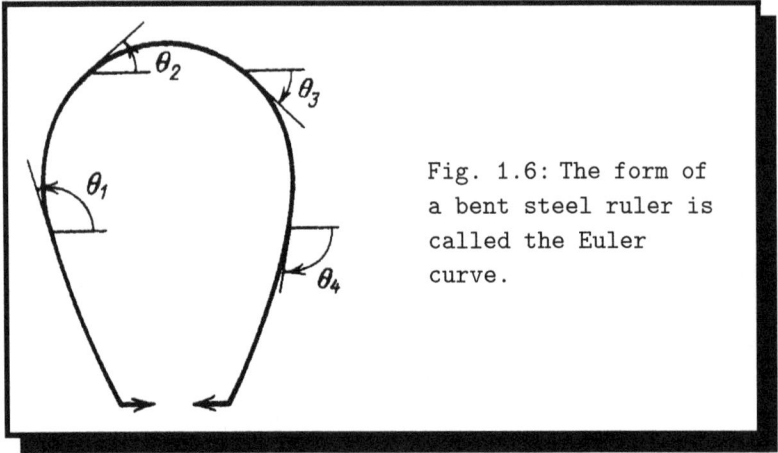

Fig. 1.6: The form of
a bent steel ruler is
called the Euler
curve.

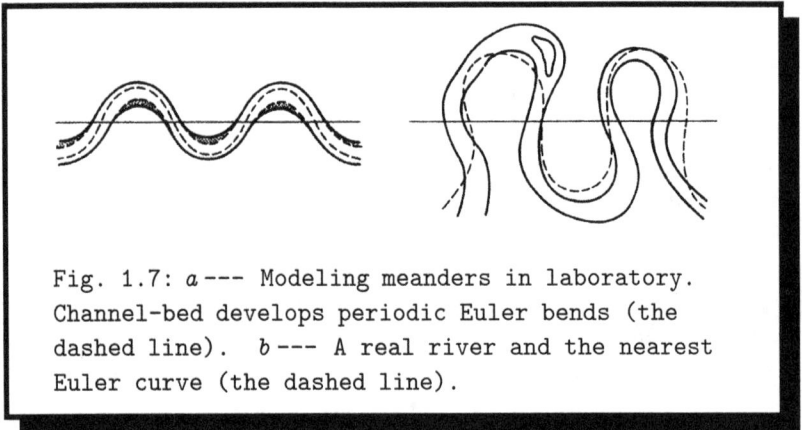

Fig. 1.7: *a* --- Modeling meanders in laboratory.
Channel-bed develops periodic Euler bends (the
dashed line). *b* --- A real river and the nearest
Euler curve (the dashed line).

By the way, the word "meander" itself is of ancient origin. It comes
from the *Meander*, a river in Turkey famous for its twists and turns (now
called the *Menderes*). Periodic deflections of ocean currents and of brooks
that form on surfaces of glaciers are also called meanders. In each of these
cases, random processes in a homogeneous medium give rise to periodic
structures; and though the reasons that bring meanders about may differ,
the shape of the resulting periodic curves is always the same.

Chapter 2

Rivers from lakes

...Old Baikal had more than three hundred sons and the sole daughter, beautiful Angara.

Ancient legend.

Let skeptical readers disbelieve the epigraph and search geographical atlases. Jokes aside, he will discover that 336 rivers fall into the Lake Baikal and only one, the mighty Angara, has it for a source. It turns out that Baikal is not the exception. As a rule, no matter how many rivers fall into a lake, only one comes out of it.

For example many rivers run to Ladoga lake but only the Neva escapes it; Svir is the only outflow of the Onega lake, etc. This may be explained by the fact that the outflowing water prefers the deepest river-bed and other possible exits are left above the level of the lake. It is hardly probable that openings of several would-be river-beds have the same elevation. A copious water supply in a brimming lake can send forth two streams. However such a situation is not steady and may take place only in relatively young (recently formed) lakes. Little-by-little, the deeper and faster stream will wash away the bed increasing the outflow. As a result the level of the lake will decline and the weaker flow will be gradually silted. Thus only the deepest of the outflowing rivers will survive.

In order for a lake to be the source of two rivers it is necessary that their origins lay exactly at the same level. This case is called a *bifurcation* (the term is now widely used by mathematicians to indicate the increase of the number of solutions to an equation). Bifurcations are uncommon and usually only a single river comes out of a lake.

The same laws may be applied to rivers. It is well known that rivers
readily flow together, whereas forks are relatively infrequent. Streams always prefer the steepest descending curve. The probability of splitting of
this curve is small. Nevertheless the situation changes at the river delta
where the main stream divides into many smaller channels.

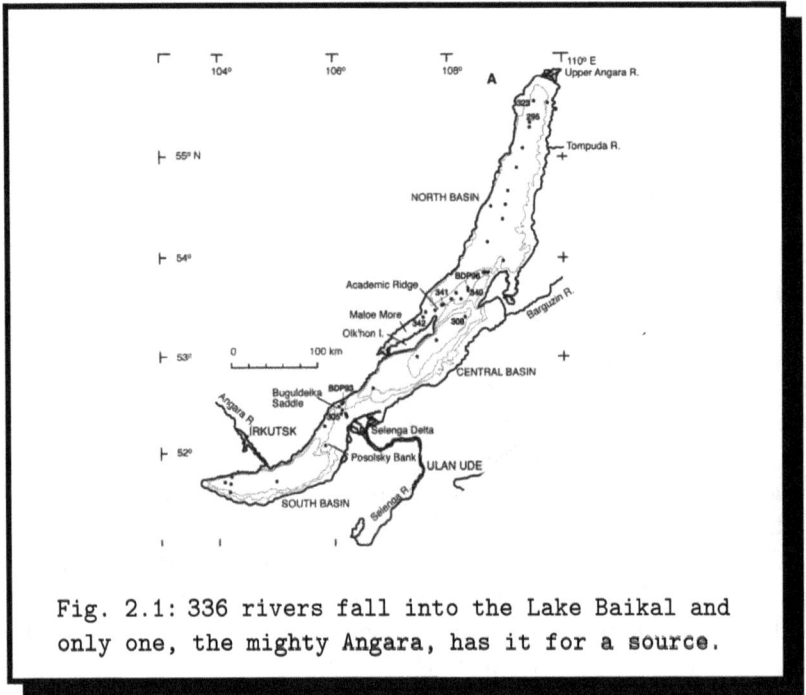

Fig. 2.1: 336 rivers fall into the Lake Baikal and
only one, the mighty Angara, has it for a source.

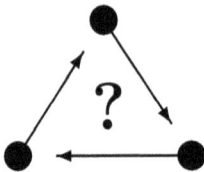

*Try to figure out why rivers behave so strangely
when approaching a big lake or the sea.*

Chapter 3

The oceanic phone booth

The walls, indeed, had ears. Or, rather, an ear. A round hole with a tube – a sort of secret telephone – conveyed every word said in the dungeon right to the chamber of Signore Tomato.

Gianni Rodari, *"The Adventures of Cipollino."*

Not so long ago – in the mid–1940s, to put a date on it – scientists from the USSR and US discovered an amazing phenomenon. Sound waves propagating in the ocean could sometimes be detected thousands of kilometers away from the source. In one of the most successful experiments, the sound from an underground explosion set off by scientists on the coast of Australia traveled halfway around the globe and was registered by another group of researchers in Bermuda, some $19,600\,km$ away (a record distance for the propagation of pulse sound signals). This means that the intensity of the sound did not change greatly as it traveled away from the source. What is the mechanism for such long-distance propagation of sound?

It looked as if the ocean contained an acoustic waveguide — that is, a channel along which sound waves traveled practically without attenuation (loss of strength). You have read about such a means in the epigraph.

Another example of an acoustic waveguide is the tube used on ships from time immemorial. The ship's captain uses a tube to give orders to the engine room from the bridge. It is interesting that the attenuation of sound traveling in air along a waveguide is so small that if we constructed a tube $750\,km$ long, it could serve as a "telephone" for calls between Pittsburgh and Detroit. But it would be inconvenient to chat over such a line, because

15

your friend at the other end would have to wait half an hour to hear your words.

We should emphasize that the reflection of a wave from a waveguide's boundaries is a crucial feature of the waveguide: it is because of this very property that the wave energy does not scatter in all directions but follows the given one.

These examples would lead us to suppose that the propagation of sound over extremely large distances in the ocean is due to some sort of waveguide mechanism. But how is such a gigantic waveguide formed? Under what conditions does it appear, and what are the reflective boundaries that make sound waves travel so far?

Since the ocean surface can reflect sound fairly well, it might serve as the upper boundary of the waveguide. The ratio of the intensity of the reflected wave to that of a wave penetrating the interface between two media strongly depends on the densities and the speeds of sound in them. If the media differ substantially then even the sound falling normally onto a flat interface will be practically completely reflected. The densities of air and water differ a thousand times, the ratio of sound velocities in them is 4.5. Therefore the intensity of the normal wave passing to air from water is only 0.01% of the intensity of the incident sound. The reflection is even stronger when the wave falls onto the interface obliquely. But, of course, the ocean surface cannot be perfectly flat because of the ever-present waves. This causes chaotic reflections of sound waves and disturbs the waveguide character of their propagation.

The results are not any better when the sound waves reflect off the ocean floor. The density of sediments at the bottom of the sea is usually within the range 1.24–2.0 g/cm^3, and the velocity of sound propagation in these sediments is only 2–3% less than that in water. So when a sound wave hits the bottom a significant amount of its energy is absorbed.

Since the ocean floor poorly reflects sound, it cannot serve as the lower boundary of the waveguide. The boundaries of the oceanic waveguide must be sought somewhere between the floor and the surface. And that is where they were found. The boundaries turned out to be the water layers at certain depths in the ocean.

How do sound waves reflect from the "walls" of the oceanic acoustic waveguide? To answer this question we will have to examine the mechanism of sound propagation in the ocean.

3.1 Sound in water

Up till now, as we have talked about waveguides, our unspoken assumption was that the speed of sound in them is constant. But the speed of sound in the ocean varies from $1.450\, m/s$ to $1.540\, m/s$ depending on temperature, salinity, hydrostatic pressure, and other factors. The increase in hydrostatic pressure $P(z)$ with depth z, for instance, adds to the speed of sound $1.6\, m/s$ per 100 meters down. An increase in temperature also adds to the speed of sound. However, the water temperature, as a rule, falls rapidly as one descends from the well-warmed upper layers to the ocean depths, where the temperature is practically constant. Due to the interplay of these two mechanisms — the hydrostatic pressure and the temperature — the dependence of the sound velocity $c(z)$ on depth in the ocean looks like that shown in Figure 3.1. Near the surface the rapidly dropping temperature takes the upper hand. Here the speed of sound decreases with depth. As we plunge deeper, the rate of decrease in temperature slows, but the hydrostatic pressure continues to grow. At some depth the two factors balance: the speed of sound reaches its minimum. Deeper down the sound velocity starts to grow due to the rise in hydrostatic pressure.

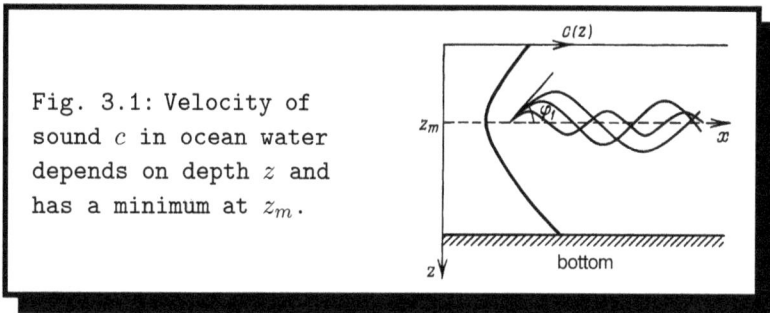

Fig. 3.1: Velocity of sound c in ocean water depends on depth z and has a minimum at z_m.

We see that the speed of sound in the ocean depends on depth and this influences the nature of sound propagation. To understand how "sound beams" travel in the ocean, we will turn to an optical analogy. We will examine how a light beam propagates in a stack of flat parallel plates with varying indices of refraction. Then we will generalize our findings to a medium with a smoothly varying refraction index.

3.2   Light in water

Let us consider a pile of flat parallel plates with varying indices of refraction $n_0, n_1, \ldots n_k$, where $n_0 < n_1 < \ldots < n_k$ (see Figure 3.2). Assume that the light beam falls onto the upper plate at an angle α_0 relative to the normal. After refraction it leaves the 0–1 boundary at an angle α_1 and that is the incidence angle for the 1–2 boundary. Upon refracting at the next interface the beam hits the 2–3 boundary at an angle α_2 and so on. According to Snell's[a] law, we have:

$$\frac{\sin \alpha_0}{\sin \alpha_1} = \frac{n_1}{n_0}, \quad \frac{\sin \alpha_1}{\sin \alpha_2} = \frac{n_2}{n_1}, \ldots \quad \frac{\sin \alpha_{k-1}}{\sin \alpha_k} = \frac{n_k}{n_{k-1}}.$$

Remembering that the ratio of the refraction indices of two media is the inverse of the ratio of the speeds of light in these media, we may rewrite these equations in the following form:

$$\frac{\sin \alpha_0}{\sin \alpha_1} = \frac{c_0}{c_1}, \quad \frac{\sin \alpha_1}{\sin \alpha_2} = \frac{c_1}{c_2}, \ldots \quad \frac{\sin \alpha_{k-1}}{\sin \alpha_k} = \frac{c_{k-1}}{c_k}.$$

Multiplying the series of equations by one another, we get

$$\frac{\sin \alpha_0}{\sin \alpha_k} = \frac{c_0}{c_k}.$$

Reducing thickness of plates to zero while increasing their number to infinity, we approach the generalized law of refraction (Snell's law):

$$c(z) \sin \alpha(0) = c(0) \sin \alpha(z),$$

where $c(0)$ is the speed of light at the point where the beam enters the medium and $c(z)$ is the speed of light at a distance z from the boundary. Thus, as a light beam propagates through an optically nonuniform medium with decreasing index of refraction, it increasingly deflects from the normal line. As the speed of light in the medium decreases (and the refraction index increases) the beam gradually turns parallel to the interface.

If we know how the speed of light varies in an optically nonuniform medium, we can use the Snell's law to find the trajectory of any beam in it. Sound beams propagating in an acoustically nonuniform medium, where the speed of sound varies, get deflected exactly in the same way. An example of such a medium is the ocean.

---

[a] W. Snell van Royen, (died 1626), Dutch mathematician.

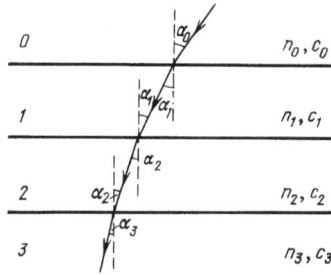

Fig. 3.2: Optically nonuniform medium may be modeled with a pile of glass plates with different refraction indices.

3.3 Water waveguides

Now let us get back to the question of sound propagation in the oceanic acoustic waveguide. Imagine that the sound source is located at the depth z_m corresponding to the minimum sound velocity (Figure 3.3). How do the sound beams travel as they leave the source? The beam propagating along a horizontal line is straight. But the beams leaving the source at an angle to the horizontal will bend because of sonic refraction (analogously to the refraction of the electromagnetic waves). Since the speed of sound increases both up and down from the level z_m, sound beams will bend towards the horizontal. At a certain point the beam will become parallel to the horizontal and after being reflected it will turn back again toward the line $z = z_m$, Figure 3.3.

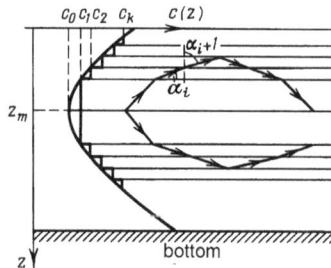

Fig. 3.3: Refraction of sound in acoustically nonuniform medium (the sound velocity $c(z)$ is minimal at the plane $z = z_m$).

Thus, the refraction of sound in the ocean allows a portion of the sonic energy emitted by the source to propagate through the water without rising up to the surface or dropping down to the ocean floor. This means that we have an oceanic acoustic waveguide. The role of the "walls" of the waveguide is played by the layers of water at the depths where the sound beams reflect.

The level z_m where the speed of sound reaches the minimum is called the axis of the waveguide. Usually the depths z_m range from $1,000\ m$ to $1,200\ m$, but in low latitudes, where the water is warmer deeper down, the axis can descend down to $2,000\ m$. On the other hand, in high latitudes the influence of temperature on the distribution of the sound speed is noticeable only close to the surface, and therefore the axis rises to a depth of $200-500\ m$. In the polar latitudes it lies still closer to the surface.

There are two different types of waveguides in the ocean. The first type occurs when the speed of sound near the surface (c_0) is less than that at the ocean floor (c_f). This usually occurs in deep waters, where the pressure at the bottom reaches hundreds of atmospheres. As we mentioned above, sound reflects well from the water-air interface. So if the ocean surface is smooth (dead calm), it can serve as the upper boundary of a waveguide. The channel then spreads through the entire layer of water, from the surface to the floor (see Figure 3.4).

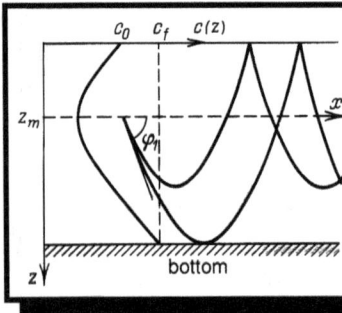

Fig. 3.4: Acoustic waveguide of the first type: dead calm, $c_f > c_0$. Sound is reflected from the surface and refracted at the bottom.

Let us see what portion of the sound beam is "captured" by the channel. We start by rewriting Snell's law as

$$c(z)\cos\varphi_1 = c_1\cos\varphi(z).$$

where φ_1 and $\varphi(z)$ are the angles formed by sound rays with the horizontal at depths z_1 and z, respectively. It is clear that $\varphi_1 = \frac{\pi}{2} - \alpha_1$, $\varphi(z) = \frac{\pi}{2} - \alpha(z)$. If the source of sound is located on the axis of the channel $(c_1 = c_m)$ then the last sound ray captured by the channel is the one tangent to the ocean floor, $\varphi(z) = 0$, as shown in Figure 3.4. Therefore all rays that leave the source at angles satisfying the condition

$$\cos \varphi_1 \geq \frac{c_m}{c_f},$$

enter the channel.

When the water surface is rough, all sound beams will scatter from it. The rays leaving the surface at angles larger than φ_1 will reach the floor and be absorbed there. Yet even in this case, thanks to refraction, the channel can capture those rays that do not quite reach the rough surface (Figure 3.5). Then the channel spreads from the surface to a depth z_k which can be determined from the condition $c(z_k) = c_0$. It is clear that such a channel captures all sound rays with angles

$$\varphi_1 \leq \arccos \frac{c_m}{c_0}.$$

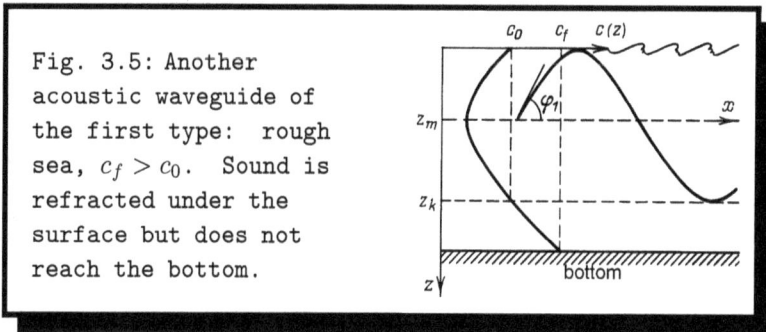

Fig. 3.5: Another acoustic waveguide of the first type: rough sea, $c_f > c_0$. Sound is refracted under the surface but does not reach the bottom.

The second type of waveguide is characteristic of shallow water. It occurs only when the speed of sound near the surface is greater than that at the floor, as in Figure 3.6. The channel occupies the water layer from the ocean floor up to the depth z_k where $c(z_k) = c_f$. It looks like a waveguide of the first type flipped upside down.

For certain types of dependencies of sound speed on depth, the waveguide focuses sound beams like a lens. If a sound source is located on the axis, the rays that leave it at different angles will periodically converge at some points on the axis. These points are called foci of the channel. So if the dependence of the speed of sound in the channel on depth is close

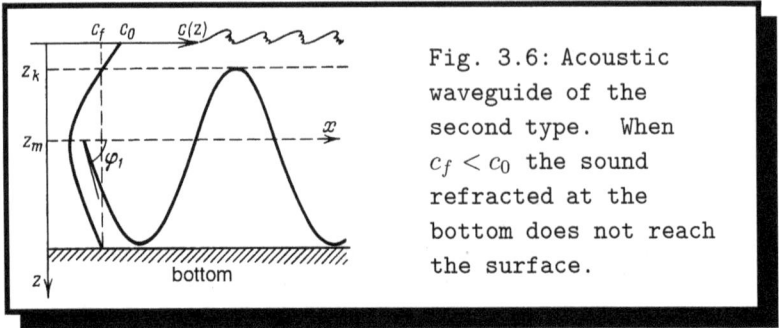

Fig. 3.6: Acoustic waveguide of the second type. When $c_f < c_0$ the sound refracted at the bottom does not reach the surface.

to parabolic: $c(z) = c_m \left(1 + \frac{1}{2}b^2 \, \Delta z^2\right)$, where $\Delta z = z - z_m$, then for rays leaving the source at small angles with the horizontal, the foci will lie at the points $x_n = x_0 + \pi \, n / b$, where $n = 1, 2, \ldots$ and b is a coefficient whose dimension is inverse to depth (m^{-1}), Figure 3.7. The parabolic function $c(z)$ is close to the actual dependence of the speed of sound on depth in deep oceanic acoustic waveguides. Deviations of $c(z)$ from a parabolic law blurs the focuses on the axis of the waveguide.[b]

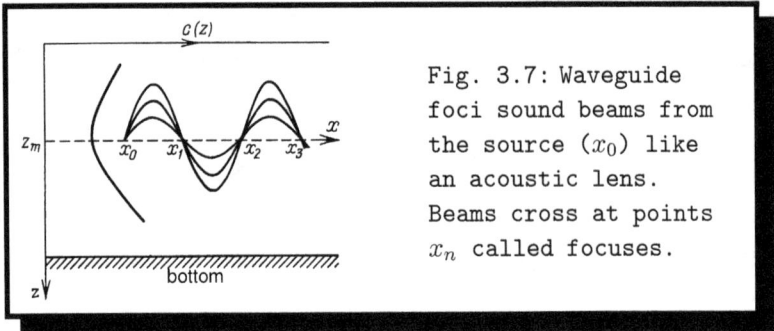

Fig. 3.7: Waveguide foci sound beams from the source (x_0) like an acoustic lens. Beams cross at points x_n called focuses.

---

[b]Like many periodic processes in nature, the propagation of beams along a parabolic waveguide obeys the wave equation. Near the axis trajectories follow the equation:

$$\frac{d^2 \, \Delta z}{dx^2} = -b^2 \, \Delta z,$$

where x is not the time but the horizontal coordinate. Obviously the trajectories are sinusoidal, $\Delta z = A \sin b \, (x - x_0)$, and cross at the zeros of the sine, $x_n - x_0 = \pi \, n / b$.

3.4 Applications?

Is it possible to send a sound signal along an oceanic acoustic waveguide and receive it at the point of origin, after it has completely circled the globe? The answer is a flat no. First and foremost, the continents present insurmountable obstacles, as well as great contrasts in depths of the world's oceans. So it is impossible to choose a direction along which there would be a continuous round-the-world waveguide. But that is not the only reason. A sound wave propagating along an oceanic acoustic waveguide differs from sound waves in the "telephone" tubes on ships that we mentioned at the outset. The sound wave traveling from the bridge to the engine room is one-dimensional, and the area of its wavefront is constant at any distance from the source. Therefore, the strength of the sound will also be constant everywhere along the tube (heat losses are not taken into account). As for the oceanic acoustic waveguide, the sound wave does not propagate along a straight line but in all directions in the plane $z = z_m$. So the wave here is a cylindrical surface. Because of this, the strength of the sound decreases with distance – that is, the sound intensity is proportional to $1/R$, where R is the distance from the source of the sound to the detector. (Try to obtain this dependence and compare it with the law of attenuation of a spherical sound wave in three-dimensional space.)

Another reason of sound attenuation is the damping of the sound wave as it travels through the waters of the ocean. Energy from the wave is transformed into thermal energy due to viscosity of water and other irreversible processes. Moreover, sound waves dissipate in the ocean because of various heterogeneities, such as suspended particles, air bubbles, plankton, and even the swim-bladders of fish.

The existence of underwater sound channels (also called the SOund Fixing And Ranging, or SOFAR, channels) is used in monitoring the temperature of oceans and other large bodies of water, since low-frequency sound pulses in them can travel thousands of kilometers. Pulse transit times can be measured using a network of stations to monitor water temperature. Comparison of these data with predictions of climate models can give some idea of whether a global warming occurs. Similar experiments can detect fluctuations of temperature in the ocean with an accuracy of up to one hundredth of a degree centigrade. Such observations under the ice of the North Pole indicates a warming of 0.5 degrees during the past 10 years.

The treaty on the prohibition of underwater nuclear tests is monitored with the help of a global hydroacoustic network in the world's oceans. This network of hydrophones detects acoustic waves generated by underwater explosions or, in some cases, appearing as a consequence of explosions in the low layers of the atmosphere.

Finally we should point out that the underwater sound channel is not the only example of waveguides in nature. Long-distance broadcasting from radio stations is possible only because of the propagation of radio waves through the atmosphere along giant waveguides. And we're sure you've heard of mirages, even if you've never seen one. Under certain atmospheric conditions, waveguide channels for electromagnetic waves in the visible range can form. This explains the appearance of a ship in the middle of desert, or a city that springs to life in the middle of the ocean.

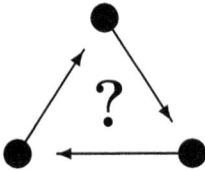

Prove that the trajectories of sound beams in a parabolic waveguide obey the wave equation.

Chapter 4

In the blue

Now the wild white horses play,
Champ and chafe and toss in the spray.

Matthew Arnold, *"The Forsaken Merman."*

Artists are endowed with sharp professional vision. This makes the world depicted in realistic landscapes outstandingly bright and colorful and features some natural events. Even the best painters do not need to understand the reasons of the portrayed processes which may often be hidden deeply. However with the help of the landscapes of a good painter one can study ambient nature even better than in the physical world. Suspending the moment in the picture the author intuitively focuses on the principals and omits inessential and passing details.

Look at the picture "In the Blue" by the Russian painter Arkadii Rylov[a] in the insert just after the Contents pages. "White birds soar among white clouds bathing in the blue sky. And the ship under the sails drifts peacefully over the gently rolling blue waves like a white bird in the ocean". This description belongs to the notable Russian art critic A. A. Fedorov-Davydov. Watching the wonderful canvas in the Moscow Tretyakov Gallery one forgets being in a museum and feels like a guest at this feast of nature.

But let us leave the aesthetic side and examine the picture with the eyes of researchers. First of all, from where did the author paint the landscape? Was he on a rocky ledge on the coast or aboard a ship?

Most probably he was on a ship since there is no surf in the foreground,

---

[a]A. A. Rylov, (1870–1939), Russian Soviet painter of epic romantic manner.

the distribution of waves is symmetrical and not deformed by the neighboring shore.

Let us try to estimate the velocity of the wind filling the sails of the
ship floating at a distance. We are not the first interested by the question
of evaluation of the speed of wind on the basis of the height of waves and
other natural evidence. It was already in 1806 that Sir Francis Beaufort[b]
introduced his approximate twelve-step scale. He related the force of the
wind to the effects it had on land objects and to sea choppiness (see Table 4.1 on pp. 34–35). This scale has been approved by the International
Meteorological Committee and is still in use.

Turning back to the picture we see that the sea is rather quiet. There
are only a few white horses on the surface. According to the Beaufort scale
this corresponds to a gentle breeze with the a speed of about 10 *mph*.

One may judge the speed of wind not only by means of the Beaufort
scale but from the brightness contrast between the sea and sky. Usually
the horizon in the open sea looks like a clear-cut border. The brightness of
the sea becomes equal to that of the sky only at dead calm. In this case the
contrast disappears and the sea becomes indiscernible from the sky. This
happens rarely because the calm must be almost absolute, the Beaufort
number being zero. Even the slightest wind disturbs the sea surface. The
coefficient of reflection from the oblique elements of the surface is no longer
equal to unity and the contrast appears between the sea and sky. It may
be studied experimentally. The dependence of the sea sky contrast on the
wind force was measured during an expedition of the Russian research ship
"Dmitrii Mendeleev". The results of the measurements are depicted in
Figure 4.1 by crosses, while the solid line represents the theoretical relation
found by A. V. Byalko and V. N. Pelevin.

By the way, why are white watercaps so much unlike the blue-green sea
water?

The color of the sea is defined by many factors. Among the most important are the position of the sun, the color of the sky, the form of the
sea surface, and the depth. In shallow waters the presence of sea-weeds
and pollution by solid particles are also relevant. All these factors affect reflection of light from the surface, its submarine scattering, and absorption.
This makes unambiguous prediction of sea color impossible. Nevertheless

---

[b]F. Beaufort, (1774–1857), English admiral, hydrographer, and cartographer, head of
the English hydrographic service.

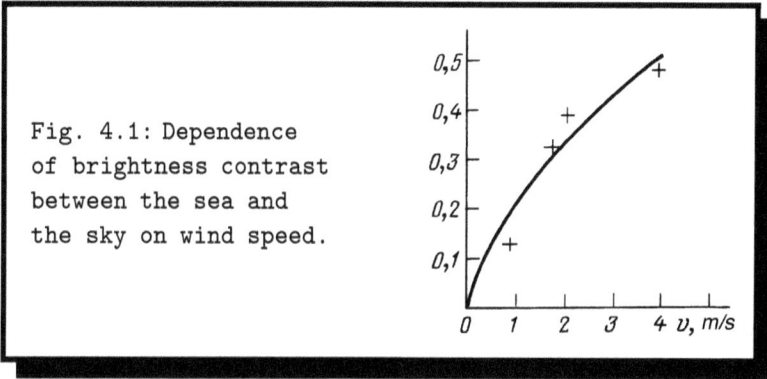

Fig. 4.1: Dependence
of brightness contrast
between the sea and
the sky on wind speed.

some details can be understood. For example, we can explain why the foreground waves that were closer to the painter were much darker in hue than others and why the sea became lighter near the horizon.

The extent of reflection of a light wave from the interface of two media with different optical densities is determined by the angle of incidence α (see Figure 1 in the insert just after the Contents pages) and the relative refraction index of the media. It is quantitatively characterized by the coefficient of reflection that is equal to the ratio of intensities of the incident and reflected rays.[c] The coefficient of reflection depends on incidence. In order to check this you may watch reflection of light from a polished table. The transparent varnish will act as the more dense optical medium. You will see that tangent rays ($\alpha \to \pi/2$) are almost completely reflected, but the smaller the angle of incidence becomes the more light penetrates the optically dense medium and the less is reflected from the boundary. The coefficient of reflection falls with decreasing of the incidence angle.

Let us consider the schematic image of a wave in Figure 1 (in the insert just after the Contents pages). It is obvious that the incidence angles α_1 and α_2 of the rays coming to the observer from the front and back sides of a wave are different, $\alpha_2 > \alpha_1$. Hence more light comes to the eyes of the observer from the distant areas and the front faces of waves look darker than the plane sea far away. Certainly, in a troubled sea the angle α varies. However from far away the angular size of the darker crests decreases rapidly even though the angle α_2 remains bigger than α_1. Near the horizon the observer

---

[c]Intensity of light is the averaged-over-time value of the light flux through a unit area perpendicular to the direction of the light propagation.

sees not solitary waves but the averaged pattern, the troughs between waves
are hidden and gradually the darker sides of the waves disappear. Because
of this the sea near the horizon looks lighter than the sea in the foreground
of the picture.

Now we can explain why watercaps are white. The seething water in
the cap swarms with endlessly moving, deforming and bursting bubbles.
Reflection angles vary from point to point and with time. As a result
sunrays are almost fully reflected by the froth and watercaps appear white.

The color of the sea is greatly affected by the color of the sky. We have
already said that the first is practically unpredictable. However the second
can be understood from physical principles. Obviously the color of the sky
is determined by the scattering of solar rays in the atmosphere of the Earth.
The spectrum of sunlight is continuous and contains all wavelengths. Why
does the scattering make the sky blue whereas the Sun seems yellow? We
shall explain that with the help of the Rayleigh law of light scattering.

In 1898 the English physicist Lord Rayleigh[d] developed the theory of
scattering of light by particles much smaller than the wavelength of the
light. He found the law that states that *the intensity of the scattered light
varies inversely with the fourth power of the wavelength.* In order to explain
the color of the sky Rayleigh applied his law to the scattering of sunlight
in the atmosphere. (For this reason the statement above is sometimes
called the "law of the blue sky".)

Let us try to understand the meaning of the Rayleigh law qualitatively.
Remember that light consists of electromagnetic waves. Molecules are built
of charged particles, i.e. of electrons and nuclei. In the field of an electromagnetic wave these particles start moving. Displacement of bound
electrons in the atom may be considered as harmonical: $x(t) = A_0 \sin \omega t$,
where A_0 is the amplitude of oscillations and ω is the wave frequency. The
acceleration of the particles is $a = -\omega^2 A_0 \sin \omega t$. Accelerated charged particles become sources of electromagnetic radiation themselves and emit the
so-called secondary waves. The amplitude of the secondary wave is proportional to the acceleration of the emitting particle. (As you know uniformly
moving charged particles make electric current but do not generate electromagnetic waves.) Therefore the intensity of the secondary emission is
proportional to the square of the acceleration of electrons in the field of the

---

[d]J. W. S. Rayleigh, (1842–1919), English physicist, chairman of the London Royal Society; Nobel Prize 1904 in physics.

primary wave (one may neglect the motion of heavy nuclei) and therefore to the fourth power of the frequency, ($I \propto a^2 \propto (x_t'')^2 \propto \omega^4$).

Now return to the sky. The ratio of the wavelength of red to that of blue is $650\,nm/450\,nm = 1.44$ ($1\,nm$ (nanometer) $= 10^{-9}\,m$). The fourth power of this number is 4.3. Thus according to the Rayleigh law the intensity of the blue light scattered by the atmosphere exceeds that of the red four times. As a result, the air layer ten miles thick acquires a blue tint. On the other hand, the blue component of the sunlight that reaches us through the atmospheric "filter" is seriously depleted. Hence the sun rays penetrating the atmosphere get a yellowish tone. The coloration may enhance getting red and orange during sunset when the rays have to pass along a longer path through the air. (Obviously the colors change in the reverse order when the sun rises.)

It is interesting that despite the claims of the Rayleigh law, the wavelength of the scattered light to be much bigger than the characteristic size of scattering particles, the intensity of scattering does not depend on the latter. At first Rayleigh believed that the color of the sky was due to the tiniest dust polluting the atmosphere. But then he decided that sunrays are scattered by molecules of gases that make up the air. Ten years later in 1908 the Polish theoretical physicist M. Smoluchowski[e] proposed the idea that scattering is affected by rather unexpected objects, namely by inhomogeneities of the density of particles. With the help of this hypothesis Smoluchowski managed to explain the long-known phenomenon of *critical opalescence* – that is the strong scattering of light in liquids and gases that occurs near a critical point. Finally in 1910 Albert Einstein[f] formulated the consistent quantitative theory of molecular scattering of light that leaned upon the ideas of Smoluchowski. In the case of gases, the intensity of scattered light exactly coincides with the earlier result of Rayleigh.

It seems that everything is in order now. But what is the origin of inhomogeneities in the air? Supposedly air must be in thermodynamic equilibrium. Gigantic inhomogeneities that make wind blow are incomparable with wavelengths of light and cannot affect the scattering.

In order to clear up the origin of the inhomogeneities we must discuss the concept of thermodynamic equilibrium in more detail. For simplicity let us consider a macroscopic amount of gas confined in a closed volume.

---

[e]M. von Smolan-Smoluchowski, (1872–1917), Polish physicist, classic works on fluctuation theory, and the theory of Brownian motion.
[f]See footnote in Chapter 1.

Physical systems consist of an enormous numbers of particles. This makes a statistical description the only possible approach. A statistical treatment means that instead of following separate molecules we calculate average values of physical characteristics of the whole system. It is not at all necessary that the corresponding value is the same for all molecules. The most probable distribution of molecules in the macroscopic gas volume would be the uniform one. But because of thermal motion there is a nonzero probability that the concentration of molecules in some part of our container will be enhanced (and as a result it will fall somewhere else). Theoretically it is even possible that all molecules will assemble in one half of the container leaving another half absolutely empty. However the probability of such an event is so small that there is little hope of realizing it even once in the $13.8 \cdot 10^9$ years that is currently believed to be the lifetime of the Universe.

Small deviations of physical quantities from their averages are not only allowed but happen perpetually due to the thermal motion of molecules. These deviations are called *fluctuations* (in Latin *fluctus* means wave). Because of fluctuations the gas density may be greater here and less there and as a result the refraction index will vary from place to place.

Let us now return to the scattering of light. All the reasoning applied to the closed container will hold in the atmosphere. Light is scattered by inhomogeneities of the refraction index that come from density fluctuations. Moreover, air is a mixture of several gases. Distinctions in thermal motion of different molecules provide another source of inhomogeneities.

The typical scale of inhomogeneity of the refraction index (and of those of density) depends on temperature. The sunlight is mainly scattered in the atmospheric layers where inhomogeneities are much less than wavelengths of visible light but much greater than the molecules of gases that compose the air. This means that scattering is effected by inhomogeneities but not by molecules, as it was presumed by Rayleigh.

Nevertheless the sky is blue and not violet contrary to the prediction of the Rayleigh law. There are two reasons for this discrepancy. First, the spectrum of the Sun contains much fewer violet rays than blue ones. The second thing responsible for the seeming disagreement of the theory with practice is our "registering device", the ordinary human eye. It turns out that visual perception markedly depends on the wavelength of light. The experimental curve characterizing this dependence is plotted in Figure 2 (inset after Contents pages). It is clear that our eye responds to violet rays much more weakly than to blue and green ones. This conceals the violet component of the scattered sunrays from the human eye.

Well, then why are the clouds that we see in the sky unmistakably white? Maybe they are composed of particles that violate the Rayleigh law and our conclusions are no longer true?

Clouds consist either of water drops or of ice crystals that in spite of being microscopic are much larger than the visible wavelengths. The Rayleigh law is not valid for these particles and the intensity of the scattered light is nearly the same for all wavelengths. This makes clouds like those painted in the picture.

Now turn your attention to the form of the clouds in Figure 3 (inset after Contents pages). The tops of the clouds in the picture are drawn fluffy and wreathing (these are the so-called heap clouds) but the lower surfaces are plainly outlined. For what reason? It is known that heap clouds (in distinction to sheet ones) are formed by uprising convective flows of humid air. The temperature of air decreases with height above the sea (as well as above the land). As long as the altitude is much less than the Earth's radius and the distance to the nearest shore surfaces of constant temperature (the isotherms) are next to horizontal plains parallel to the sea surface, the temperature drop is sufficiently fast near the sea level, being about $1°\,C$ per hundred meters (this makes about $1.6°\,F$ per 100 yd). (Generally speaking the dependence of the air temperature on the altitude is far from linear but for elevations less than several miles these numbers are correct.)

Now, what happens to the upward flow of humid air? As soon as it reaches the altitude where the temperature of air corresponds to the dew point of the vapor, the water it carries begins condensing to tiny drops. The isotherm where this happens contours the bottom of the cloud. Irregularities of the surface where condensation takes place stay within some tens of meters whereas clouds stretch for hundreds and thousands of meters. Hence their lower bounds are almost flat. This is confirmed by the row of clouds in the distance hovering horizontally above the sea.

However the rise does not stop at the bottom of a cloud. The air continues the upward motion and cools rapidly. The remaining vapor suffers condensation, first into droplets and then into little ice crystals. These crystals usually form the top of heap clouds. After having lost the vapor and cooling down, the air slows down and turns back. It flows sideways and around the cloud. Convective flows lead to the formation of the characteristic curls at the tops of heap clouds and the descending cold air keeps clouds apart. Thanks to that they do not merge into a heavy grey mass, being interspaced by blue intervals.

You see in the foreground a gaggle of white swans. Let us estimate the frequency of wing flaps of a medium-sized bird (say with the mass $m \approx 5\,kg$ and the area of wings $S \approx 0.5\,m^2$) when it flies without gliding. Let the average speed of the wings be $\bar{v}$. Then in the time Δt the bird wings will render speed $\bar{v}$ to the mass $\Delta m = \rho S \bar{v} \Delta t$ of the air (here ρ is the air density, which at sea level and at $15°\,C$ is equal approximately $1.225\,kg/m^3$). The total momentum passed by the wings to the air will be $\Delta \bar{p} = \rho S \bar{v}^2 \Delta t$. In order to keep at the same height we must compensate the weight of the bird. This means that $\Delta \bar{p} = m g \Delta t$. We may conclude that

$$m g = \rho S \bar{v}^2,$$

and the mean speed of the moving wings is $\bar{v} = \sqrt{m g/\rho S}$. This speed can be related to the frequency ν of flaps and the length of the wings L in the usual way:

$$\bar{v} = \omega L = 2\pi \nu L.$$

Assuming that $L \sim \sqrt{S}$ we find:

$$\nu \approx \frac{1}{2\pi S} \sqrt{\frac{m g}{\rho}} \approx 2\,s^{-1}. \tag{4.1}$$

Thus according to our calculation the bird should flap its wings twice per second. This order of magnitude represents quite a reasonable estimate.

It is interesting to discuss the analytical formula in more detail. Let us suppose that all birds have roughly the same body form regardless to the size and species. Then one may link the area of the wings to the mass of the bird by the relation $S \propto m^{2/3}$. Substituting this into the earlier found expression for the frequency of the wing flaps we obtain that

$$\nu \propto \frac{1}{m^{1/6}}.$$

From this we conclude that the frequency of flaps grows with decreasing bird mass. This absolutely agrees with the common sense explanation. Certainly the assumption that all birds have wings of the same form is extremely rough since the wings of most big birds are relatively bigger than those of small ones. Nevertheless this only supports the trend.

Let us note that the same formula (4.1) could be derived using dimensional analysis (except the important factor 2π). It is clear that the frequency of wing flaps depends on the bird's weight, the area of the wings S, and the density ρ of the ambient air. Let us search for a relation between the four of them. Suppose that $\nu = \rho^\alpha S^\beta (m\,g)^\gamma$ with α, β, γ being unknown numbers. Comparison of the dimensions of the quantities in both sides of the relation gives $\alpha = -\gamma = -1/2$ and $\beta = -1$. From here it follows that

$$\nu \sim \frac{1}{S}\sqrt{\frac{m\,g}{\rho}}.$$

Questions and answers that can be found *"out of the blue"* are far from being exhausted. Curious and observant readers will perhaps find in the picture other more instructive aspects. However why should we limit ourselves by the frame of the picture? There are plenty of interesting questions and problems in the everyday world all around us.

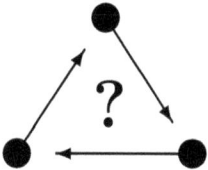

Can you explain why astronauts say that from outer space the Earth looks blue?

Table 4.1: BEAUFORT WIND SCALE

Beaufort Force Number	State of Air	Wind velocity		Description of sea surface
		(knots)	(mph)	
0	calm	0–1	< 1.15	sea like a mirror
1	light airs	1–3	1.2–3.5	ripples with appearance of scales are formed, without foam crests
2	slight breeze	4–6	4.6–6.9	small wavelets still short but more pronounced; crests have a glassy appearance but do not break
3	gentle breeze	7–10	8.0–12	large wavelets; crests begin to break; foam of glassy appearance; perhaps scattered white horses
4	moderate breeze	11–16	13–18	small waves becoming longer; fairly frequent white horses
5	fresh breeze	17–21	20–24	moderate waves taking a more pronounced long form; many white horses are formed; chance of some spray
6	strong breeze	22–27	25–31	large waves begin to form; the white foam crests are more extensive everywhere; probably some spray
7	moderate gale	28–33	32–38	sea heaps up and white foam from breaking waves begins to be blown in streaks along the direction of the wind; spindrift begins to be seen
8	fresh gale	34–40	39–46	moderately high waves of greater length; edges of crests break into spindrift; foam is blown in well-marked streaks along the direction of the wind

Table 4.1: BEAUFORT WIND SCALE *(continued)*

Beaufort Force Number	State of Air	Wind velocity (knots)	(mph)	Description of sea surface
9	strong gale	41–47	47–54	high waves; dense streaks of foam along the direction of the wind; sea begins to roll; spray affects visibility
10	whole gale	48–55	55–63	very high waves with long overhanging crests; resulting foam in great patches is blown in dense white streaks along the direction of the wind; on the whole the surface of the sea takes on a white appearance; rolling of the sea becomes heavy; visibility affected
11	storm	56–65	64–75	exceptionally high waves; small- and medium-sized ships might be for a long time lost to view behind the waves; sea is covered with long white patches of foam; everywhere the edges of the wave crests are blown into foam; visibility affected
> 12	hurricane	> 65	> 75	the air is filled with foam and spray; sea is completely white with driving spray; visibility very seriously affected

Chapter 5

The moon-glades

Heaven was full of silent stars, and there was a
moonglade on the water that stretched almost from him
to Rose.

Edgar Rice Burroughs, *The Efficiency Expert.*

Reflections of various light sources from the surface of water often look
like long shimmering lanes leading from the source to our eye. Remember
the setting sun reflected by sea or street lights along a night river quay.
The glittering of the moon adorns the sea or lake with a wide strip of light.

Fig. 5.1: The sun-glades. The wind speeds are
(from left to right): $12\,m/s$; $12\,m/s$; $5\,m/s$;
$2\,m/s$. The altitudes of the sun above horizon
are: $30°$; $20°$; $13°$; $7°$.

All this happens because every wavelet on the surface gives a separate image of the source. Let us try to understand why reflections from thousands of illuminated ripples make a glade, that is, an oblong figure directed from the light source to the observer.

As you already know, wavelets are formed at Beaufort numbers between 1 and 3. At weaker winds water is calm and the surface reflects like a plain mirror. Stronger winds bring on foam and white horses and the contour of the glade becomes vague. One may visualize ripples as scores of wavelets running chaotically in all directions. Slopes of their surfaces do not exceed some limiting value α which depends on the wind and can reach $20°$–$30°$.

We assume that the observer is on a cliff at an altitude of h above the water surface, and the angle between the average position of the latter (in complete calm) and the direction to the moon is ω. It is important that, since the Moon is at a distance of $384,000\,km$ from the Earth, and the length of the moon glade consist in the best case a few kilometers, this angle remains the same for both points at the beginning and at the end of the moon glade (see Fig. 5.2, where, because of the lack of space, we have slightly distorted the picture and denoted by means of ω two different angles).

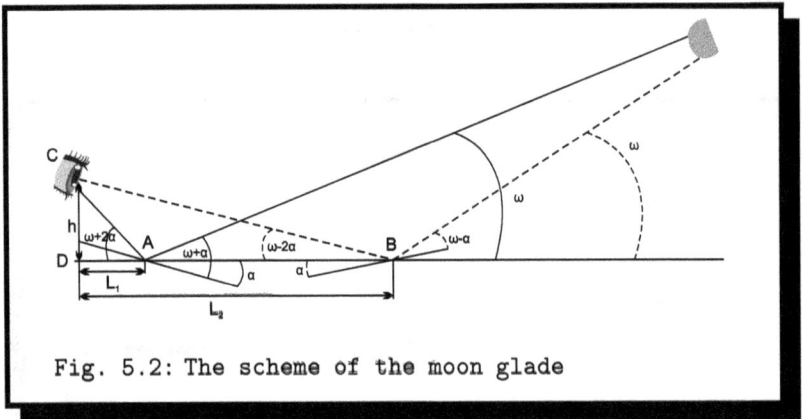

Fig. 5.2: The scheme of the moon glade

Now let us determine the position of the point of the moon glade closest to the observer. Let the steepness of the slope of the wave reach its maximum value (angle α) at the point A where the wave runs to the observer, whose eye is at the point C. It is clear that the angle of incidence of light from the Moon at this "mirror" will be $\omega + \alpha$, and the light will be reflected from this wave at the same angle. The reflected ray will make an angle $\omega + 2\alpha$ with respect to the horizontal plane, as seen from Figure 5.2. If it still hits the observer's eye, then the point A is at the distance $L_1 = h \cdot \cot(\omega + 2\alpha)$ from the point D as seen from the same figure. The luminous points will be to the right of A when the angles of inclination of the wave are less than α. The last one we can see will correspond not to the rising but to the falling slope at a limiting angle α (point B). Here, the mirror is tilted at the same angle α as at the point A, but down, not up. It is easy to see that the angle of incidence of moonlight on it is $\omega - \alpha$. At the same angle, the light will be reflected from the wave; with respect to the horizontal plane it will propagate at the angle $\omega - 2\alpha$. The condition that the ray still hits the observer's eye at the point C can be written as $L_2 = h \cdot \cot(\omega - 2\alpha)$. Thus, the length of the moon glade is $L = L_2 - L_1 = h[\cot(\omega - 2\alpha) - \cot(\omega + 2\alpha)]$, which can be transformed to

$$L = \frac{h \cdot \sin 4\alpha}{\sin^2 \omega - \sin^2 2\alpha}. \tag{5.1}$$

This expression shows that for $\alpha = 0$ (calm) $L = 0$, and we are dealing simply with the image of the Moon. For $\alpha \to \omega/2$, the moon glade becomes infinite. If the Moon is high ($\omega > 2\alpha$), then there cannot be any moonglade at all.

Chapter 6

The Foucault pendulum and the Baer law

> ... There was something, however, in the appearance of this machine which caused me to regard it more attentively.
>
> Edgar Allan Poe, *"The Pit and the Pendulum."*

Those lucky enough to have been to St. Petersburg will remember the famous pendulum in St. Isaac's Cathedral.[a] Others might have heard about it (Figure 6.1). The swings of the pendulum are accompanied by the slow rotation of the plane of oscillations. This observation was first made in 1851 by the French scientist J. Foucault.[b] The experiment was carried out in the spacious hall of the Pantheon in Paris, the ball of the pendulum had a mass of 28 *kg*, (62 *lb*) and the string was 67 meters (73 *yd*) long. Since then this sort of pendulum is named after Foucault. How can one explain its motion?

You know from textbooks that if Newton's laws[c] were true on the Earth then the pendulum would keep to the plane of oscillations. This means that in the reference frame rigidly bound with the Earth the laws of Newton must be "corrected". In order to do this one has to introduce special forces called *inertia forces*.

---

[a]St. Petersburg, the city and seaport on the Gulf of Finland of the Baltic Sea was during 1712–1917 the capital of the Russian Empire. In 1917, after the revolution, it was renamed Leningrad but now bears the original name.

[b]J. B. L. Foucault, (1819–1868), French physicist, foreign member of Russian Academy of Sciences.

[c]Isaac Newton, (1642–1727), English philosopher and mathematician; formulator of the laws of classical mechanics and gravity.

Fig. 6.1: In March 1931 the Foucault pendulum was first presented in St. Isaac's Cathedral occupied at that time by the Leningrad Antireligious Museum.

6.1　Inertia forces in the rotating reference frame

Inertia forces must be introduced in any reference frame that accelerates with respect to the Sun (or, to be precise, to the so-called stationary stars). These are called non-inertial reference frames in distinction to inertial ones that move uniformly with respect to the Sun and stationary stars.

Strictly speaking, the Earth does not present an inertial frame of reference because it orbits the Sun and revolves itself. Usually one may neglect accelerations arising from these motions and apply Newton's laws. However, this is not the case with a Foucault pendulum. Precession of the oscillation plane is explained by the action of the special force of inertia called the Coriolis force.[d] Let us examine it.

Here is a simple example of a rotating reference frame where inertia forces reveal themselves clearly.

Imagine a man riding a merry-go-round in Gorky Park.[e] Let the radius of the circle be r and the angular velocity of rotation ω. Suppose that the man tries to jump from his seat to the one in front of him, Figure 6.2, moving with velocity v_0 with respect to the platform.

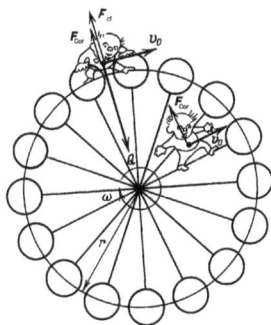

Fig. 6.2: Inertia forces in revolving reference frame.

⚠ Warning! The experiment is purely imaginary being strictly forbidden by safety regulations.

First let us consider the motion of our hero in a stationary reference frame. Obviously the motion is circular with the linear velocity v which

---

[d]G. G. Coriolis, (1792–1843), French civil engineer.
[e]See the novel by M. Cruz-Smith for further reference.

equals the linear velocity $\omega\, r$ of the merry-go-round and his relative velocity:

$$v = \omega\, r + v_0.$$

The centripetal acceleration is defined by the common formula,

$$a_{cp} = \frac{v^2}{r} = \frac{v_0^2}{r} + \omega^2\, r + 2\, v_0\, \omega.$$

According to Newton's second law, the acceleration is due to the horizontal component of the force exerted on the man by the rotating platform, the seat, handles etc.,

$$m\, a_{cp} = Q.$$

Now consider the motion in the reference frame bound to the merry-go-round. Here the linear velocity is v_0 and the centripetal acceleration is $a'_{cp} = \frac{v_0^2}{r}$. With the help of the two previous equalities we may write:

$$m\, a_{cp} = \frac{m\, v_0^2}{r} = Q - m\, \omega^2\, r - 2m\, v_0\, \omega.$$

In order to apply the second Newton law in the rotating frame of reference we must introduce the force of inertia:

$$F_{in} = -(m\, \omega^2\, r + 2m\, v_0\, \omega) = -(F_{cf} + F_{Cor}),$$

where the minus sign indicates that it is directed away from the axis. In the non-inertial frame the equation of motion will be:

$$m\, a'_{cp} = Q + F_{in} = Q - (F_{cf} + F_{Cor}).$$

It seems that the inertia force throws you from the center of the merry-go-round. However the word "seems" is not a slip. No new interactions between the bodies appear in the rotating reference frame. The only real forces acting on the man are the same reactions of the seat and bars. Their net horizontal component Q is directed towards the center. In the stationary reference frame the force Q resulted in the centripetal acceleration a_{cp}. In the rotating frame due to kinematical reasons the acceleration changed to the smaller value a'_{cp}. In order to restore the balance between the two sides of the equation we had to introduce the force of inertia.

In our case the force F_{in} consists of the two addends. The first is the centrifugal force F_{cf} that increases with the frequency of rotation and with the distance from the center. The second is the Coriolis force F_{Cor} named after the person who first calculated it. This force has to be introduced only when the body moves relative to the rotating frame. It does not depend on the position of the body but on only its velocity and the angular velocity of rotation.

If the body in the rotating frame moves not along a circle but radially, Figure 6.2, then, just the same, one must introduce the Coriolis force. Now it is perpendicular to the radius unlike the previous case. One of the basic features of the Coriolis force is that it is always perpendicular both to the axis of rotation and to the direction of the velocity vector. It may look strange but in the rotating frame inertia forces not only push a body away from the center but tend to swerve it astray.

We must emphasize that the Coriolis force, like all other inertia forces, is kinematic in origin and cannot be related to any physical objects.[f] Here is an explicit example.

Imagine a cannon set at the North pole and pointed along a meridian (the pole is chosen for simplicity). Let the target lie on the same meridian. Is it possible that the projectile hits the target? From the point of view of an external observer which uses the inertial frame bound to the Sun, the answer is obvious: the trajectory of the projectile lies in the initial meridional plane whereas the cannon revolves with the Earth. Thus the projectile will never reach the target (unless a whole number of days will elapse). But how could one explain this fact in the reference frame bound to the Earth? What causes the projectile to stray from the initial vertical plane? In order to restore consistency one has to introduce the Coriolis force, which is perpendicular to the rotation axis and to the velocity of a body. This force pulls the projectile away from the meridional plane and it misses the target.

Now let us return to the precession of the oscillation plane of the Foucault pendulum from which we have started. It is caused by a quite similar reason. Suppose again that the pendulum is situated at the pole. Then for a stationary observer the oscillation plane is at rest and it is the Earth that rotates. A denizen of the North pole will see the opposite. For him the meridional plane looks fixed whereas the oscillation plane of the pendulum performs a full revolution every 24 hours. The only way to explain this is with the help of the Coriolis force. Unfortunately, in general, the picture is not so transparent as at the pole.[g]

---

[f]Even though the force of inertia is not produced by any real bodies, observers feel it as a real force, akin to gravity. Remember the centrifugal force in the turning car.

[g]Oscillation plane of a Foucault pendulum located elsewhere turns $2\pi \sin \alpha$ radians per day, where α is the latitude of the place.

6.2 Interesting consequences

The Coriolis force which appears due to the Earth's rotation leads to a number of important effects. But before discussing those let us establish the direction of the Coriolis force. We have said that it is always perpendicular to the rotation axis and to the velocity of motion. However this leaves two possibilities depicted in Figure 6.3. Remember that an analogous situation emerges when defining the direction of the Lorentz[h] force exerted on a moving charge by a magnetic field. You may remember from textbooks that it is perpendicular to the velocity of the charge and to the magnetic induction. Still in order to define it unambiguously one has to resort to the *left-hand rule*.

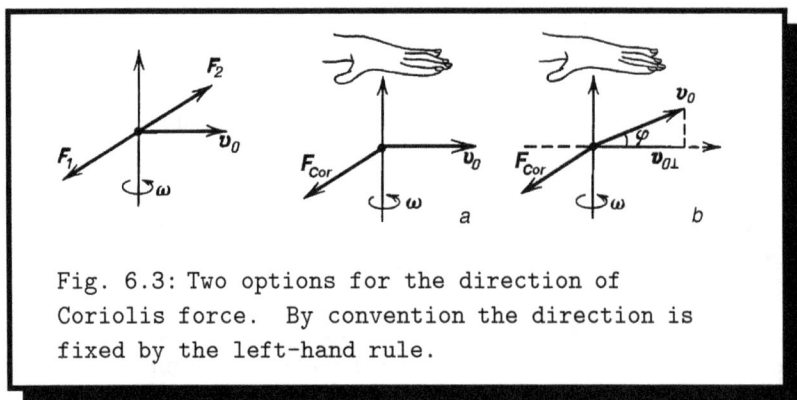

Fig. 6.3: Two options for the direction of Coriolis force. By convention the direction is fixed by the left-hand rule.

The direction of the Coriolis force can be determined by means of a similar rule elucidated in Figure 6.3, *a*. First of all we must assign a direction to the axis. By convention looking in this direction one sees the clockwise rotation.[i] Now let us pose the left hand with the four fingers pointing in

---

[h]H. L. Lorentz, (1853–1928), the Dutch physicist; Nobel Prize 1902 in physics.

[i]The *gimlet* rule states that this is where a gimlet rigidly attached to the frame would move.

the direction of the velocity so that the axis of rotation pierces through the palm. The thumb positioned at a right angle will show the direction of the Coriolis force.

The alternatives in defining the directions of Coriolis and Lorentz forces correspond to the two kinds of symmetry encountered in nature, the left and right symmetries. In order to classify the symmetry one has to use "standards" such as hand, gimlet, cork-screw, etc. Certainly nature does not care about your hand or gimlet. These are simply tools that help to fix the direction of the force.

This completes the discussion of the Coriolis force for the case when the velocity of a body in the rotating frame is perpendicular to the axis. The magnitude of the force is $2m\,\omega\,v_0$ and the direction is defined by the left-hand rule. But what happens in the general case?

It turns out that if the velocity v_0 makes an arbitrary angle with the rotation axis, Figure 6.3, *b,* then only the projection of v_0 onto the plane perpendicular to the axis is important. The value of the Coriolis force is given by the following formula:

$$F_{\mathrm{Cor}} = 2m\,\omega\,v_{\perp} = 2m\,\omega\,v_0\cos\phi.$$

The direction of the force is determined by the same left-hand rule although now the fingers must be parallel not to the velocity, but to its projection onto the plane perpendicular to the axis, Figure 6.3, *b.*

Now we have learned everything about the Coriolis force: both how to calculate its value and to define its direction. Armed with this knowledge we may explicate a number of interesting effects.

It is well known that trade winds which blow from the tropics to the equator are always deflected westward. This effect is explained in Figure 6.4. First let us consider the Northern hemisphere where trade winds blow from north to south. Position the left hand above a globe, palm down. The axis of Earth's rotation enters the palm and is being perpendicular to the four fingers. The Coriolis force is perpendicular to the page being leveled at you, that is to the west. Trade winds of the Southern hemisphere in their own turn blow from the tropic of Capricorn north to the equator. However, neither the direction of rotation nor the projection of wind velocity onto the equatorial plane change. Therefore the direction of the Coriolis force does not change either and in both cases the winds are diverted to the west.

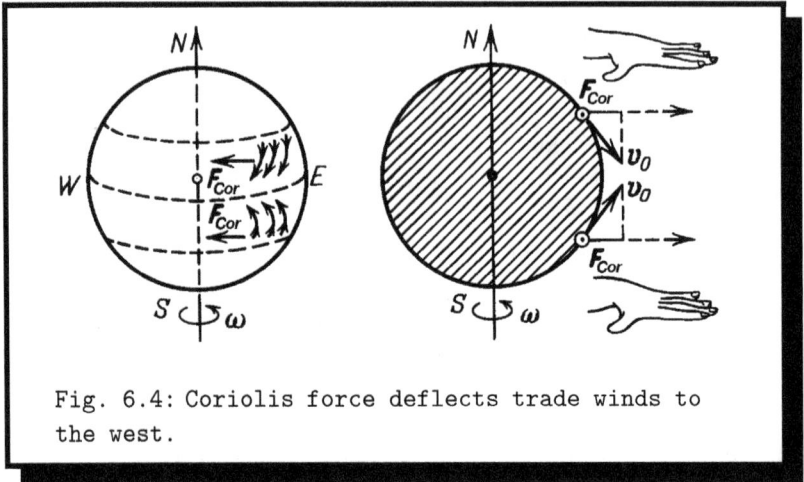

Fig. 6.4: Coriolis force deflects trade winds to the west.

Figure 6.5 illustrates Baer's law.[j] The right banks of rivers in the Northern hemisphere are steeper and more undermined than the left ones (and vice versa in the Southern hemisphere). The reason is again the Coriolis force that pushes flowing water to the right. Because of friction the surface velocity of a stream is bigger than that at the bottom; hence the Coriolis force is also bigger. This gives rise to the circulation of water shown in Figure 6.6 by arrows. The soil of the right bank is washed away and settles at the left side. This strongly resembles the erosion of the bank at river turns which was described in the chapter dedicated to meanders.

The Coriolis force always leads to eastward deviation of falling bodies. (Tackle this yourself.) In 1833 the German physicist Ferdinand Reich carried out precision experiments in the Freiburg mine. He found that the average (over 106 measurements) deflection of bodies which were dropped from the height of 158 m (173 yd) was 28.3 mm (1.11 in). This was one of the first experimental proofs of the Coriolis theory.

---

[j]K. E. von Baer, (1792–1876), Estonian zoologist and pioneer embryologist who was among the founders of the Russian Geographical Society.

Fig. 6.5: Coriolis force drives water flows to the right in the Northern hemisphere and to the left in the Southern.

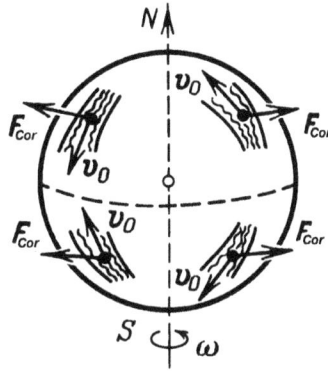

Fig. 6.6: The right banks of rivers in the Northern hemisphere are steeper and more undermined than the left ones.

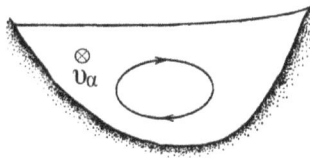

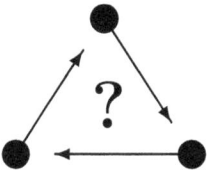

Try to estimate the difference of water levels at the right and left sides of the Volga river.

Does the Baer law apply to rivers streaming along parallels or the equator? What changes if a river crosses the equator, like the Congo?

Chapter 7

The moon-brake

Time and tide wait for no man.

A proverb.

Long ago people identified the Moon as the reason for tides. The Moon attracts the water of the ocean and forms a water "hump" in the ocean. The hump keeps its place on the moon side as the Earth rotates about the axis. When the high water advances to a coast the tide rises and when it retreats the ebb starts. The theory looks rather natural but it leads to a contradiction. This would mean that tides must be a daily event but instead, they happen every twelve hours.

The first explanation was given by Newton's theory of tides which appeared shortly after the discovery of the law of gravity. We shall study this question using the idea of inertia forces. According to the previous chapter, one may apply Newton's laws of mechanics in a rotating reference frame after adding to interactions between physical bodies the forces of inertia.

The Earth rotates around its axis, around the Sun and around...the Moon. Usually one forgets the latter, but it is this rotation that makes possible the theory of tides. Imagine that two balls, one light and one heavy, linked by a string are placed on a smooth surface, Figure 7.1.

Rotations of the tied balls are interrelated. Each one follows a circle of its own radius but the common center of the two is at the center of mass of the system. Of course the bigger ball traces the smaller circle but it moves! Similarly, the Moon and the Earth are attracted to each other, according to the law of gravity, and are orbiting their common spatial center of inertia C,

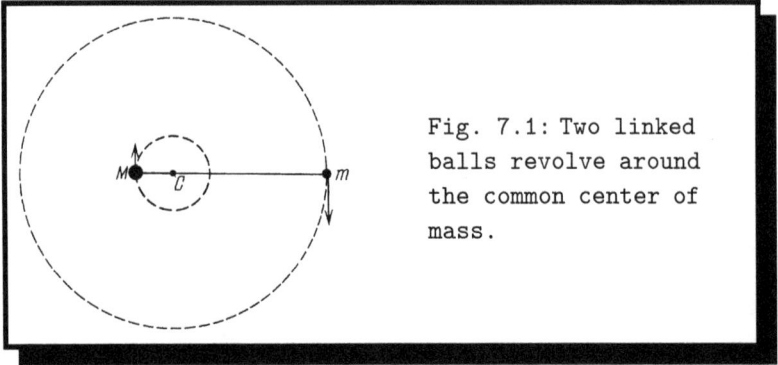

Fig. 7.1: Two linked balls revolve around the common center of mass.

Figure 7.2. Because of the large mass of the Earth, this point lies inside the globe being shifted with respect to the center O. The angular velocities of rotation around C of both the planet and the satellite are evidently the same.

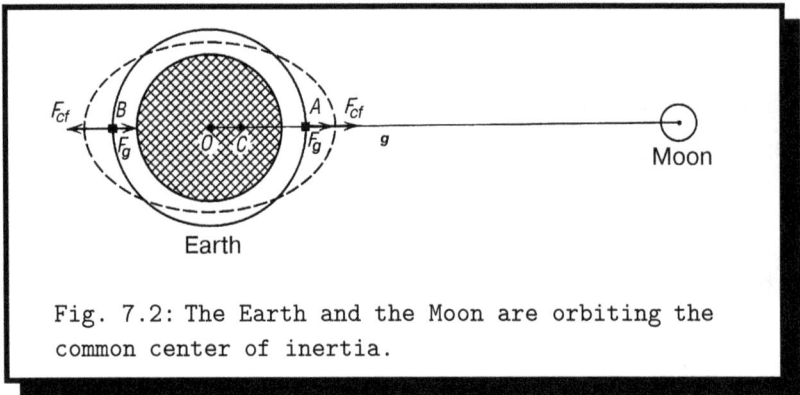

Fig. 7.2: The Earth and the Moon are orbiting the common center of inertia.

Consider now the rotating reference frame where both the Earth and the Moon stay at rest. Inasmuch as the reference frame is noninertial every mass element experiences not only the force of gravity F_g but a centrifugal force F_{cf} as well. The farther from the center C the stronger this force becomes.

Let us imagine for simplicity that water is evenly spread over the entire surface of the globe. May this be in equilibrium? Obviously not. Gravitational attraction to the Moon and centrifugal forces will destabilize the state. On the moon side, the two forces are directed away from the Earth's center and give rise to the water hump A, Figure 7.2. But the situation on the remote side is quite alike. As we step away from the common center of mass, the centrifugal force increases whereas the attraction to the Moon falls down. The resultant force is again directed away from the center of the Earth and is the origin of the second hump B. The equilibrium configuration is represented in Figure 7.2.

Of course this explanation of tides is much simplified. It does not take into account the nonuniform distribution of water on the Earth, effects of attraction to the Sun and many other factors that may essentially influence the picture. Still the theory answers the chief question. Once the humps do not move (with respect to the rotating frame of reference) but the planet revolves around its own axis the tides must occur twice a day.

Now we should explain the principle of the moon-brake. It turns out that the humps actually lie not on the line connecting the centers of the Earth and the Moon (as it was shown for simplicity in Figure 7.2) but are a little displaced, Figure 7.3. The reason is that because of friction, the ocean rotates together with the Earth. Therefore the mass of water in the humps is continuously renewed. However the deformation is always retarded with respect to the force that brings it on. (The force gives rise to an acceleration but it takes time for particles to gain speed and reach the place.) Thus the top of the hump that is the point of the highest tide is not at the point of the strongest attraction to the Moon which lies on the line connecting the centers. The hump is formed with a lag and it is shifted in the direction of the daily rotation of the Earth. According to Figure 7.3 this implies that the force of gravitational attraction to the Moon F_g does not pass through the center of the Earth and brings about a torque that slows down the gyration. The duration of the revolution is enhanced daily! This phenomenon was first recognized by the wonderful English physicist Lord Kelvin.[a]

The "moon-brake" has operated flawlessly for many millions of years and has the capability to notably change the length of the day. Scientists discovered in corals that have lived in the ocean about 400 million years ago, structures called "diurnal" and "annual" rings. When the diurnal rings were counted it turned out that there were 395 of them per year! The length of a year, that is the period of the circumvolution of the Earth around the

---

[a]W. Thomson, (1824–1907), 1st Baron Kelvin since 1892; English physicist and mathematician, chairman of the London Royal Society.

Fig. 7.3: The water humps are shifted from the
line connecting the centers of the Earth and the
Moon.

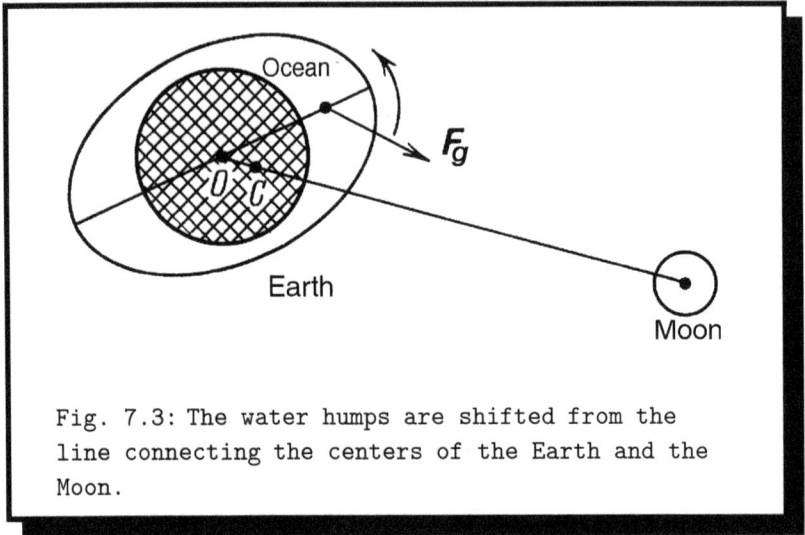

Sun, has probably not changed since then. Hence in those times the day
lasted only 22 hours!

Now the moon-brake keeps working on making days and nights longer.
At the end of the story, the period of the Earth's daily rotation will become
equal to that of the Moon orbiting and the impeding will cease. The Earth
will forever remain turned to the Moon by the same side, as presently the
Moon is. The increase of the day will effect the climate. The extended day
on the sunny side of the globe will be opposed by the prolonged night on
the rear. Cold air from the night side will rush to the warm hemisphere.
Winds and dust-storms will break out... But this prospect is so far off that
mankind will definitely discover how to prevent these calamities.

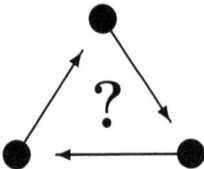

*What is the effect of the moon-brake on the du-
ration of the lunar month (that is the period of
orbiting of the Earth by the Moon)?*

PART 2

Saturday night physics

We have become so accustomed to our circumstances that we do not even notice the many wonders nor think about the actual causes behind them. Yet at closer inspection, one discovers a great many grounds for contemplation.

"When tossing pebbles into water, focus on the circles they produce, otherwise your tossing will be a mere time-frittering," — wrote the great Koz'ma Prutkov.[a]

We shall try to convince you that even the most improbable phenomena in the world around us can be explained by ordinary physics.

---

[a] Koz'ma Prutkov was the great Russian writer, poet, and philosopher of the nineteenth century. His selected works along with those by Alexander Pushkin and Mikhail Lomonosov may be found on the desk of any Russian scholar.

Chapter 8

Why the violin sings

The violin had no color, but sound it had.

N. Panchenko, *"The poem about a violin."*

Whenever an object moves through a medium, there always arise resistive forces trying to slow the motion. When a body slides mechanically along a rigid surface it will be the force of dry friction; in a liquid or in a gas it will be liquid friction (viscosity) and in the air it will be the aerodynamic resistance.

The interaction between a body and the surrounding medium is a rather complicated process leading usually to the work-to-thermal energy conversion of mechanical energy of the body. However, the reverse situation, when the medium is in fact supplying the body with energy, is possible too. And this usually leads to some sort of oscillation. For example, the dry friction force between a wardrobe being moved and the floor will slow down the motion. The same friction between the bow and string of a violin will make the string reverberate. As we will see later, the cause of vibration in the latter case is the decreasing dependence of friction on the velocity of the motion. The vibration indeed occurs when friction decreases with augmenting velocity.

Let us illustrate the generation of mechanical oscillations using as an example a concerto violin. The sound of a violin is caused by the moving bow, right? It is impossible, of course, to explain here all the complicated phenomena involved in formation of a particular musical tone, yet let us try to understand, in principle, why the string starts vibrating when the bow is being smoothly pulled against it.

The friction force between the bow and string is dry friction. We can easily distinguish two different kinds of friction – friction of rest and sliding friction. The first acts between the touching surfaces when contacting bodies are at rest with respect to each other; the second – when one body is actually sliding along the surface of the other.

As is well known, in the former case (no sliding), friction will balance an external force (being equal in magnitude and opposite in direction) up to a certain maximal value, called F_{fr}^0.

The sliding friction, in its turn, depends on the material and condition of the contacting surfaces, as well as on the relative velocity of the bodies. It is the latter circumstance that we will discuss in more detail. The character of relation between sliding friction and velocity varies for different bodies: as velocity rises the sliding friction at first often drops down and then begins to go up too. Such dependence of the dry friction force on velocity is illustrated by the graph in Figure 8.1. The friction force between the hair of the bow and the string behaves in this way too. When v, the relative velocity of the bow and string, is zero the friction between them does not exceed F_{fr}^0. Then, for the descending part of the curve, $0 < v < v_0$, any slight increase of the relative speed, say, by Δv, leads to the corresponding decline of friction force and, vice versa, when velocity is going down the change of force will be positive (see Figure 8.1). And, as we are going to see, it is exactly due to this not so evident at first glance feature that the energy of the string can grow at the expense of mechanical work done by the force of dry friction.

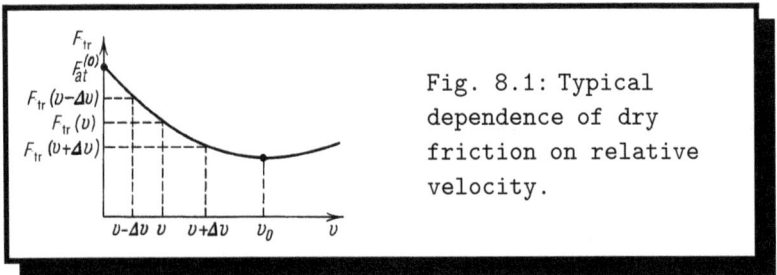

Fig. 8.1: Typical dependence of dry friction on relative velocity.

When the bow starts its motion, the string is getting drawn along with it, and the friction is compensated by the tension of the string, Figure 8.2. The resultant of the tension forces is proportional to the deviation x of the string from equilibrium:

$$F = 2T_0 \sin \alpha \approx \frac{4T_0}{l} x,$$

where l is the length of the string and T_0 is the tension force, which for small stretches, x, may be taken to be constant. Thus, when the string is being pulled along with the bow, the force F is growing until it reaches the maximal value of friction, F_{fr}^0. Then the string begins sliding against the bow.

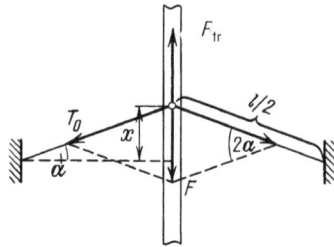

Fig. 8.2: When string follows the bow without sliding the friction of rest compensates the resultant of the two tension forces.

For the sake of simplicity, let us assume that at the beginning of slipping the friction force drops abruptly from the maximum rest value F_{fr}^0 down to the relatively weak force of sliding friction. In other words, we can approximately consider the slipping of the string as an almost free motion.

At the exact instant when the string takes off from clinging to the bow to sliding, its velocity is equal to the that of the bow, and, therefore, it keeps moving in the same direction. Yet now, the net tension force, not compensated by anything, will start slowing the motion of the string down. Consequently, at a certain moment the velocity will drop to zero, the string will stop and then it will reverse its motion and go back against the bow. Further, after the maximal swing to the other side, the string will again have to start moving in the same direction with the bow. Throughout this time the bow continues to move with the same constant velocity u and at some point the speeds of string and bow will match both in their magnitude and, this time, in the direction too. The slipping between the string and the bow will disappear and, again, the friction will balance the string tension. Now, as the string approaches the neutral position, the tension force subsides causing the corresponding waning of friction that counteracts it. And then, after the string passes the equilibrium position, everything happens again.

The ensuing graph for the string deviation versus time is shown in Figure 8.3, *a*. The periodic motion of the string is composed of two different parts for each period. That is, for $0 < t < t_1$ the string follows the bow with constant speed u and the deviation x is linearly proportional to the elapsed time ($\tan \alpha = u$). At t_1 the "take-off" occurs and for the interval $t_1 < t < t_2$ the dependence of x on time becomes a sinusoid. At the instant t_2, when the tangent to the sinusoid has the same slope α as the starting linear piece of the curve (and, hence, the velocities of the string and the bow are equal), the string is captured by the bow again.

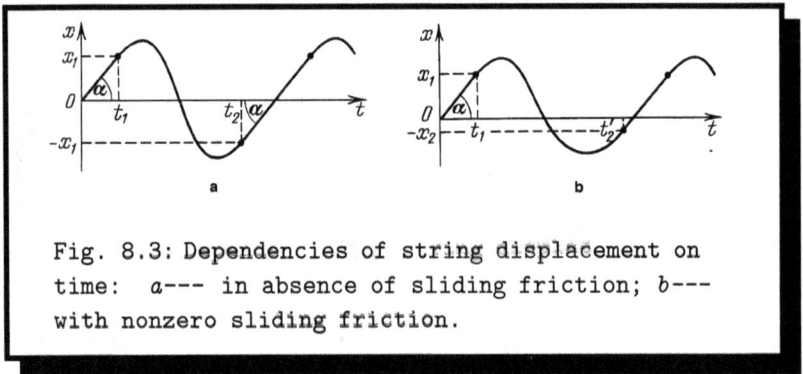

Fig. 8.3: Dependencies of string displacement on time: *a*--- in absence of sliding friction; *b*--- with nonzero sliding friction.

Figure 8.3, *a* illustrates an ideal case when there is no sliding friction between the bow and the string and, consequently, there is no energy loss as the string moves freely. The total work performed by friction forces (in the intervals without sliding) during the complete cycle of oscillation equals zero because for negative x the friction force acts against the motion and the mechanical work is negative, whereas for $x > 0$ the work is the same in magnitude yet positive in sign.

Now let us try to figure out what happens if the sliding friction force is not negligibly small any more. Then it should cause energy loss. The graph for the string motion with sliding is presented in Figure 8.3, *b*. For positive values of x the curve is actually steeper than for negative ones. Hence, now the clinging of string to the bow happens at a smaller negative deviation ($-x_2$ in the picture) than the positive x_1 at which the string first starts slipping against the bow: $x_2 < x_1$. This results in positive mechanical work A done by friction in the intervals when the bow and the string go along together:

$$A = \frac{k\left(x_1^2 - x_2^2\right)}{2}.$$

Here $k = \frac{4T}{l}$ is the coefficient of proportionality which relates the value of friction of rest that pulls the string away from the equilibrium and the string swing.[a]

This positive fraction of the total work compensates for energy losses due to the sliding friction and makes the string oscillate without damping.

Generally speaking, to replenish the energy, it is not at all necessary for the string to cling to the bow. It is enough for their relative velocity v to stay within the falling part of the sliding friction dependence on relative speed (see Figure 8.1). Now let us take a closer look at the vibration of the string in this case.

Suppose the bow is being pulled with constant speed u and the string is driven away from its neutral equilibrium position by x_0 so that the net tension force $F(x_0)$ is again compensated by the sliding friction force $F_{\mathrm{fr}}(u)$. If, by chance, the string deviates in the direction the bow moves, their relative velocity shall decrease. This will cause friction to rise. (Note that we are on the dropping part of the $F_{\mathrm{fr}}(u)$ curve!) This, in turn, makes the string stretch even more. Of course, at some point the elastic force will exceed the friction (remember that the vector sum of tensions is directly proportional to the deviation of string from the neutral position, whereas the friction is limited by F_{fr}^0). There the string will start slowing down, then it will reverse its motion and go in the opposite direction. On the way back the string will pass the point of equilibrium, then stop at the utmost position on the other side and repeat everything again... Thus oscillations will be amplified.

It is important to notice that the described oscillations, once started, will proceed without damping. Indeed, when the string moves with velocity Δv in the direction of the bow and $u > \Delta v > 0$, then the friction performs positive work. On the other hand, when going back the work of friction will be negative. The relative velocity $v_1 = u - \Delta v$ in the former case is less than that, $v_2 = u + \Delta v$, in the latter one. The friction, $F_{\mathrm{fr}}(u - \Delta v)$ will, on the contrary, be greater in the first situation than that, $F_{\mathrm{fr}}(u + \Delta v)$, in the

---

[a]Remember that at the linear piece of the curve, Figure 8.3, the force of friction is equal in magnitude to the resultant of the tension forces, Figure 8.2.

second one. Thus, the positive mechanical work done by friction when the string and bow are moving together surpasses the negative work performed when the string moves back. This results in positive net work during the vibration cycle. Consequently, the amplitude of vibrations increases with each successive oscillation. And this keeps going up until a certain limit. If $v > v_0$ so that the relative velocity of bow and string v finally leaves the descending part of the plot $F_{\text{fr}}(v)$ (Figure 8.1) then the negative work of friction can overcome the positive one, forcing the amplitude of oscillations to wane.

As a result, a stable vibration with some equilibrium amplitude will finally be attained, for which the total work done by friction will be exactly equal to zero. (To be precise, the positive work during the cycle compensates the energy loss due to the air resistance, nonelastic character of deformation, etc.). These steady oscillations of the violin string will proceed without damping.

It is quite common that sound vibrations are excited when one body moves along the surface of another: dry friction in a door hinge causes it to screech; and so do our shoes, floor tiles, and so on. You can produce screeching by just pressing and pulling your finger along, say, a smooth and firm enough surface.[b] And phenomena which occur in these examples may be very similar to the excitation of vibrations of the violin string. At first there is no sliding, then an elastic deformation develops up to the point when the "take-off" happens and "their majestic" oscillations commence. And once having started, they do not subside abruptly but continue without noticeable damping. Because of the same dropping character, the friction forces do positive mechanical work that procures the energy required for the oscillations.

If the dependence of friction on relative velocity of the surfaces changes its character, the screech goes away. Everyone knows, for instance, that you can simply lubricate the surfaces to get rid of an irritating screak. And the physical reason behind it is trivial: for slow motion the liquid friction is proportional to velocity and, hence, the conditions required to bring about and then to sustain oscillations disappear when you substitute dry friction with the liquid one. Conversely, when vibrations are desirable, the participating surfaces are often specially treated in order to reach a sharper decrease of friction force with increasing velocity. For instance, for this exact reason rosin is applied on violin bows.

---

[b]Chapter 9, "The chiming and silent goblets" gives a less trifling example of the kind.

No surprise that understanding the laws of friction often helps in solving various practical and industrial problems. For example, while machining a metal piece, undesired vibrations of the cutter can develop. These vibrations are caused by the force of dry friction between the tool and metal shaving slithering along its surface, Figure 8.4. Here again the dependence of friction on speed may have for high quality steels the familiar "dropping" character. Which, as we know by now, is the principal condition for exiting oscillations. A common way to preclude such vibration (which can turn out to be quite detrimental for both the cutter and the piece in work) will be to use along with the naive lubrication a special sharpening of the cutter, basically to hone it to the correct angle, so that no slithering will occur and there will be no reason for oscillations.

Fig. 8.4: Vibrations of cutter of machine tool may be eliminated by correct choice of the sharpening angle.

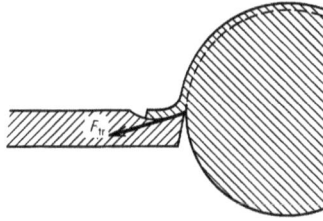

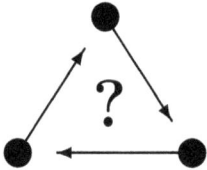

Can you describe (or even write a formula for) the motion of a string subject to constant sliding friction?

Chapter 9

The chiming and silent goblets

The carriage resembled an open shell made of glittering crystal; its two large wheels seemed to be built of the same material. When they were turning, they produced marvelous sounds: Full, yet still growing and approaching, these chords reminded the tones of glass harmonica, yet of amazing strength and power.

E. T. A. Hoffmann, *"Genannt Zinnober."*

It is not a novel idea that one can make simple wine-glasses sing. However, it turns out that there is a very peculiar way to do this. How peculiar? Well, judge for yourself.

If you dip a finger in water and start circling it carefully along the edge of a glass keeping the rim wet, it makes, at first, a rather screeching sound. But then, after the water has thoroughly and uniformly covered the glass edge, the tone should turn into something more melodic. By varying the pressure of the finger you can easily change the pitch of the produced sound. The height of the pitch will also depend on the size of the glass and thickness of its walls.[a]

Notice, by the way, that not every single glass is capable of making these pleasing chiming tunes, so the search for a suitable one may turn out to be quite a meticulous affair and take a while. The best "singers" turn out to be very thin-walled goblets, those having the shape of the paraboloid of revolution, with a long slim stem. Another critical parameter

---

[a]The mechanism of exciting sound is the same as in the bow-instruments, see Chapter 8, "Why the violin sings."

that determines the resonant tone of the glass is the level of liquid in it: generally, the fuller the glass the lower its pitch. When the water level passes the midline of the glass, waves will develop on the surface of liquid, because of the wall shaking. The maximal disturbance marks the position of the finger inducing the sound at the moment.

Fig. 9.1: Benjamin Franklin playing one of his inventions, a mechanical form of musical glasses called the ''armonica''.

A famous American scientist (as well as one of the greatest statesmen in the history of his country – a rare yet proven possible, at least back in those days, combination) Benjamin Franklin, who is mostly known for his experiments with atmospheric electricity,[b]had employed the phenomenon discussed in the previous paragraph to create a peculiar musical instrument, very similar to that described in Hoffmann's "Genannt Zinnober". That was a series of perfectly polished glass cups, each with a drilled orifice in the middle, arranged equidistantly on the same shaft. There was also a pedal under the box where the cups on the axle were situated, like in an old-fashioned sewing machine, to make the shaft rotate. And just by simple touching with wetted fingers, one could change the tone of the system from a sound forte down to meager whistling. Now it is hard to believe, but the people who had heard this "goblet organ" playing assured that the harmony of its sounds was amazingly appeasing on both the listeners and the performer. In 1763 Franklin had given his own instrument as a present

---

[b]B. Franklin, (1709–1790), American public official, writer, and scientist.

to an Englishwoman, Ms. Davis. She demonstrated it for several years traveling around Europe, and then the famous instrument disappeared without a trace. Probably memories of that true story had reached the German writer E. T. A. Hoffmann. He was a talented musician himself, and used them in his "Genannt Zinnober".

And since we are talking glasses, it seems also worth mentioning another interesting fact. It may sound iconoclastic but it is not accepted as proper etiquette to clink champagne glasses. Really. And the deal here is that for some, of course, purely physical reason, goblets filled with champagne or any fizzy carbonated drink make, when clinked, an inexpressive muffled sound. So what is the matter here? Why do goblets with champagne not ring?

Physically speaking the melodic ear-pleasing tingle that we hear after clinking glasses comes from high-frequency sound waves. Our glasses act like resonators in the high-frequency sound (that is $10-20\,KHz$) and even ultrasound (higher than $20\,KHz$) range. When we clink either empty glasses or glasses filled with noncarbonated beverage, these oscillations keep ringing for a rather long period of time once induced. On the other hand, this automatically suggests a likely cause for muffling the sound produced by goblets with champagne. Those are the tiny pinching bubbles of carbon dioxide stampeding from the opened bottle in their delectable effervescent rush. They may lead to strong scattering of short sound waves in the goblet. Similar processes take place in the atmosphere, where fluctuations of molecular density scatter the light in the short wave part of the spectrum (see Chapter 4, "In the blue").

Even for the highest frequencies perceived by human ear ($\nu \sim 20\,KHz$) the wavelength of sound in water, $\lambda = c/\nu \sim 10\,cm$ ($c = 1450\,m/s$ is the velocity of sound in water) is considerably larger than the size of CO_2 bubbles in champagne (say, less than $1\,mm$), and, consequently, the latter seem to be quite legitimate candidates to cause a Rayleigh-type scattering of sound. Yet, let us look at the problem a bit closer. What does for instance our estimate for $\lambda_{\min}$ really mean? Just for simplicity, let us forget for now about the complicated shape of a real goblet and think of a rectangular box with a plane one-dimensional expansion–compression sound wave in it. For the excess of air pressure in the box we can write:

$$P_e\,(x,\,t) = P_0\,\cos\left(\frac{2\pi\,x}{\lambda} - \omega\,t\right), \qquad (9.1)$$

where P_0 is the amplitude of pressure oscillations, ω is the sound frequency, λ is the corresponding wave length and x is the coordinate along the axis of propagation.

Since even the minimal value of λ is well over the glass dimensions, $x \ll \lambda_{\min}$, the function $P_e(x, t_0)$ (called also the *pressure field* of the sound wave) at any given instant t_0 varies only slightly over the volume of the glass. Thus, the first term in (9.1) is vanishingly small and the space-time distribution of the excess pressure is determined mainly by the second term in the argument of the cosine. This actually shows that because of the negligible value of $x \ll \lambda$ the excess pressure field in the glass is almost uniform but rapidly changing:

$$P_e\left(t\right) = P_0 \cos \omega\, t. \tag{9.2}$$

Note the difference between the pressure field (9.2) and a conventional standing wave in a rigid box of length l. The resonance condition for the latter would read: $l = \frac{n\lambda}{2}$, where $n = 1, 2, \ldots$ This long sound wave simply could not fit in the glass. However walls of a real glass are elastic and take part in the oscillations of the content. Vibrations of the walls transfer sound to the ambient air making it audible.

The total pressure of liquid in the glass is the sum of $P_e\left(t\right)$ and the atmospheric pressure:

$$P_e\left(t\right) = P_{\text{atm}} + P_0 \cos \omega\, t.$$

We are now just one step away from understanding the true reason behind the observed fast damping of sound in champagne goblets. The answer is hiding in the fact that gas-saturated liquid is a so-called *nonlinear acoustic medium*. This piece of "scientific" vernacular means in reality the following. It is known that the solubility of a gas in liquid depends on pressure. The higher the pressure the greater the volume of gas that can be dissolved in a unit volume of liquid. But, as we have already established, in the presence of sound oscillations the pressure field in the glass varies. At the moments when the pressure of liquid drops below the atmospheric one, the outgassing increases. The release of gas bubbles distorts the simple harmonic time dependence of pressure. It is in this particular sense that one calls the gas-saturated liquid a nonlinear acoustic medium.[c] The outgassing inevitably takes energy from sound oscillations making them wane much faster. After

---

[c]Remember nonlinear sound distortions as the nightmare of Hi-Fi fans.

the glasses have been clinked, all kinds of sound frequencies are excited in them; then, however, due to the mechanism sketched above, high pitch-modes will die away much quicker than low-pitch ones. In the absence of the high frequencies a pure high-pitched ringing melody of crystal will turn into a miserable muffled thump.

Yet, it turns out that gas bubbles not only damp sound waves in liquid but in certain circumstances can also generate them. Indeed, it has been found that sound oscillations can be excited by wee air bubbles in water hit by a high-power laser pulse. The effect is caused by the impact of the laser beam upon the bubble surface. There the light may experience total internal reflection. After such a "hit" the bubble quivers for some time (until the vibration is damped) exciting sound waves in the surrounding medium. We may try to evaluate the frequency of these oscillations.

There are lots of important phenomena in nature, which, however different they may seem, can be described by the same equation, the equation of the harmonic oscillator. These are different kinds of oscillations – a weight bouncing on a spring, atoms vibrating in molecules and crystals, electrical charge flowing back and forth from plate to plate of the capacitor in LC-circuits, and many others. The imperative physical feature that unites all the previous examples is the presence of a "restoring force". The latter occurs if the system has been driven away from equilibrium by some external perturbation. It always tends to bring the system back to equilibrium and linearly depends on displacement. And gas bubbles vibrating in a liquid are just one more example of such oscillatory systems. Hence, we can try to use the well-known relation obtained for a mass on a spring to estimate the typical frequency of oscillations of the bubbles. Of course to do so we have to figure out what plays the role of the "coefficient of elastic force" in this case.[d]

The first candidate could be the surface tension σ of the liquid: $k_1 \sim \sigma$; at least it has the desired dimension (N/m). It seems reasonable to substitute into the formula for the oscillator's frequency the mass of the liquid involved in the oscillations instead of mass of the weight. Clearly this mass by the order of magnitude should be about the mass of water

---

[d]Coefficient of elastic force k of a spring characterizes the proportionality between the value of restoring force F and displacement x from equilibrium:

$$F = -kx.$$

expelled by the bubble. This gives the volume of the bubble times the density of the liquid: $m \sim \rho r_0^3$. So, one can write the following expression for the natural frequency of the bubble oscillations:

$$\nu_1 \sim \sqrt{\frac{k_1}{m}} \sim \frac{\sigma^{\frac{1}{2}}}{\rho^{\frac{1}{2}} r_0^{\frac{3}{2}}}. \tag{9.3}$$

However it turns out not to be the only possible solution. We have not yet utilized another important parameter, that is the air pressure inside our bubble, P_0. When multiplied by the radius of the bubble, it also gives the same (N/m) dimension of "elasticity coefficient", $k_2 \sim P_0 r_0$. And after having plugged this new coefficient into the same relation for the natural frequency of the oscillator, we obtain an entirely different value for the frequency of the bubble's vibration:

$$\nu_2 \sim \sqrt{\frac{k_2}{m}} \sim \frac{P_0^{\frac{1}{2}}}{\rho^{\frac{1}{2}} r_0}. \tag{9.4}$$

Which of these two values is the true one? It may sound surprising, yet, actually both of them are correct. They just correspond to two different types of oscillation of the air bubble. The first one represents oscillations occurring when the bubble was originally squashed, say by the laser impulse. In such a motion, the shape and, therefore, the surface area of the bubble are constantly changing, while the volume remains the same. In this process the "restoring force" is determined by the surface tension.[e] The second type, on the contrary, takes place when the bubble has been squeezed uniformly from all directions and then released. In this case it starts throbbing due to the pressure forces. Our second frequency ν_2 corresponds to radial oscillations of this kind.

Obviously the effect of laser beam impact on the bubbles is asymmetrical. Therefore, the sound waves produced by the bubbles after this type of excitation are likely to belong to the first type. Furthermore, if, for

---

[e]Notice there is a whole bunch of different kinds of oscillation, in which the bubble's volume does not change. Those vary from trivial alternated squashing and squeezing of the bubble in various directions to much more outlandish transformations when the bubble turns into something like, say, a doughnut. The frequencies of these oscillations may vary quantitatively yet by the order of magnitude all of them are determined by Eq. (9.3).

instance, the size of bubbles is known, one could determine the type of vibration from the frequency of the sound generated by the bubbles. In the experiments that we are talking about, this frequency was found to be $3 \cdot 10^4\ Hz$. Unfortunately, the dimensions of tiny air bubbles in water are hard to measure with a sufficient degree of accuracy. It is clear though, that they should be of the order of some fractions of a millimeter. After plugging $\nu_0 = 3 \cdot 10^4\ Hz$, $\sigma = 0.07\ N/m$, $P_0 = 10^5\ Pa$, $\rho = 10^3\ kg/m^3$ into the corresponding formulae one may find the characteristic dimensions of bubbles that produce sound for both types of oscillations:

$$r_1 \quad \sim \quad \frac{\sigma^{\frac{1}{3}}}{\rho^{\frac{1}{3}} \nu_0^{\frac{2}{3}}} = 0.02\ mm;$$

$$r_2 \quad \sim \quad \frac{P_0^{\frac{1}{2}}}{\rho^{\frac{1}{2}} \nu_0} = 0.3\ mm.$$

It turns out that the size of bubbles does not differ much. Obviously, the difference is not enough to determine what type of oscillations were generated in the experiment. However, the estimates of the radii of bubbles are in perfect agreement with what we would expect from our day-to-day observations. This is evidence in favor of our way of reasoning, based mainly on dimensional arguments.

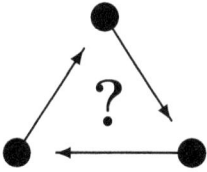

Do you have any idea why in the champagne glass the high-pitch overtones die out far quicker than the principal mode?

Chapter 10

The bubble and the droplet

A boy, with bowl and straw, sits and blows,
Filling with breath the bubbles from the bowl.
Each praises like a hymn, and each one glows;
Into the filmy beads he blows his soul.

Old man, student, boy, all these three
Out of the Maya-foam of the u' niverse
Create illusions. None is better or worse.
But in each of them the Light of Eternity
Sees its reflection, and burns more joyfully.

Hermann Hesse, *"The glass bead game."*

The numerous guises of surface tension in the natural and technological world around us are pervasive and amazingly varied. It gathers water into droplets and allows one to blow up a soap-bubble shimmering with rainbow colors or to write with an ordinary pen. Surface tension also plays a significant role in the physiology of the human body. It has been utilized in space technology too. Why, after all, does the surface of a liquid behave in the way it does, like a stretched elastic membrane?

The molecules in the narrow layer, really close to the liquid surface, could be considered as "dwelling" in very special circumstances. They happen to have neighbors, identical to them as molecules, only on one side, whereas the "inner" denizens are completely surrounded by their twin relatives that look and act identically.

Because of the attractive interaction between these closely lying molecules, the potential energy of each of them is negative. The abso-

lute value of the latter, on the other hand, could at first approximation
be premised as proportional to the number of the nearest neighbors. It is
clear then, that surface molecules, each having fewer neighbors right next
to them, must have a higher potential energy than the ones in the volume
of the liquid. Another factor raising the potential energy of the molecules
in the surface layer is that the concentration of molecules in the liquid
decreases near the surface.

Of course, molecules of liquid are moving with incessant thermal motion — some of them dive inside, leaving the surface, and others go up to
take their place. However, one can always speak of the average surplus
potential energy of the surface layer.

The reasoning above shows that in order to extract a molecule from
inside the liquid up to the surface, external forces must perform positive
work. Quantitatively this work is expressed by the surface tension σ, which
is equal to the additional potential energy of molecules occupying a unit
surface area (compared to the potential energy these molecules would have
if they remained inside).

We know that the most stable state, among all possible states of a
system, is that with the lowest potential energy. In particular liquid will
always try to assume the shape corresponding to the minimal surface energy
for the given conditions. This is the origin of surface tension, which actually
always tries to shrink the surface of liquid.

10.1   Soap-bubbles

Paraphrasing the great English physicist Lord Kelvin,[a] you can simply blow
up a soap-bubble, stare at it, study it all your life long, and still be able
to extract more lessons of physics from it. For instance, the soap film is an
excellent object for exploring the various effects of surface tension.

Gravitational forces do not play any noticeable part in the considered
case, for the soap film is very thin and, therefore, its mass is negligible. So
the protagonist here will be the surface tension force which, as we have just
shown, will try to make the surface area of the film as small as possible,
within the given circumstances, of course.

But why necessarily soap films? Why cannot we, for example, study,

---

[a]See page 53.

say, films of distilled water? Especially considering the fact that its surface tension is several times that of the aqueous soap solution (just a fancier name for the soapy water).

It turns out that the answer does not depend so much on the value of the surface tension coefficient, but lies rather in the structure of the soap film itself. Indeed, any soap is abundant with so-called surface-active agents *(surfactants)*, or long organic molecules with two ends that have a completely opposite affinity to the water. That is, while one end (called "the head") clings to water avidly, the other one ("the tail") stays completely indifferent to the water. This leads to the rather complex structure of soap film in which the soapy solution is armored by a fence made of those densely packed highly oriented layers of surface-active agents,[b] Figure 10.1.

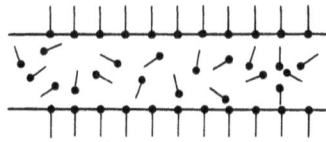

Fig. 10.1: Stability of soap film is guaranteed by presence of the surface-active organic molecules.

But for a moment, let us go back to our soap-bubbles. Most of us not only just marveled at these gorgeous creations of nature at one occasion or another, but were making them ourselves. They are so perfectly spherical in their shape and can hover in the air for so long, before rupturing finally against an obstacle. The pressure inside them appears to be higher than the atmospheric one. This additional pressure is due to the fact that the soap film of the bubble attempts to minimize the surface area and gives the air inside an extra squeeze. Moreover, the smaller the bubble radius R is, the higher the additional pressure inside. Now we shall try to find the magnitude of this addition, ΔP_{sph}.

First we conduct a so-called mental experiment. Suppose that the surface tension of the film of the bubble drops a tiny bit, and its radius increases, consequently, by a certain value, $\delta R \ll R$, Figure 10.2. This, in

---

[b]Surfactants are mainly used in order to reduce the surface tension and improve wetting properties of detergents. In the meantime they help to stabilize the film and prolong the lifetime of soap-bubbles.

turn, causes the following increase of the external surface area:

$$\delta S = 4\pi \, (R + \delta R)^2 - 4\pi \, R^2 \approx 8\pi \, R \, \delta R;$$

($S = 4\pi \, R^2$ stands for the surface area of the sphere). And, therefore, for the incremental surface energy, one can write:

$$\delta E = \sigma \, (2 \, \delta S) = 16\pi \, \sigma \, R \, \delta R, \tag{10.1}$$

(since δE is proportional to the tiny δR, the surface tension coefficient, σ, can be assumed to be constant).

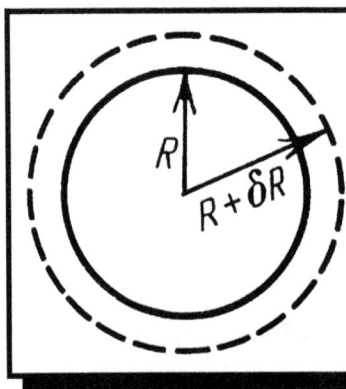

Fig. 10.2:
Infinitesimal
inflation of
soap-bubble.

By the way, notice the extra "2" factor showing in (10.1), although absent from the original definition of surface energy. That is because now we have taken into account both surfaces of the bubble, its internal surface as well as the external one; when the bubble radius grows by δR each of its surfaces stretches by an additional $8\pi \, R \, \delta R$.

This fictitious increment of surface energy is thanks to the mechanical work of the compressed air trapped inside the bubble. The pressure in the bubble remains almost the same when its volume grows by a small amount δV so one can equate this work to the enhancement of the surface:

$$\delta A_{air} = \Delta P_{sph} \, \delta V = \delta E.$$

The volume change here, on the other hand, is equal to the volume of the thin-walled spherical shell, Figure 10.2:

$$\delta V = \frac{4\pi}{3}(R + \delta R)^3 - \frac{4\pi}{3} \, R^3 \approx 4\pi \, R^2 \, \delta R,$$

entailing,

$$\delta E = 4\pi\, R^2\, \Delta P_{sph}\, \delta R.$$

Now compare this expression to the earlier established formula (10.1). This gives for the additional pressure inside the spherical soap-bubble that balances the surface tension forces:

$$\delta P_{sph} = \frac{4\sigma}{R} = 2\sigma'\, \rho; \qquad (10.2)$$

(we denoted by $\sigma' = 2\sigma$ the doubled coefficient of surface tension of the liquid). Obviously, in the case of a single curved surface (for instance, that of a spherical droplet), this additional pressure would be $\delta P_{sph} = 2\sigma\,/\,R$. This relation is called the *Laplace formula*.[c] The reciprocal of radius is conventionally called the *curvature* of the sphere: $\rho = 1/R$.

Thus we have arrived at the important conclusion that the incremental pressure is proportional to the sphere's curvature. Yet a sphere is not the only shape a soap-bubble can take. Indeed, having placed the bubble between two hoops, for example, one can easily stretch it in a cylinder crowned by round spherical "caps", on its top and bottom as in Figure 10.3.

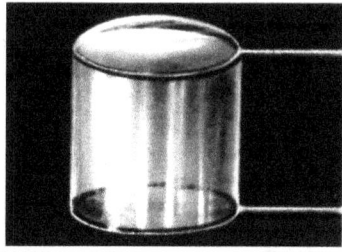

Fig. 10.3: With the help of wire frames one can make a cylindrical soap-bubble.

What will the value of the additional pressure be for such an "unorthodox" bubble? It is clear that the curvature[d] of the cylindrical surface varies

---

[c]Pierre-Simon Laplace, (1749–1827). A great French mathematician, one of that splendid constellation of French mathematicians, contemporaries of the Great French Revolution – J. L. Lagrange, L. Carnot, A. M. Legendre, G. Monge, etc. Laplace became best known for his contribution to the theory of probability and celestial mechanics, as well as for the famous quote from Napoleon that he "carried the spirit of the infinitely small into the management of affairs", when the scholar had miserably failed in his short-lived assignment as "Minister of the Interior".

[d]What is the curvature of a plane (two-dimensional) curve? For a circumference it is defined in the same way as for the sphere: $\rho = 1\,/\,R$, where R is the radius. Each

in different directions: it is zero along the generating line (for cylinders that are straight), however for the cross section perpendicular to the axis, the curvature equals $1/R$, where R is the radius of the cylinder. Well, then what value of ρ should we substitute into the previously derived formula? It turns out that the difference in pressures on the two opposite sides of an arbitrary surface is defined by the average curvature. Let us try to figure out what it will be for our right cylinder.

First, erect a normal[e] to the cylinder surface at a point A, and then construct a set of planes passing through the normal. The resulting cross sections of the cylinder by these planes, (called the *normal sections*), can be circular, elliptical or even degenerate into two parallel straight lines, (Figure 10.4). Surely, their curvature at the given point is different: it is maximal for the circle and minimal (nil actually) for the longitudinal section. The average curvature is defined then as the half sum of minimal and maximal values of the curvature of the normal sections at this point:

$$\bar{\rho} = \frac{\rho_{max} + \rho_{min}}{2}.$$

This definition applies not only to a cylinder. In principle, the average curvature at a given point can always be calculated in this manner.

For the lateral surface of a cylinder, its maximal curvature at any point is $\rho_{max} = 1/R$, where R is the radius of the cylinder and the minimal value $\rho_{min} = 0$. Thus, the average curvature of the cylinder is $\bar{\rho} = 1/2R$, and the additional pressure inside the cylindrical bubble is:

$$\Delta P_{cyl} = \frac{\sigma'}{R}.$$

So it turns out that the additional pressure in the cylindrical bubble is equal to that of a spherical bubble with double the radius. That makes the radii of the spherical caps of such a cylindrical bubble twice the radius of the cylinder itself. Hence the caps are just spherical segments rather than the full hemispheres.

And what would happen if one eliminates the additional pressure in the bubble completely by, say, pricking its caps? The first solution popping to

---

tiny piece of any other curve can, just the same, be considered as an arc of a certain radius. The reciprocal of this radius is called the curvature of the plane curve at the given point.
[e]That is, the line perpendicular to the tangential plane at the point A.

mind would be that, since there is no additional pressure, the surface should not have any curvature at all. Yet surprisingly, the walls of the cylinder are actually bending inside taking the shape of a *catenoid* (from the Latin *catena* for "chain"). This shape can be generated by rotating the so-called *catenary line,* around its *X*-axis.[f] So what is the matter here?

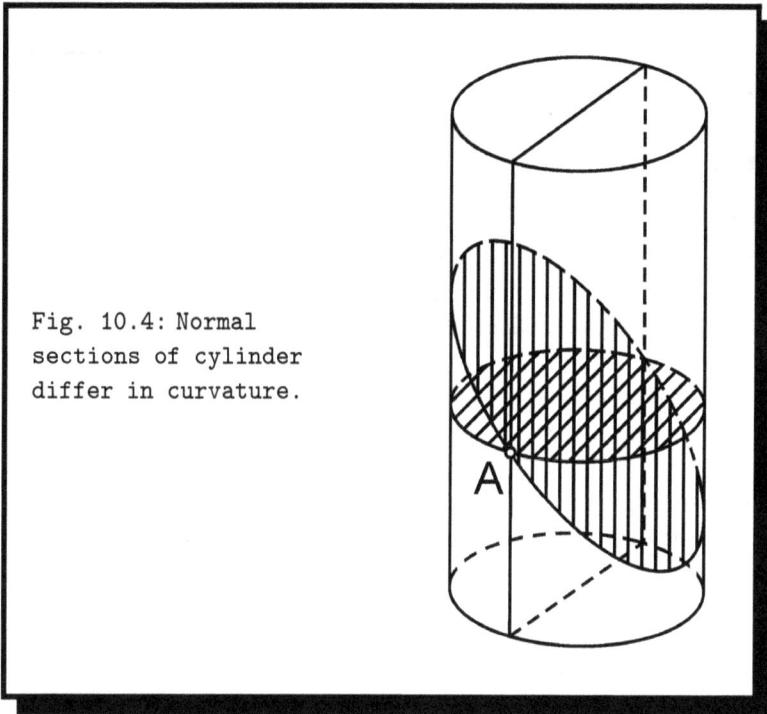

Fig. 10.4: Normal
sections of cylinder
differ in curvature.

Let us examine this surface closer, in Figure 10.5. It is easy to notice that its narrowest, or waist, portion, also called the *saddle,* is concave as well as convex. Its cross section, perpendicular to the rotation axis, is a circle; on the other hand, dissecting along the axis gives, by definition, the

[f]Catenary is the curve formed by a perfectly flexible uniform chain suspended by the endpoints. The form of the curve (up to similarity transformations $a \rightarrow \alpha a$) is given by the equation:
$$y = \frac{a}{2}(e^{\frac{x}{a}} + e^{-\frac{x}{a}}).$$

catenary. The inward curvature should raise pressure inside the bubble, but the opposite curvature would lower it. (Pressure under a concave surface is higher than the pressure above it.) In the case of catenoid, the two curvatures are equal in magnitude but have the opposite directions, therefore, negating each other. The average curvature of this surface is zero. Hence, there is no additional pressure inside such a bubble.

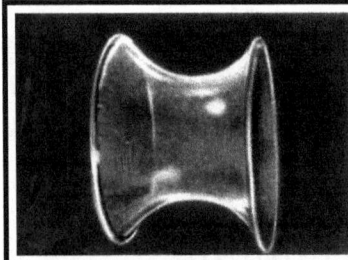

Fig. 10.5: Left to itself the soap film takes the form of catenoid. This surface has zero average curvature.

A catenoid is not unique though, and there are a bunch of other surfaces, seemingly "badly" curved in all possible directions, yet having their average curvature equal to zero, and consequently, not exerting any extra pressure. To generate these surfaces, it is enough to immerse a wire frame into soapy water. While lifting the frame back from the solution, one can immediately see various surfaces of zero curvature, formed depending on the frame shape. However, a catenoid is the only surface of revolution[g] (besides the plane, of course) with zero curvature. Surfaces of zero curvature bounded by a given closed curve may be found with the help of methods of a special branch of mathematics, called *differential geometry*. An exact mathematical theorem claims that surfaces of zero curvature have the minimal area among all the surfaces with the same boundary — a statement which seems pretty natural and obvious for us now.

A plethora of combined soap-bubbles makes froth. In spite of seeming disorder there is an indisputable rule held in the formation of soap films in a foam: the films intersect one another only at equal angles, (Figure 10.6). Indeed, look, for example, at the two joint bubbles partitioned by the common wall, (Figure 10.7). The additional (with respect to atmospheric) pressures inside the bubbles will be different. According to the Laplace

---

[g]That is the surface which may be generated by rotation of a curve.

formula, (10.2):

$$\Delta P_1 = \frac{2\sigma'}{R_1}, \qquad \Delta P_2 = \frac{2\sigma'}{R_2}. \tag{10.3}$$

So the common wall must be bent in order to compensate the pressure difference in the bubbles. Its radius of curvature, hence, is determined by the difference $\Delta P_2 - \Delta P_1$ from Eq. (10.3):

$$\frac{2\sigma'}{R_3} = \frac{2\sigma'}{R_2} - \frac{2\sigma'}{R_1},$$

which gives after regrouping

$$R_3 = \frac{R_1 R_2}{R_1 - R_2}.$$

Again, Figure 10.7 depicts a cross section of these two bubbles by a plane passing through their centers. The points A and B here mark the intersections of the plane of the picture with the circumference where the bubbles are touching. At any point on this circumference there are three films coming together. As long as their surface tension is the same, the tension forces can balance each other only if the angles between the crossing surfaces are the same. Therefore, each of them is equal to 120°.

10.2   On different kinds of droplets

We now discuss the shapes of droplets. Here things become a bit more complicated. Now the surface tension which, as always, tries to minimize the area of the surface is counteracted by other forces. For example, a liquid droplet is almost never spherical, although it is a sphere that has the smallest surface area among all shapes for a given volume. When sitting on a flat steady surface, droplets look rather squashed; when in free fall, their shapes are even more complex; only in the absence of gravity in space they finally assume the form of perfect spheres.

   The Belgian scientist J. Plateau,[h] in the middle of the nineteenth century, was the first to come up with a successful solution of how to eliminate the effects of gravitation when studying the surface tension of liquids.

---

[h] Joseph Antoine Ferdinand Plateau, (1801–1883), Belgian physicist; worked in the fields of physiological optics, molecular physics, surface tension. Plateau was the first to put forward the idea of a stroboscope.

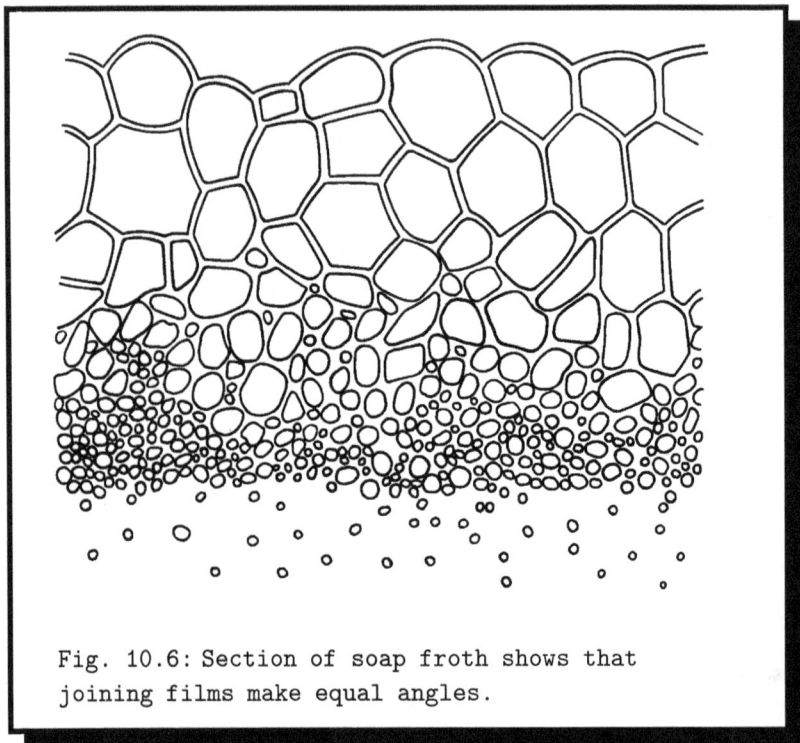

Fig. 10.6: Section of soap froth shows that
joining films make equal angles.

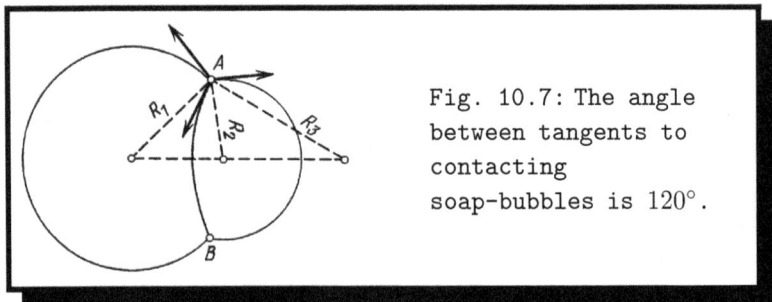

Fig. 10.7: The angle
between tangents to
contacting
soap-bubbles is 120°.

Surely, in those days researchers did not even dream of having genuine
weightlessness, and yet J. Plateau simply suggested compensating the gravitational forces with the Archimedean buoyancy force. He submerged his
subject liquid (oil) in a solution with the exact same density, and, as his

biographer tells us, was utterly surprised to see that the oil droplet had developed a spherical shape. He used his golden rule to "become surprised at the right time," and then experimented and contemplated upon this peculiar phenomenon for a long while.

He had his method to study a variety of entailing effects. For instance, he meticulously investigated the process of droplet formation at the end of a tube.

Normally, no matter how slow a droplet is being made, it separates from the tip of the tube so fast that the human eye cannot follow the details of this event. So Plateau had to dip the tip of the tube he was using into a liquid, with a density just slightly less than that of the droplets themselves. The gravitational force influence was, by doing so, substantially diminished and as a result, really large droplets could be formed and the process of their taking off from the end of the tube could be clearly seen.

In Figure 10.8 you see the different stages droplets undergo in their formation and separation (of course, these pictures were taken using the modern high-speed film technique). Let us try now to explain the observed sequence. During the slow growth stage, the droplet is in equilibrium at each particular instant. For a given volume, the droplet shape is determined by the condition of a minimal sum of its surface and potential energies, the latter of which is the result of gravitational forces, of course. The surface tension is trying to shape droplets spherically, whereas gravity, on the contrary, tends to situate the droplet's center of mass as low as possible. The interplay of these two yields the resulting vertically stretched form (the first shot).

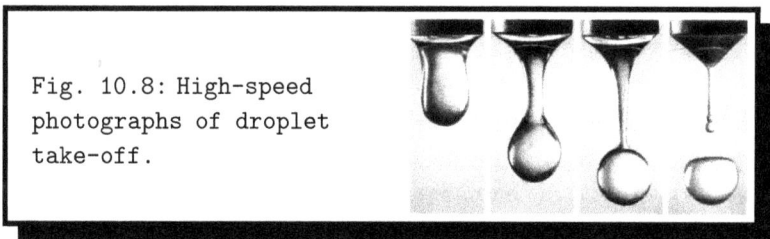

Fig. 10.8: High-speed photographs of droplet take-off.

As the droplet continues to grow gravitation becomes more prominent. Now most of the mass accumulates in the lower part of the droplet, and the droplet begins to develop a characteristic neck (the second shot of Fig 10.8). The surface tension forces are directed vertically along the tangent to the neck, and for some time they manage to balance the droplet's weight. Not for long, however, and at a certain moment just a slight increase in the droplet's size is enough for gravity to overtake the surface tension and to break this balance. The droplet's neck then narrows promptly (the third shot), and off the droplet finally goes (the fourth shot). During this last stage, an additional tiny droplet forms on the neck and follows the big "maternal" one. This secondary little droplet (called Plateau's bead) is always there, however, because of the extreme swiftness of the droplet leaving the tap, we basically never notice it.

We will not be going into the details of the formation of Plateau's bead as it is a pretty complicated physical phenomenon. We will instead try to find an explanation of the observed shape of the primary droplets in their free fall. Instant photographs of the falling droplets show clearly the almost spherical shape of the little secondary drops, while the big primary ones look rather flat, something like a bun. Let us estimate the radius at which the droplets start losing their spherical shape.

When a droplet is moving uniformly (at a constant speed), the force of gravity acting, say, on the narrow central cylinder AB of the droplet, Figure 10.9, must be balanced by forces of surface tension. And this automatically requires that the radii of curvature of the droplet at A and B should differ. Indeed, the surface tension produces the additional pressure, defined by the Laplace formula: $\Delta P_L = \sigma' / R$, and, if the curvature of the droplet surface at the point A is greater than that at the point B, then the difference of these Laplace pressures could compensate for the hydrostatic pressure of liquid:

$$\rho g h = \frac{\sigma'}{R_A} - \frac{\sigma'}{R_B}.$$

We can check how much R_A and R_B should really differ to satisfy the previous relation. For tiny little droplets with radii of about $1\,\mu$ ($10^{-6}\,m$), the value of $\rho g h$ is of the order of $10^{-2}\,Pa$, whereas $\Delta P_L = \sigma' / R \approx 1.6 \cdot 10^5\,Pa$! So, in this case, the hydrostatic pressure is so small when compared to the Laplacian that one can safely disregard it, and the resulting droplet will be very close to an ideal sphere.

But it is a completely different story for a drop of, say, $4\,mm$ radius. Then the hydrostatic pressure is $\rho g h \approx 80\,Pa$, but the Laplacian one is $\Delta P_L = 78\,Pa$. These values are of the same order of magnitude, and, consequently, the deviation of such a droplet from the ideal spherical shape

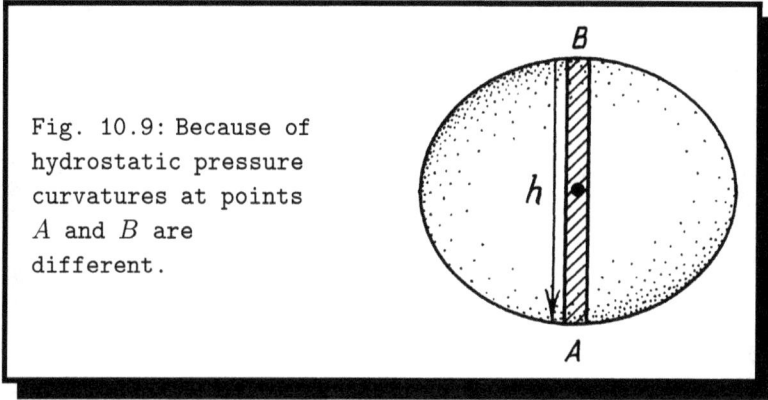

Fig. 10.9: Because of hydrostatic pressure curvatures at points A and B are different.

becomes quite noticeable. Assuming $R_B = R_A + \delta R$ and $R_A + R_B = h = 4\,mm$, one finds $\delta R \sim h \left[\sqrt{\left(\frac{\Delta P_L}{\rho g h}\right)^2 + 1} - \frac{\Delta P_L}{\rho g h} \right] \approx 1,6\,mm$, and the difference of the radii of curvature at A and B now turns out to be of the same order as the size of the droplet itself.

The previous calculations show us which droplets should deviate substantially from the sphere. However, the predicted asymmetry turns out to be the opposite of that observed in the experiment (indeed, real droplets in Figure 10.10 are flattened from the bottom!). What is the matter here? Well, the thing is that we assumed the air pressure to be the same over and under the droplet. And this is certainly true for slow moving drops. But when the speed of the droplet is sufficiently high, the surrounding air does not have enough time to smoothly flow around. A region of pressure appears before the droplet and an area of lower pressure forms right behind it (where real turbulent vortices are formed). The difference of the front and back pressures can actually exceed the hydrostatic pressure, and the Laplace pressure now must compensate for this difference. In such circumstances, the value of $\frac{\sigma'}{R_A} - \frac{\sigma'}{R_B}$ turns negative, meaning that R_A is now greater than R_B. That (to our final satisfaction) is exactly what has been seen in the experimental pictures.

And finally, a short quiz about the giants and whoppers. Have you ever seen them among droplets? Not many. They simply do not survive under normal circumstances. And for good reason: droplets of large radii turn out to be unstable and spatter into a bunch of little ones almost

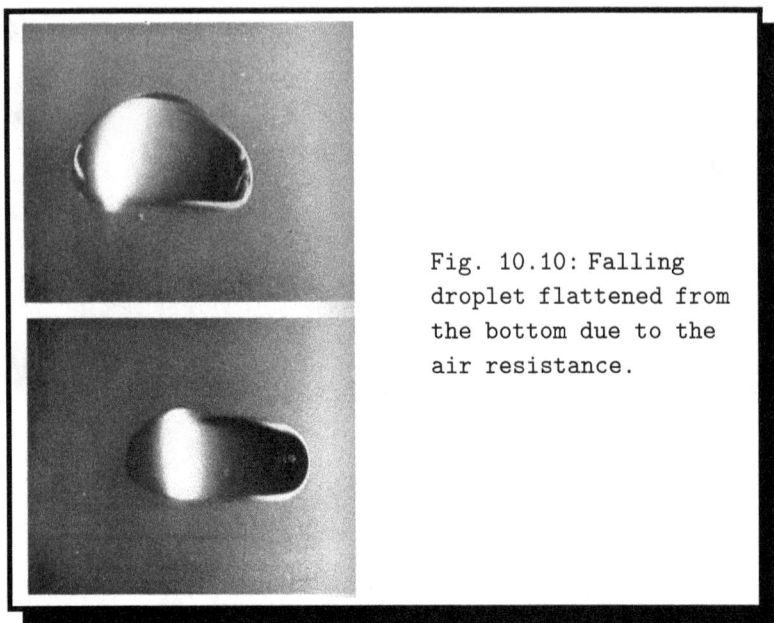

Fig. 10.10: Falling
droplet flattened from
the bottom due to the
air resistance.

instantly. It is the surface tension that assures longevity of a droplet on a
hydrophobic[i] surface. Yet once the hydrostatic pressure becomes greater
than the Laplacian, the droplet spreads over the surface and breaks into
smaller ones. One can use the following relation to estimate the maximal
radius of a still stable droplet: $\rho g h \gg \frac{\sigma'}{R_A}$, where $h \sim R$. From which, one
can find:

$$R_{max} \sim \sqrt{\frac{\sigma'}{\rho g}}.$$

For water, for instance, $R_{max} \approx 0.3\,cm$ (of course, this is just an order
of magnitude estimate of the maximal size of droplets). This is why we
never see, for example, really gigantic droplets on the leaves of trees or
other unwettable surfaces.

---

[i]In chemistry, *hydrophobicity* is the physical property of a molecule (known as a hy-
drophobe) that is seemingly repelled from a mass of water. (Strictly speaking, there is
no repulsive force involved; it is an absence of attraction). In contrast, *hydrophiles* are
attracted to water.

Chapter 11

The mysteries of the magic lamp

... "Symmetriads" appear spontaneously. Their birth resembles eruption. All of a sudden, the ocean starts coruscating as if tens of square kilometers of its surface were covered with glass. A while later, this glassy envelope pops up and bursts outwards in a shape of a monstrous bubble, in which, distorted and refracted, arise the reflected images of the whole firmament, the sun, clouds, the horizon. . .

Stanislaw Lem, *"Solaris."*

The series of pictures, shown on the cover, have not been taken either on Solaris, or from a spaceship diving in the recondite abyss of Jupiter's atmosphere. Neither have they been taken from the window of a bathyscaphe having dared to approach an erupting underwater volcano. Not even close. They are just photographs of a working Lava-lamp, a gadget anyone could find without much trouble, for example, in a "hands-on" toy shop or sometimes in a big department store. And yet, it turns out that this seemingly simple device conceals plenty of beautiful and subtle physical phenomena.

The design of the lantern is not very complicated. It consists of a cylinder with transparent walls, in the base of which under its glass bottom a regular electrical bulb is mounted. The glass in the lower part is covered with a multicolor light filter, and a coiled metal wire is wrapped around the bottom perimeter as in Figure 11.1. One sixth of the cylinder is filled with a wax-like substance (which we will call from now on *substance A*), and the rest of its volume is filled with a transparent liquid (say, *liquid B*). The particular criteria for choosing these substances, as well as their properties, will be discussed a bit later, when we will be closely studying the physical processes taking place in the lamp.

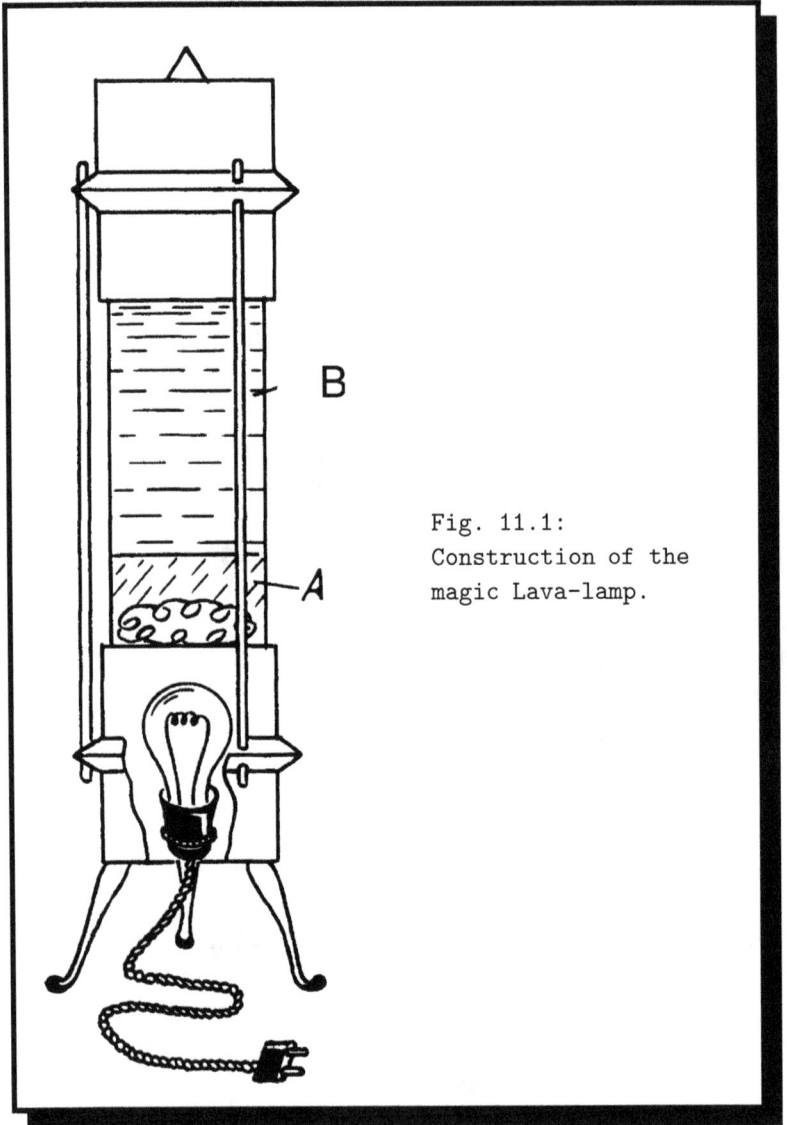

Fig. 11.1:
Construction of the
magic Lava-lamp.

It appears more convenient to conduct observations of the Lava-lamp in the dark, having it as the only source of light. So, let us turn it on and prepare to wait. As we will see, the events taking place inside the lamp could be separated into several stages. We will call the first one *"the phase of rest and accumulating strength"*.

The substance A is amorphous and, therefore, does not have a strict, well-ordered internal structure.[a] As its temperature goes up, it becomes more and more malleable, softens, and gradually turns liquid. It is worth recalling at this time a principal difference between the crystalline and amorphous substances. For the former ones, this solid-to-liquid transition (melting, in ordinary words) happens at a certain temperature point and requires a particular amount of energy (*the heat of melting*), which is expended for breaking the material's crystal structure. On the other hand, solid and liquid states of an amorphous substance are not critically different. When the temperature is rising, amorphous materials simply soften and become liquid-like.

When turned on, the bulb of the lantern, illuminating the cylinder from the underneath through the color filter with a kind of red-green glow, serves also as the heater. In the bottom floor, close to the bulb, there consequently develops a "hot spot" (the area of elevated temperature). Substance A in this hot region becomes softer, whereas the same time neither the upper crust of A, nor the liquid B have had enough time to warm up remaining relatively cold. As the larger part of A softens, the top solid crust becomes thinner. At the same time, due to thermal expansion, the volume of the lower, now liquid, part of A tends to increase, raising the pressure underneath the crust. Eventually, A finally breaks the crust and lunges bubbling upwards. It is like a tiny volcano is born. The quiescent phase of "rest and accumulating" is over and the new period of "volcanic activity" kicks in (see top-right figure on the cover).

The substances A and B are chosen in such a way that the density of the warmed A, rushing up from the crack, is slightly lower than that of the still rather cold B, causing the new portions of A, successively leaving the rift, to surface one after another.[b] On their way up, these pieces start cooling

---

[a] We will dwell on the difference between crystalline and amorphous substances later in Chapter 20.

[b] This resembles the famous experiment, in which a droplet of aniline, that at first peacefully rests at the bottom of a tall glass cylinder with water, immediately starts up to the surface as soon as the temperature reaches about $70°$ C and the aniline's density becomes less than that of water.

down and, when reaching surface, become solid again, assuming various and quite peculiar shapes. Yet their density is back to the initial value, which is higher than the density of B, and these smithereens begin to sink slowly. Some of them, usually the smaller ones, however, continue to hover near the surface for a long time. And the reason for this recalcitrant behavior is, of course, our old acquaintance, surface tension. Indeed, A and B are chosen so that the B-liquid does not wet the A-solid. Hence the surface tension force acting on fragments of A is directed up, trying to push them out from the liquid. It is exactly the same reason why, for example, the water-striders can freely stay (and quite audaciously run) on the surface, or an oiled metal needle does not sink.

Meanwhile, the excessive pressure in the lower part of the cylinder, under the crust, has been relieved, the edges of the crack have become molten, and new portions of melted A are continuously trickling out of the crater. However, now they do not sever in the form of bubbles, instead they stretch leisurely as an extended upward narrow stream. The outer surface of this stream, in contact with the cold B, quickly cools down and stiffens, producing a sort of trunk. And if one tries to look through this trunk, one may well be surprised, for the trunk turns out to be a hollow, narrow walled tube, filled with liquid B. The explanation would be that when the stream of melted A leaves the crater and runs upwards, at some point it simply does not have enough material to continue growing. Then, pressure inside the trunk decreases and, resulting from that, a crack develops somewhere in the junction between the crater edge and the trunk, and then, sure enough, the cold liquid B starts pouring into the cleft. The top of the A tube in the meantime keeps going up, and the liquid B fills the tube inside, cooling down and shaping the tube's inner walls, finally causing them to completely solidify.

As the vine of the volcanic plant makes its way to the surface, on the bottom of the lantern the melting continues and the next ball of the "hot" liquid A leaves the crater. It goes up, but now it goes up inside of the developed tube, and when it gets to the top of the tunnel, the ball, still being warm enough, extends the tube by another increment. So the plant keeps on growing, adding one by one these successive blocks (see second-top figure on the cover). Soon, shoving off the crusty chips of the preceding "volcanic activity", there starts protruding another stem near the first one, and maybe another, after. These esoteric underwater plants swirl and intertwine, like exotic shoots of jungle verdure, among the falling rocks, continuing at leisure to descend from the surface. The picture halts for a time. We could name this stage *the phase of the rocky forest.*

If at this point one turns the lantern off, the petrified thicket will remain there "forever" and the lamp will not be able to return to its original, clearly segregated two-phase state. Yet, surprisingly, after the kaleidoscope of the described events, we have not reached the working regime of our magic lamp. So let us keep it on and continue to watch.

In time, the liquid B is still warming up, and the boulders resting on the floor are starting to melt again, the tangled vines of the magnificent plants are wilting down. An interesting fact: there are no really squeezed shapes among the droplets formed from the liquefied rocks. They all are turning out quite spherical. Under normal conditions, the force squeezing water droplets on a hydrophobic surface is their weight. And it is balanced by the force of surface tension, tending to make the drops ideally spherical, for the sphere has the minimal surface area for a given volume. In the Lava-lamp's flask, besides gravity and the surface tension, there is an Archimedean buoyancy force also acting on the droplets, and, because of the closeness of the densities of A and B, almost equalizing the gravitational force. So the droplets happen to be in a kind of nearly weightless situation, with nothing to preclude them from donning their predestined round shape (we have already discussed that topic in Chapter 10).

For a single droplet, in the absence of gravity, the ideal spherical form is the most energetically favorable one. For two, or several drops, touching each other, from the same logic, it would be more beneficial to merge into one, because the surface area of a single large ball is less than the total area of surfaces of several smaller ones, of the same aggregate mass (we will allow the reader to check this statement on their own). However, looking at a Lava-lamp, one notices that those almost spherical droplets of A tend to linger together without actually coupling. It seems especially striking if one remembers how promptly, almost momentarily, mercury or water droplets, for example, couple on an unwettable surface. What determines the time it takes for, say, two droplets to join?

Interestingly enough, this question attracted the attention of different researchers and engineers for a long time. Not only from the point of simple scientific curiosity, but also because of its critical importance for under-standing physical processes in some very practical areas. For instance, in powder metallurgy, where the preliminary metals are powdered into grains, pressed and *baked* together to produce new alloys with desired physical properties. Back in 1944, the bright Soviet physicist Yakov Frenkel[c] had proposed a simple and quite useful model of such a merging process in his pioneering work, which became fundamental in establishing the theoretical basis for this important branch of modern metallurgical technology. And now we are going to use the underlining idea of his work to estimate the time it would take for two droplets in the magic Lava-lamp to couple.

Let us consider two identical droplets in close proximity, so that they start touching each other. At the point of their contact a connecting isthmus develops, Figure 11.2, which continuously grows as the two droplets merge. We will use the energy considerations (for it is the simplest and the shortest way) to estimate the coupling time. The total energy available for the system of the two drops ΔE_s results from the difference between the surface energies of its initial and final states, that is the summed surface energy of the two separate droplets of radii r_0 and the big "unified" droplet of the radius r:

$$\Delta E_s = 8\pi\,\sigma\,r_0^2 - 4\pi\,\sigma\,r^2.$$

Since after merging the total volume of the drops does not change, one can write the following equality: $\frac{4\pi}{3}r^3 = 2 \cdot \frac{4\pi}{3}r_0^3$, and find $r = r_0\sqrt[3]{2}$, so that

$$\Delta E_s = 4\pi\,\sigma\,(2 - 2^{\frac{2}{3}})\,r_0^2. \tag{11.1}$$

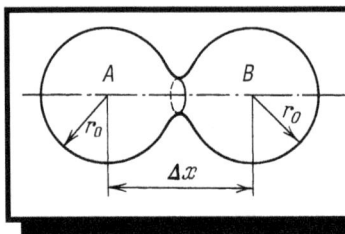

Fig. 11.2: Initial stage of the droplet coalescence.

According to the Frenkel idea, this additional energy is spent for work against forces of liquid friction, appearing in the process of redistributing droplet material as well as the surrounding medium, during the droplet merging. We can estimate this work within an order of magnitude. To find

[c]Ya. I. Frenkel, (1894–1952), specialist in solid-state physics, physics of liquids, nuclear physics, etc.

the liquid friction force, we will apply the famous Stokes' formula[d] for a spherical ball of radius R moving with velocity $\vec{v}$ in a liquid of viscosity η: $\vec{F} = -6\pi\,\eta\,R\vec{v}$. We suppose further that the viscosity of the droplet material is significantly higher than that of the surrounding liquid $\mathcal{B}$, which allows us to leave η_A in the Stokes' expression as the only viscosity coefficient.[e] Also, we may plug r_0 in the place of R. And then, noticing that the same quantity Δx characterizes the scale of the mutual displacement of the droplets when they merge, one can assume that $\Delta x \sim r_0$. So finally, one could write for the work of the liquid friction forces:

$$\Delta A \sim 6\pi\,\eta_A\,r_0^2\,v.$$

From this expression, it is clear that the faster the droplets are merging the greater is the amount of energy required (because the liquid friction force increases with speed). However, the available energy resource is limited by ΔE_s, (11.1). So these two relations will provide us with the required merging time τ_F (called the *Frenkel time* of merging). Assuming $v \sim r_0\,/\,\tau_F$ to be the process speed, we find:

$$\Delta A \sim \frac{6\pi\,\eta_A\,r_0^3}{\tau_F} \sim 4\pi\,\sigma\,(2 - 2^{\frac{2}{3}})\,r_0^2,$$

and finally:

$$\tau_F \sim \frac{r_0\,\eta_A}{\sigma}. \tag{11.2}$$

For water droplets of, say, $r_0 \sim 1\,cm$, $\sigma \sim 0.1\,N/m$ and $\eta \sim 10^{-3}\,kg/(m \cdot s)$ this time turns out to be around just evanescent, $\tau_F \sim 10^{-4}\,s$. Yet, for example, for the much more viscous glycerin ($\sigma_{gl} \sim 0.01\,N/m$, $\eta_{gl} \sim 1\,kg/(m \cdot s)$ at $20°\,C$), the corresponding time is $\sim 1\,s$, proving the fact that for different liquids, depending on their viscosity and surface tension coefficient, τ_F can vary within a rather broad range.

---

[d]George Gabriel Stokes, (1819–1903), renowned British physicist and mathematician, mostly famous for the theorem and the formula, both having commemorated his name.
[e]The Stokes' expression was derived for a different situation, when a spherical body moved in viscous liquid. However, it is pretty obvious that in the case of two merging droplets, the liquid friction force can only depend on viscosity, droplet size, and the speed of the process. Hence, from the dimensional considerations, the Stokes' formula turns out to be the only combination of these three physical quantities with the dimension of force (and we do not care about the exact proportionality coefficient, for our estimate is within an order of magnitude only).

It is worth emphasizing at this point that even for the same liquid, due to the strong temperature dependence of viscosity, Frenkel time can vary quite a bit. Going back to glycerin, for instance, its viscosity drops 2.5 times when the temperature rises from 20° to 30° C. The surface tension coefficient, on the other hand, stays pretty indifferent to the temperature variation — in the considered temperature range, σ_{gl} does not change by more than a couple of percent. This allows us to safely assume that the temperature dependence for Frenkel time is purely determined by the viscosity temperature dependence.

Now, let us look again at the balls of $\mathcal{A}$ still lying peacefully on the floor of the lamp through the derived estimation for the Frenkel merging time. As long as the liquid $\mathcal{B}$ stays rather cold, $\mathcal{A}$'s viscosity remains high, and it will prevent the balls from rushing to merge. It is the same reason that two touching wax balls do not couple into one at room temperature. Although, if one heats them enough, the viscosity of the wax plummets and the balls merge expeditiously. One more thing that plays an important role in the process is the state of the surfaces of the potential partners — if they are rough and contaminated it is difficult for the initial bridge to develop.

The merging of the $\mathcal{A}$ droplets is absolutely critical for the lantern's working cycle to go on. This explains the presence of a special means, in order to facilitate this redistribution of $\mathcal{A}$, from numerous drops into a uniform, melted mass. It is the metal coil, wired along the lantern's bottom perimeter. This coil is warmed up by now, and when approaching and touching it, the droplets receive that needed heat, lowering their viscosity and, by doing so, warming greatly their desire to join back in the prime body of the liquefied $\mathcal{A}$. Soon, after all droplets finally disappear into the maternal mass, one, unified liquid phase of $\mathcal{A}$ remains in the bottom of the lantern cylinder. And, because it is still continuously being heated, the liquid $\mathcal{A}$ cannot remain motionless. A new stage of the Lava-lamp life commences. We will call it *the phase of protuberances*.

Formed in the surface layers of $\mathcal{A}$, such protuberances languidly take off for their upward trek to the surface of $\mathcal{B}$, pulled, of course, by the buoyancy force, and gradually assuming, as they go, an increasingly spherical shape (see middle figure on the cover). Having reached the upper layers of $\mathcal{B}$ (where $\mathcal{B}$, due to its low thermal conductivity, still remains cold), the protuberances cool down a little, nerveless, remaining liquid this time, and begin to drown slowly, landing on the $\mathcal{A}$'s slightly popped up surface. Because of their relatively high viscosity, it is quite difficult for them to dive into the $\mathcal{A}$ medium right away. So they bounce on its surface for a while, drifting to the periphery, where the "surgical" metal coil opens up their surface and they end their life cycle exactly where they started it.

The bulb in the cylinder base keeps heating the system, creating new protuberances. As the temperature continuously rises, the rate of their birth goes up as well. When taking off from the $\mathcal{A}$ surface, protuberances leave behind them smaller droplets,[f] which kind of freeze perplexed in space, hesitating whether they really should go up into the unknown following their parent, or maybe just return to safety, into the original medium. In time, a dozen of such orphaned liquid balls are hovering in the cylinder, some of which do finally dare to continue upwards, whereas the coy ones descend back (see second last figure on the cover): a new stage of *collisions and calamities* emerges. And this turns out to be the longest and the most impressive phase of the lantern's activity.

The spheres are colliding, veering in various directions, however managing to avoid merging in the process. It seems like it would be advantageous, energetically, for the striking drops to couple (for the same reason we have mentioned just some paragraphs previously). Yet, once again, they happen to run into the time problem. The duration of collision t, is all they have, and if τ_F turns out to be much longer than t, then there is not enough time for the droplets to join and having collided, they will just simply bounce apart. Let us try to give an estimate of the collision time. Most of the collisions in the lamp are glancing ones (Figure 11.3), during which the soft liquid balls slightly deform and slide along each other. The characteristic time of such an encounter must be about $t \sim r_0 / v$. The velocity of the balls flowing in $\mathcal{B}$, v, is just several centimeters per second, and the ball radii are of the order of a couple of centimeters too. It makes $t \sim 1\,s$, which is too short for them to unite, leaving then no other option except to continue roaming aloofly and unattached in the lamp cylinder. At times they loiter by the bottom, then they wander though the bulk of $\mathcal{B}$, colliding with each other, yet not merging.

This *"phase of collisions and calamities"* can go on for hours. The manual usually recommends switching the lantern off after 5–7 hours of operating. But under certain circumstances, when the temperature of the ambient air is sufficiently high (say, you happen to marvel at the magic gadget on a sultry Austin or squashingly hot Tucson summer evening),

---

[f]By the way, these are the same Plateau balls that have been mentioned in Chapter 10.

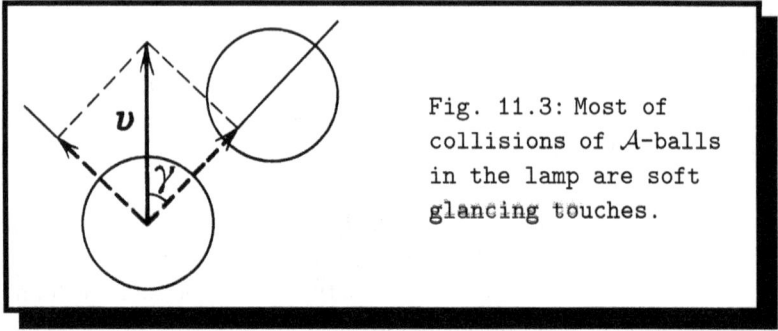

Fig. 11.3: Most of collisions of $\mathcal{A}$-balls in the lamp are soft glancing touches.

the described collision stage turns out not to be the last one. Finally, after a stationary temperature distribution along the cylinder height has been attained (and the whole liquid $\mathcal{B}$ has warmed up), the densities of $\mathcal{A}$ and $\mathcal{B}$ become practically the same, and the entire $\mathcal{A}$ congregates into a single gargantuan ball. At first, the whopper is hanging in the bottom part, bouncing at times against the cylinder walls. Then, because of these "colder wall" contacts, it cools down a little, becomes a bit denser, and, consequently, sinks down to the floor. After touching the bottom though, the ball gains an extra portion of heat, its density drops again, and it returns to its previous position, where it stays until it cools down again. Then the cycle starts all over. This phase, unmentioned in the lantern's instructions, we could name *the super-ball time* (see bottom-right figure on the cover).

Finally, after we, along with our magic Lava-lamp, have gone through the numerous stages of its work, gaining some understanding of the mechanism of the described processes, let us take a concluding look at these phenomena in a general manner. The first question coming to mind is why do these successive, often repeating, events of the birth, life, and death of the spheres occur at all? It is clear, from all our previous discourse, that the driving force behind the processes is the temperature difference between the top and bottom ends of the lantern (in thermodynamical terms, between the the *heat sink and* heat source). If one supposes that the flux of heat propagates in the system only because of the heat conduction by the $\mathcal{B}$-liquid, the temperature of $\mathcal{B}$ will be simply changing gradually along the height, and nothing unusually amusing would happen. The birth of spheres, on the other hand, as well as ordinary convection, is a consequence of instabilities which sometimes develop in systems, where a thermal flow due to the variation of temperature along the boundary occurs. The study of the behavior and properties of these systems is the subject of the branch of physics called *Synergetics*.

Chapter 12

The water mic or about one invention of Alexander Bell's

A pool among the rock
If there were the sound of water only
Not the cicada
And dry grass singing
But sound of water over a rock
Where the hermit-thrush sings in the pine trees
Drip drop drip drop drop drop drop
But there is no water

T. S. Eliot. *"The Waste Land."*

These days, everybody knows what the microphone is. Right? We see it on TV: those nifty pin-looking things clinging to the anchorman's lapels, or the old-fashioned, a ball on a handle type that reporters stick into people's faces in feverish anticipation of a story. Radio-interviewers often ask their guests to speak directly into their microphone, hinting to us the presence of the latter; and movie makers, no matter how sophisticated and outlandish their sound effects are contrived to be, end up registering them with some kind of a microphone. One can easily buy a decent mic in any home electronics store and use it with their tape or CD recorder, computer or telephone. The design of the gadget is described in most of today's high school physics textbooks. However, we can assure our revered reader that there are only a few who are aware of the existence of the so-called water microphone. Do not be surprised. Indeed it turns out that one can rather efficiently amplify different sounds with the help of a simple water stream. The device employing such a principle of sound amplification was invented by the American engineer Alexander Graham Bell, who is mostly known as one of the inventors of another gadget we cannot imagine our daily life without – the telephone.[a]

But first, let us pay attention to that "water stream" amplifier thing.

If there is a hole, say a little round orifice, drilled in the bottom of a reservoir of water, one may notice that the stream flowing downwards through the hole consists of two parts, differing in properties. The upper is transparent and steady, looking as if made of glass; yet as it goes further from the outtake, the stream becomes thinner and finally at the point of the minimal cross section, it turns into the second part, which is rather opaque and jittery. At first glance, it still looks continuous, without interruptions, like the upper region. However, it is possible at times to swiftly pass one's finger across this part of the stream without getting wet. The French physicist Felix Savart[b] after having meticulously investigated properties of liquid streams, arrived at the conclusion that at the narrowest point of the stream it breaks its continuity and splits into a series of separate droplets. Today, over a century after the discovery, one can easily prove this by taking photographs of the trickle with a flash, or by looking at the stream in stroboscopic lighting, see Figure 12.1; in Savart's day, however, researchers had to study the trickles in the dark, observing them with light from electrical sparks.

Look at the momentary image of the lower portion of the stream, Figure 12.1. It is composed of successive, alternatively bigger and smaller, droplets. As the picture clearly shows, the bigger ones are actually oscillating, gradually changing their shape from a flattened, horizontally stretched ellipsoid (droplets 1 and 2 in the picture), to round balls (3), then to the ellipsoid again (4, 5, 6), squeezed and stretched vertically, and then back to a sphere (7), etc. Each droplet, pulsing rapidly in its free fall,[c] produces

---

[a]Alexander G. Bell, (1847–1922), Scottish-born American inventor. The first public demonstration of speech transmission using his electrical apparatus took place in 1876.
[b]F. Savart, (1791–1841), French physicist; worked in the fields of acoustics, electromagnetism and optics.
[c]The frequency of pulsations can be estimated similarly to that of air bubbles in liquid, Chapter 9, "The chiming and silent goblets":

$$\nu \sim \sigma^{1/2} \rho^{-1/2} r^{-3/2}.$$

Putting $\sigma = 0.07\,N/m$, $\rho = 10^3\,kg/m^3$, $r = 3 \cdot 10^{-3}\,m$, one finds that $\nu \approx 50\,Hz$. Worth noting here is that the "shooting rate" of a typical movie camera is 24 pictures per second. This is enough for the human eye to take the film for continuous motion.

Fig. 12.1: Sequence of alternating big and small droplets after the splitting of a water stream.

different images in one's eye at different instances of time. This causes us to perceive the lower part of the stream as kind of misty, widening in the regions where the droplet-ellipsoids are stretched latitudinally, and narrowing where they are elongated vertically.

Another interesting finding of Savart's was that sounds have a strong effect on the upper transparent part of the water jet: if a sound of a certain pitch (meaning frequency) is excited nearby, the transparent region of the stream turns instantly opaque. Savart gave the following explanation. The droplets, which the stream finally breaks into, actually start developing from the very beginning of the fall, right by the outlet. At first they are outlined only by circular notches that become more prominent as the liquid falls, until the point in the stream where they split completely. These notches are so close together that they make a slight sound. Thus a music tone, in unison with this "natural" pitch, will make the continuous stream of liquid break into separate droplets earlier turning the transparent flow bleary.

The English physicist John Tyndall[d] later repeated F. Savart's experiments in his laboratory. He managed to produce a water stream, transparent and uninterrupted, of about ten feet long. And then, by using the sound of appropriate tone and volume from one of the pipes of an organ, he was able to break this stream into countless separate drops, transforming it by doing so into a misty unsteady trickle. In one of his articles, he related his observation of the water stream falling into a basin. He observed that the water stream falling into a basin sounds like that when the falling stream crossed the surface above the point of transition from the transparent to opaque part at moderate pressure, then the stream entered the liquid silently; but when it crossed the surface below the rupture point, a murmuring started and one saw numerous bubbles formed. In the former case, not only was there no serious sputtering, but liquid rather piled up around the basis of the stream in the basin where the direction of the liquid motion was actually reversed.

These features of water streams were used by A. Bell in his design of the water mic, depicted in Figure 12.2. It consisted of a metal tube and a branch pipe with a funnel soldered on its side; the bottom of the tube was mounted on a massive support, whereas its top was covered with a membrane-like rubber piece, fastened to the tube by a lace. As we know already from Tyndall's experiments, the lower part of the water stream, split into separate droplets, makes a murmuring noise as it reaches the water in the basin. On the other hand, when the upper, still continuous, portion of the stream enters the liquid in the reservoir, the flow remains soundless. One can perform a similar demonstration with a piece of cardboard, placed across a stream of water. As one gradually pulls the cardboard sheet up, the drumming generated by hitting droplets becomes lighter and, after the "transition point" is passed, the noise ceases.

The membrane in Bell's microphone plays exactly the same role as the cardboard piece in the preceding example. Yet, because of the resonator (tube) and the side pipe with the trumpet, the slightest tap of a droplet is amplified and echoed much louder. Thus, the tiny droplets that hit the rubber membrane will make a rapping equal to that of a hammer against an anvil.

---

[d] J. Tyndall, (1820–1893), English physicist, specialized in optics, acoustics and magnetism.

Fig. 12.2: The water microphone by Alexander Bell amplified sound at the expense of energy of falling droplets.

One could easily use this apparatus to illustrate the sensitivity of a water stream to different musical tones, the fact described by Savart and Tyndall. That is, if we touch a faucet from which a slim stream of water runs by a vibrating tuning fork, Figure 12.2, the flow will momentarily break into drops which "commence" their rather ear-splitting chorus. This amplification of the original sound which was pretty weak at the expense of the energy of the falling stream does indeed constitute the physical principle of the water microphone. If one substitutes the tuning fork with, say, a wrist watch, it will make its ticking audible to the entire audience in the room. One of the renowned popularizers of science at the end of the nineteenth century claimed that he had tried to transmit the sound of his voice by connecting a funnel to a glass tube from which water was running. Although the water stream in his utensil had presumably started "talking", the voice was so roaring, horrifying and indiscernible, that according to the legend, the spectators just scooted away.[e] While reading these lines, the authors feel that it is quite a blessing that Bell's main invention, the telephone with the electrical mic in its receiver, happened to be free of such a disadvantage.

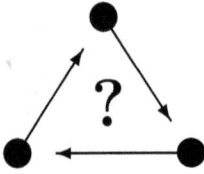

Returning to Chapter 10, "The bubble and the droplet", try to understand why droplets in the lower part of the stream are pulsing periodically. Are the small ones throbbing too?

---

[e]See footnote on page 70.

Chapter 13

How waves transmit information

It is monstrous, monstrous:
Methought the billows spoke and told me of it;
The winds did sing it to me, and the thunder,
That deep and dreadful organ-pipe, pronounced
The name of Prosper: it did bass my trespass.

W. Shakespeare, *"The Tempest."*

We have grown so accustomed to television, radio, cellular phones, and Internet that we are no longer even a tiny bit surprised by the fact that we can so easily receive information from pretty much any corner of the world. Yet it was not always like that (actually not even close) and not a very long time ago:

As Russian authors, we may just as well use an example from Russian history. In order to send a message to St. Petersburg about the coronation ceremony of Empress of Russia Elizabeth which took place in Moscow in 1741, a human chain of soldiers with signal flags in their hands was mustered all along the road from St. Petersburg to Moscow. When the crown was put on the new empress' head, the first of the soldiers waved his flag, so did the second when he had seen his neighbor signaling, then the third, the fourth, and so on. So the news about the crowning event had made its way to the Northern Russian Capital, where a fired cannon notified the crowds of anticipating subjects.

Now, let us ask ourselves a legitimate question: what was actually "moving" along this peculiar chain? Although each soldier remained in his original place, at a certain moment he changed his *state,* raising the flag. And this change of state was exactly the thing which moved along the chain. In situations of this sort physicists say that a *wave was running (propagating)* along it.

There are a great variety of waves, depending on what physical properties vary when they propagate. For sound (acoustic) waves, the density of the matter through which they travel fluctuates, whereas, for instance, in the electromagnetic waves (light, radio, television, etc.), the intensities of electric and magnetic fields oscillate. There are temperature waves, waves of concentration in chemical reactions, epidemic waves, and so on and so forth. In a poetic way, one could say that waves penetrate the entire edifice of contemporary science.

The simplest possible type of waves are the monochromatic ones, where states change at each point according to a simple harmonic law, with a certain constant frequency (the sine or cosine law). The monochromatic sound waves are what we call the sound tones, or pitches. One can excite such waves using, for example, a tuning fork. Monochromatic light waves are generated by lasers. With a simple stick, while dipping it periodically up and down in water, one can make ripples that are pretty close to being monochromatic. A similar wave could be produced in our live chain on the road to St. Petersburg as well.

Imagine that each soldier not just simply raises his flag, but starts continuously and periodically waving it from side to side; and the next soldier follows the one before him with a certain fixed delay or *phase shift*. A wave takes off running along the chain. We bet our dear reader has seen something like that at a stadium during competitions, when the ardent (or simply bored) spectators start doing what is called "Mexican wave".

These monochromatic events are quite pleasing to the eye, yet can they actually transmit information? – Obviously, no. Periodic oscillations do not tell us anything new, so no information is transmitted. Whereas, with the single toss of a hand the diligent servicemen did manage to communicate to St. Petersburg (more than $600\,km$ away from Moscow) the important news. What is the difference between these two kinds of waves? If we take an instant picture of the motion for both situations, we will see that in the first case all the soldiers were involved in the motion, whereas in the second, only one of them. In other words, when a signal is being transmitted, the wave (whatever kind it was) at each moment is localized in space. One could imagine two, or three, or even several near-standing participants raise their hands at the same time. In such a case, the length of conveyed signal would increase. Having been able to generate signals of different lengths, one could send not only a message about a single event (such as the coronation having been accomplished, for example) but, in principle, any kind of information. Take, for instance, the famous Morse code (which was patented much later by the way, in 1854).

Besides regiments of soldiers, there are, of course, other information-transmitting signals: light, sound, electric current, etc. It is interesting to note that any signal can be presented as a sum of monochromatic waves with different frequencies. This possibility is ensured by the so-called *super-position principle*, stating that the oscillations of overlapping (interfering) waves at each point of a medium, should be just simply summed up. And, therefore, depending on the phase shift, oscillations can either amplify each other (for example, two identical waves with no phase shift will result in an oscillation of doubled amplitude, Figure 13.1, *a*) or extenuate each other (again: two identical waves being out of phase will cancel each other completely, Figure 13.1, *b*). It turns out also, that one can select (or tune) the amplitudes and frequencies of combined monochromatic waves in such a manner that they will nullify each other in most places, except for a certain region where, conversely, they will amplify each other.

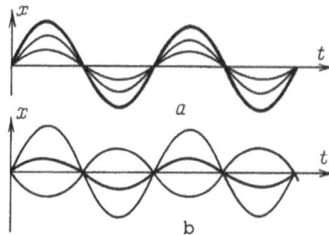

Fig. 13.1:
Monochromatic waves can amplify or attenuate each other depending on the phase shift.

The result of the summation of a large number N of waves with identical amplitude A_0, and frequencies lying within a small interval $2\,\Delta\omega$ around the basic frequency ω_0 is shown in Figure 13.2. It is like an instant photograph of a wave, showing variation of a fluctuating quantity A in different points of space at a fixed moment of time. There is a central maximum with $N\,A_0$ amplitude, and also numerous secondary peaks, though with quickly waning amplitudes, indicating that, indeed, the overlapping waves mostly extenuate themselves, having been noticeably amplified only in the vicinity of the central maximum.

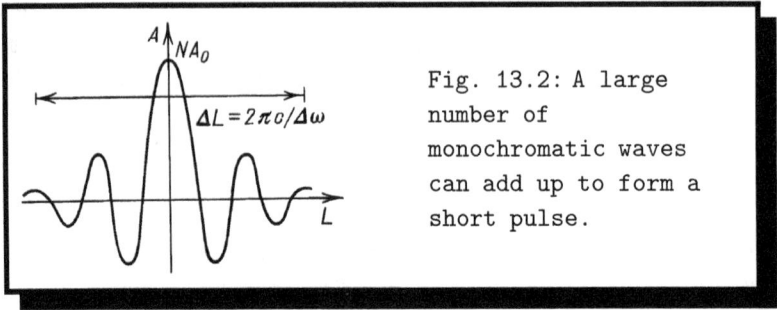

Fig. 13.2: A large number of monochromatic waves can add up to form a short pulse.

It is worth noticing that the central peak does not stand still, but moves with the wave's propagation speed. If the monochromatic component waves all move with the same speed c (as in the case of electro-magnetic waves in a vacuum, for instance), the central maximum also moves with the same speed c, keeping its constant width $\Delta L = \frac{2\pi c}{\Delta \omega}$. Hence, the time length of the running signal will be equal to $\Delta t = \frac{2\pi}{\Delta \omega}$.

Now one could easily write a simple yet quite fundamental relation:

$$\Delta \omega \cdot \Delta t \sim 2\pi.$$

So, the signal length and the width of the range where the frequencies of its components lie are inversely proportional to each other. Qualitatively, such a relationship seems quite natural: if there is a segment of sinusoid, corresponding to a signal of rather long duration (Δt is large), then it must be an almost monochromatic wave and $\Delta \omega$ is small. However, if a short signal is required, it should be combined of many waves with different frequencies. Everyone, we believe, noticed glitches and noises generated on the radio on all bands when lightning strikes nearby.

Thus, each signal can be made of a set of monochromatic waves, or, just the same, each signal can be decomposed into such waves. The amplitude versus frequency dependence for monochromatic waves, composing a signal, is called the signal *spectrum*.[a] In this situation, for example, it will be a rectangle of altitude A_0 and width $2\,\Delta\omega$, depicted in Figure 13.3. This, of course, is a trivial spectrum; signal spectra, just like signals themselves, can have various, at times most peculiar shapes.

---

[a] Although by spectrum physicists sometimes just mean the set of frequencies of monochromatic waves, making a signal, we stick to the more particular definition that takes into account the waves' amplitudes as well.

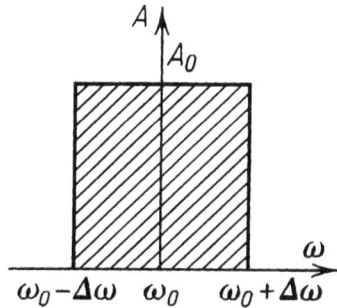

Fig. 13.3: The spectrum of the pulse depicted in Figure 13.2.

When we pronounce sounds, for example, we make the air vibrate in a certain way, so these vibrations propagate as sound signals of a certain shape. Their spectra strongly vary depending on whether we utter vowels or consonants. The vowels have spectra with two distinct characteristic peaks at certain frequencies (they are called *formants*). The spectra of consonants, on the other hand, are more "smeared", spread over the whole audible sound frequency range: Figure 13.4 represents the spectrum of the letter S sound. An entire method called *harmonic analysis* has been developed allowing us to find spectra of registered signals, as well as restore signals from their spectra.

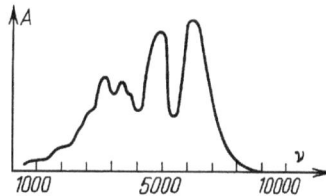

Fig. 13.4: Spectrum of the consonant S sound.

Peculiar though it may sound, solid bodies are capable of "yelling" too. Thermal motion makes atoms in the crystal lattice oscillate, generating, thereby, elastic waves propagating inside the crystal. These oscillations are sound waves too. The position of their spectral maximum depends on temperature, however, it lies in the range of extremely high frequencies. For example, at an absolute temperature of about $5\,K$ the maximum stays already around $3 \cdot 10^{11}\,Hz$, while at room temperatures it moves to

$1,8 \cdot 10^{13}$ Hz. In the audible frequency range, though, the amplitude of these oscillations is negligible; so that in order to hear "what the solids are talking about", one must use some special devices. And by listening to this chatting (studying the sound signal spectra, actually), researchers have already discovered plenty of very important secrets hidden in the solid-state.

But what kind of signals are used in practice to transmit information? For short distances sound signals have worked just fine, for as far back as the human history goes. The limitation though is that these types of waves tend to dissipate quickly. Yet, if one amplifies them at certain intervals (re-transmits them) such signals may travel quite a distance. In Africa, for example, until recently people were sending messages using the tam-tams, "drumming" a piece of information from one village to the next (similar to what the Russian soldiers did with their flags).

In the contemporary world most of the signals are transmitted in the form of electromagnetic waves, which can cover much longer distances, before dying out. For example, one can make an electromagnetic wave carry sound signals. In order to do that, the frequency of this electromagnetic wave (called the *carrier wave*) is kept constant, whereas its amplitude is varied (*modulated*) in accordance with the sound oscillation to be transmitted, Figure 13.5. This way a signal containing the required information is generated. Then, at the receiving end, the signal is "deciphered" — the envelope corresponding to the modulating sound signal is extracted. Such a sending — receiving method is called, therefore, *amplitude modulation* or *AM.* It has been employed in radio and television broadcasting.[b]

There arises a new question though: how much information per unit of time can one actually transmit, with the help of waves? To sort this one out, let us look at the following situation. It is known that any number can be presented in binary notation, as a sequence of ones and zeros. In a similar way, any information can be written encoded into a row of successive pulses and pauses of certain duration. The signals can be transmitted

---

[b]Of course, after modulation, the electromagnetic wave no longer remains monochromatic. For example, in the case of simple amplitude modulation of the carrier wave of, say, frequency ω_0 and amplitude $A(t) = A_0 (1 + \alpha \sin \Omega t)$, in Figure 13.5:

$$x(t) = A(t) \sin \omega_0 t = A_0 \sin \omega_0 t + \frac{\alpha A_0}{2} [\cos(\omega_0 - \Omega) t - \cos(\omega_0 + \Omega)t] .$$

You see that the spectrum even of this simplest modulation consists already of three different frequencies: $\omega_0 - \Omega$, ω_0 and $\omega_0 + \Omega$.

Fig. 13.5: Amplitude
of amplitude-modulated
carrier wave varies in
accordance with the
transmitted
low-frequency signal.

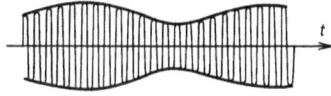

by amplitude-modulated waves, for instance, Figure 13.6. The higher the desired speed of information transmission, the shorter these signals must be. Yet for reliably transporting the information the length of the signal should not be shorter than the period of the carrier sinusoidal wave. This gives us right away the categorical limit on the maximum rate of information transmission. When one wants to raise the speed, one necessarily needs to increase the carrying frequency. This, indeed, once again gives the considered relation for the time length of a signal: $\Delta t \approx \frac{2\pi}{\Delta\omega}$, where $\Delta\omega$ becomes comparable to ω_0.

Fig. 13.6: The
simplest way to
transmit digital
information would be
simply to chop the
carrier into segments.

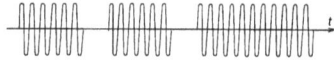

For example, to broadcast a musical program, it is sufficient to use electromagnetic waves with frequencies of around several hundred kilohertz ($1\,\mathrm{KHz} = 10^3\,\mathrm{Hz}$). The human ear is capable of sensing sound frequencies up to approximately $20\,\mathrm{KHz}$, therefore the frequencies composing the signals, in this case, will be at least an order of magnitude less than that of the carrier wave. Yet, this frequency range does not fit for broadcasting of television programs by analog transmission method. The images on the TV screen are generated 25 times per second, and they, in turn, consist of tens of thousands of separate dots (pixels). So the required modulation frequency is about $10^7\,\mathrm{Hz}$ and the corresponding carrying wave should have a frequency of several tens or hundreds of megahertz ($1\,\mathrm{MHz} = 10^6\,\mathrm{Hz}$). That is why in TV technology, engineers had to utilize the *very-high (VHF)* and *ultra-high (UHF)* frequency bands.[c] Consequently, the *very-short* radio waves, with wave length of the order of one meter, are employed even though such waves can propagate only over rather limited (basically, within direct visibility) distances.[d]

An even better candidate for fast information transmission would be ordinary light, having its frequency in the range of $10^{15}\,\mathrm{Hz}$, which would boost the transmission speed by several tens of times. Although the idea is rather old (Alexander Bell[e] was actually the first who applied light signals for sending sound messages back in 1880), its technological feasibility arrived only recently with the development of high-quality sources of monochromatic light. Lasers, coupled with fiber-optical light guides, capable of transmitting light with extremely low propagation losses, as well as all the grandeur of modern electronics, allow efficient coding–decoding of these light signals.

Now we can with all certainty state that the age of copper wires is fading away and is being replaced with the coming epoch of ultra-fast information transmission networks based on fiber-optics technology.

---

[c]The terms VHF and UHF refer to the frequency (ν) bands $30-300\,\mathrm{MHz}$ and $300-3000\,\mathrm{MHz}$ respectively. Sometimes both ranges are united into the single very-short wave band. The corresponding wave lengths are given by the usual formula $\lambda = \frac{c}{\nu}$, where $c = 3\cdot10^8\,m/s$ is the speed of electromagnetic waves in vacuum. This gives $\sim 1\,cm$–$10\,m$ for lengths of very-short waves. Actually in this band *frequency modulation (FM)* is used rather than AM.

[d]Interesting enough the first television sets (with mechanical field scanning), made back in the 1920s, worked in the medium-wave (MW) band. The quality, due to the discussed problems, was very poor and the images were almost indiscernible. This made researchers and engineers switch to very-short waves and develop the electronic field scanning techniques. However, this MW television had its own advantage — because of the longer (compared to the very-short waves) propagating range, programs from Moscow, for example, could reach, say, Berlin without any TV satellites or retranslating stations.

[e]Whom you already know from Chapter 12.

Chapter 14

Why electric power lines are droning

What is the Tacoma Narrows Bridge? The Tacoma Narrows Bridge was built in 1940. After Months of swaying up, down and side to side the Bridge collapsed. Taking with it the life of a poor dog (Tubby).

"Mr. H's World O'Physics."

Long ago, the Ancient Greeks noticed that a tightly drawn string sometimes sounded melodious in the wind, as if it were singing. Perhaps back then people knew the Aeolian harp, named so after Aeolus, the God of Winds in Greek mythology. It is made up of a frame (or an open box) with several strings stretched across, which is then placed where the wind can pass through it. Even a single string in such an instrument can generate quite a spectrum of different tones. Something of the same nature occurs, although with a far more limited tune variety, when the wind swings the cables of telegraph lines.

The puzzle of this phenomenon had long bewildered scientific pundits of the past, until the end of the seventeenth century when Sir Isaac Newton applied his newly developed analytical method to problems that we would call fluid dynamics.

According to the law, first stated by Newton, the resistance force acting on a body moving in a liquid or gas is proportional to the square of the velocity v:

$$F = K \rho v^2 S.$$

Here S is the cross-sectional area of the body perpendicular to the direction of motion, ρ is the density of liquid (or gas), and K is just a proportionality coefficient.

Later it turned out that the formula does not apply universally. When the speed of the body is low compared to the thermal velocities of molecules of the medium, the previous relation starts faltering. We have discussed already in Chapter 11 that for relatively slow-moving bodies the resistance force becomes linearly proportional to the speed (Stokes' law). This situation occurs, for example, when tiny droplets move in a rain cloud, or residue flakes precipitate onto the bottom of a glass, or the drops of substance $\mathcal{A}$ are roaming restlessly inside a Lava lamp (see the same chapter). However, in the modern world with its jet velocities, Newton's law of resistance force is much more common.

Either way, could we not find a satisfactory explanation to the phenomena of droning power lines or the tunes of the Aeolian harp just knowing these basic relations for the resistance? No, unfortunately, it is not at all that simple. Really, if the resistance force remained the same (or increased with the speed of the wind), the string would be simply drawn by the breeze without any oscillations at all.

So, where is the trick here? Well, it turns out that in order to understand the nature of the string vibration in this case, we cannot get away with just a couple of rather general ideas not touching on flow mechanism. We ought to delve a bit deeper and discuss in more detail how liquid actually flows around a body at rest. (This is simpler than considering a body moving in the static liquid but does not affect the result, of course).

The case when the current is relatively slow is depicted in Figure 14.1. The liquid flow lines are passing smoothly around and behind a cylinder (the picture shows the cross section). Flow of this kind is called *laminar*, and the resistance force arises from the internal friction (viscosity) of the liquid and indeed is proportional to the velocity of the liquid (again – our reference frame is tied to the body). Both velocity and friction force at any point in such a flow are time-independent (the flow is *stationary*), and this rather insipid situation is of no interest to our Aeolian problem. Now let us look at Figure 14.2. The flow velocity has increased and new whirling characters have appeared — the eddies or vortices, if you wish. Friction is not the definitive cause any more. Now it depends less on changes of momentum on a microscopic scale and more on scales comparable to the size of the body. The resistance force becomes proportional to the second power of velocity, v^2. Finally, see Figure 14.3, the flow velocity has grown even further, and the eddies are now aligned into neat well-ordered chains.

Fig. 14.1: Lines of slow laminar flow around long cylindrical wire.

Fig. 14.2: At higher flow velocities vortices appear behind the wire.

Fig. 14.3: In fast flows a periodic vortex trail is formed in the wake.

Here indeed lies the answer to the riddle of the string's vibrations! These nicely structured tails of vortices break loose recurrently from the surface of the string and excite the string oscillations, like plucking fingers would do.

This arrangement of vortices, tailing behind the body, was first discovered and studied experimentally at the beginning of the twentieth century. It found theoretical explanation in the works of the Hungarian scientist, Theodore von Karman.[a] Now these periodic vortex wakes are known as the *Karman trails* (or even the "Karman Vortex Street").

---

[a]T. von Karman, (1881–1963). Having greatly contributed to the field of fluid dynamics, he is known as the father of supersonic flight. Hungarian-born, he worked for the US government during World War II, and then for the NATO Aeronautical Research and Development Advisory Group.

As velocity continues to increase even further, the vortices do not have enough time to spread over a large area of liquid. The "swirling" region narrows, the eddies start mixing with each other and the flow becomes chaotic, irregular *turbulent flow*. The latest experiments, though, show that for extremely high velocities there develops another kind of periodicity, but here we are getting well beyond the scope of our chapter, and the best we can do for the curious is to refer them to, say, the exciting book by James Gleick, titled *Chaos*. It is worth noting that, although the phenomenon of Karman vortex trails may look like a rather academic example of just another nice quirk of nature, having not much practical significance, in reality it is quite the opposite. Electric power transmission lines, for example, swing in the wind when it is at a constant speed because of these periodically born and released vortices. And this inevitably creates quite powerful stress at the wire fastenings to the supporting towers, which, if neglected, could (and, unfortunately, have) led to their breakage (sometimes very dangerous). The same holds for tall industrial smoke stacks.

Probably the most renowned engineering calamity of this kind happened in Tacoma (Washington), in November 1940, when a new auto bridge (two lanes, a mile long) built just four months before, Figure 14.4, started swaying violently and collapsed, luckily not killing or injuring anybody (except for the legendary dog mentioned in the epigraph).

Fig. 14.4: Growing oscillations excited by turbulent vortices led to the collapse of Tacoma Narrows Bridge.

A special committee from the Federal Works Agency was appointed to investigate the accident, with T. von Karman as one of its members, by the way. And the committee's conclusion read that the Tacoma Narrows Bridge had crashed due to "forced vibrations excited by random action of turbulent wind". Well, a new bridge was built after a while, but this time with a completely different profile for the wind-exposed surfaces, eliminating, thereby, the cause of unruly vibrations.

Chapter 15

The footprints on the sand

You walked with me, footprints in the sand,
And helped me understand where I'm going.

Simon Cowell, *et al.*, *"Footprints on the Sand."*

Have you ever thought whether, when walking on a beach, you compress the sand under your feet? On the face of it, stepping onto sand presses the grains together. But actually things may turn out quite the opposite way. Here is the proof – the footprints left on wet sand stay dry for quite a while. The English scientist known for his works on hydrodynamics, O. Reynolds,[a] noted in his talk at the meeting of the British Association in 1885 that when the foot steps onto sand still wet after the ebb, the surrounding area immediately becomes dry... According to him the pressure of the foot loosens the surrounding sand and the stronger the pressure the more water is absorbed. This makes the sand dry until enough water comes from below.

But why does pressure widen intergranular spaces so that the available water no longer fills them? For scientists of the nineteenth century this was not an idle question. The answer was immediately related to the atomic structure of matter. This is the topic of the present chapter.

---

[a]O. Reynolds, (1842–1912), English physicist and engineer, specialist in the theories of turbulence, viscous flows, and lubrication.

15.1   The dense packing of balls

Is it possible to fill an entire space with rigid balls of the same radius? Of course not; there will always be voids in between. The fraction of space occupied by the balls is called the *packing density*. The closer the balls lie, the less space is left between them, and the higher the packing density. But when does the density reach the maximum? The answer to this question gives a clue to the mystery of the footprints on the beach.

Let us start from a simpler case and study the packing of equal circles in a plane. Dense packing of circles can be obtained by inscribing those into the cells of a mosaic *(tilings of the plane)* composed of equal regular polygons. There are only three variants: to cover the plane by equilateral triangles, by squares, and by hexagons. The packing of circles in the cells of the square and hexagonal mosaics is shown in Figure 15.1. It is easily seen that the second pattern *(b)* is more economical. The accurate calculation (that you may carry out yourselves) proves that in this case 90.7 percent of the surface is covered by circles whereas for squares, *(a)*, the portion makes only about 78 percent. The hexagonal packing is the densest possible in the plane (or, as modern physicists like to say, in two dimensions). Maybe this is the reason that bees use it for the honeycomb.

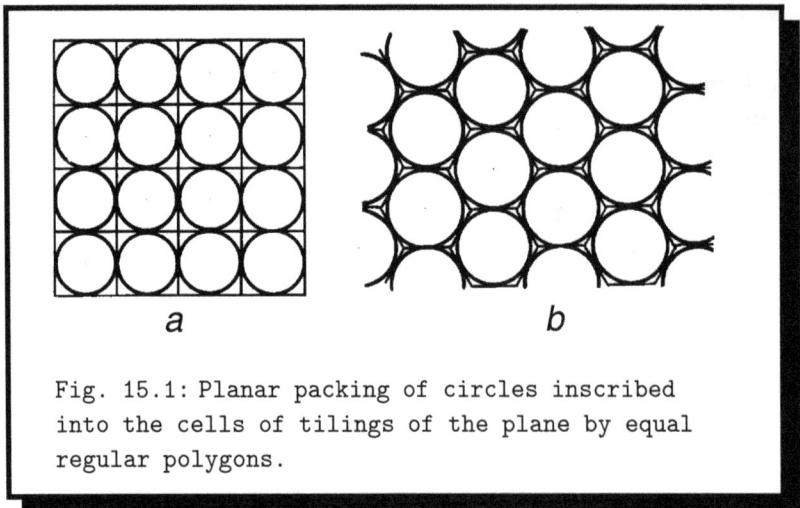

a   b

Fig. 15.1: Planar packing of circles inscribed into the cells of tilings of the plane by equal regular polygons.

A dense spatial packing of balls may be realized as follows. Let us first prepare a dense plane layer of balls placed on a flat surface in the economical order described above. We shall call it the X-layer. Now try to put a similar hexagonal second layer on top of it. We could do this so that each of the balls of the upper layer would lie right over a ball of the lower one, as if we were filling cells of an invisible honeycomb. However in this XX-packing too much empty space will be left. The volume occupied by balls when laid in this manner is only 52 percent of the whole.

To augment the density one must simply put the upper balls into the holes formed by three touching balls below. (This may be called the XY-arrangement). But it is impossible to fill all the holes at once – one of the two adjacent holes will always remain free, Figure 15.2. Therefore when putting on the third layer we face a choice. We may either put balls above the holes of the ground X-layer which were left free by the second Y-one (one of these points is denoted by A in Figure 15.2, b) and build a new Z-layer; or we can place them right over the balls of the base (the point B in Figure 15.2, b) in the X-order. Regular spatial patterns are obtained if successive layers follow periodically one of these prescriptions, that is, either $XYZXYZ\ldots$ or $XYXY\ldots$ As a result we obtain the two ways of spatially packing balls depicted in Figure 15.3. In both cases the balls fill about 74 percent of the space.

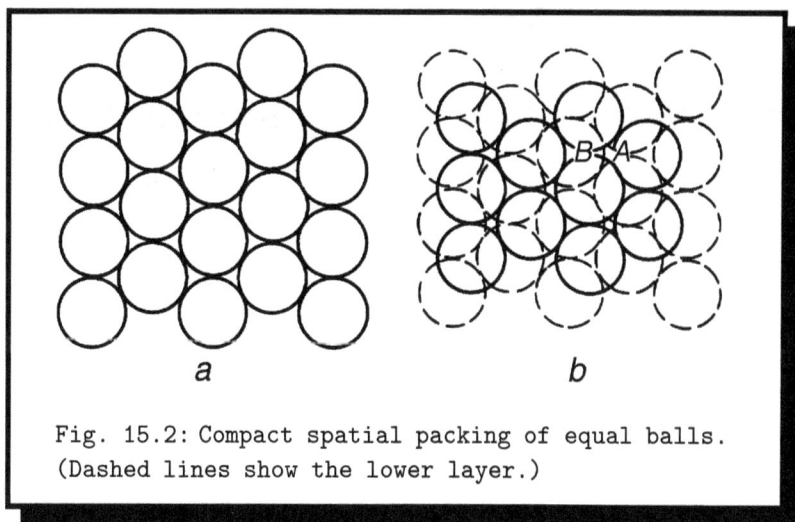

Fig. 15.2: Compact spatial packing of equal balls. (Dashed lines show the lower layer.)

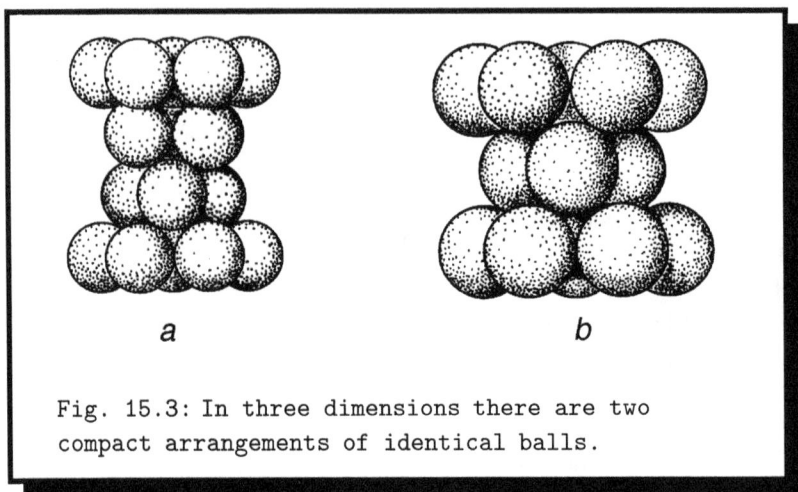

a   b

Fig. 15.3: In three dimensions there are two
compact arrangements of identical balls.

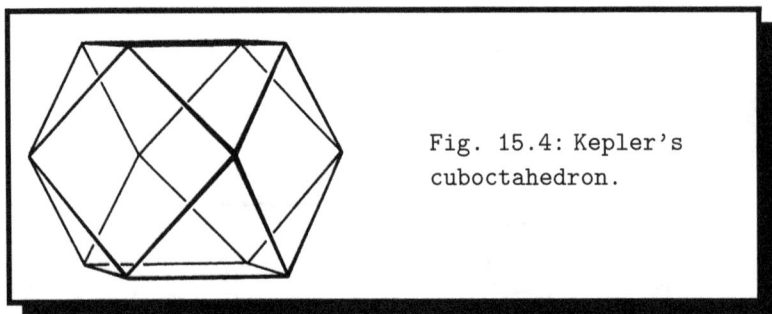

Fig. 15.4: Kepler's
cuboctahedron.

It is easy to count that every ball in these packings touches 12 others
and the points of contact are vertices of a 14-faced polyhedron. The faces
of these polyhedra are alternating squares and equilateral triangles. Say,
the first choice (Figure 15.3, *b*) produces the cuboctahedron[b] shown in
Figure 15.4.

So far we have only studied how to arrange balls so that they fitted
a periodical spatial honeycomb. But is it possible to construct a dense
packing without this condition? An example is presented in Figure 15.5.

---

[b]Cuboctahedron belongs to the so-called Archimedean solids. This class comprises 13
convex polyhedra with congruent vertices and surfaces composed of regular polygons of
two different types. The name *cuboctahedron* itself belongs to J. Kepler (see page 123).
The packing in Figure 15.3, *b* does not correspond to any of Archimedean bodies since
it generates two different types of vertices.

The balls in planes mark out the sides of concentric regular pentagons. The nearest balls of the same pentagon are in contact, but pentagons within a layer are separated. The sides of pentagons in alternating layers contain even and odd numbers of balls respectively. The packing density of this arrangement is 72 percent which is not much less than that of the dense hexagonal arrangements shown in Figure 15.3. There is a way to pack balls so that the centers do not form a lattice and the packing density reaches 74 percent. Yet the question whether denser packings exist remains open.

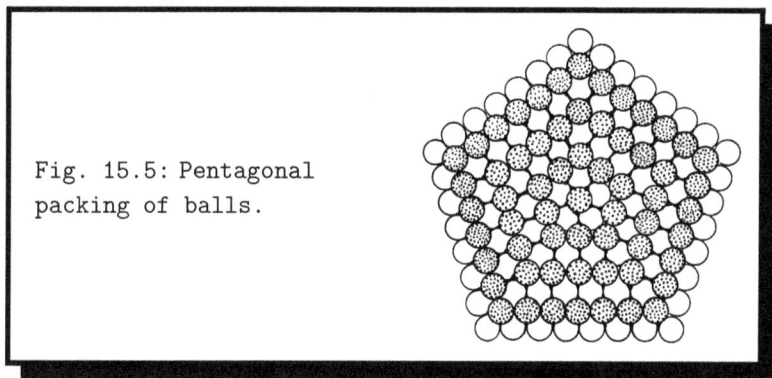

Fig. 15.5: Pentagonal packing of balls.

Let us return to the footprints on the beach. Now we know that there are special arrangements of balls that leave only a small amount of void in between. If one disturbs such an arrangement displacing, for instance, the balls of one layer away from the holes of the next one, the voids will grow. Certainly nobody cares to arrange the grains of sand in a special way. But how could we force sand to be densely packed?

Remember common sense. What do you do when pouring grain into a can? You shake the can gently in order to fill it better. Even if the grain is in clumps patting helps it fit in.

The scientific investigation of this trick was undertaken in the 1950s by the British scientist G. Scott who loaded spherical flasks of different radii with ball bearings. When they were filled without shaking, so that the balls found places randomly, the empirical dependence of packing density on the number of balls had the form:

$$\rho_1 = 0.6 - \frac{0.37}{\sqrt[3]{N}},$$

where N was the number of balls. You can see that if the number of balls is very big (it had reached several thousands in the experiments) then the density tends to become constant and corresponds to filling 60 percent of the space. But shaking the container while filling helps to increase the density:

$$\rho_2 = 0.64 - \frac{0.33}{\sqrt[3]{N}}.$$

Yet even in this case, it is much less than 74 percent which corresponded to the regular packing of balls.

The results of the experiment deserve attention. Why is the addition inversely proportional to $\sqrt[3]{N}$? The balls near the walls of the flask are in a special position with respect to those inside. This affects the density of packing. The magnitude of their contribution is proportional to the ratio of the surface ($\sim R^2$) to the volume ($\sim R^3$) of the container, being inversely proportional to the size of the system (R). By the volume of the system we mean the entire volume occupied by the balls including spaces between them. The size of the system is $R \sim \sqrt[3]{N}$ since the volume is proportional to the number of balls. Dependencies of this type often appear in physics when one has to take into account surface effects.

You see that accurate experiments are in agreement with common sense and prove that shaking granular substances helps to enlarge the density of packing. But what is the reason? Remember that positions of stable equilibrium always correspond to minima of potential energy. A ball may forever lie steadily in a hole but it will immediately roll down from a bump. Something of the sort happens here as well. Shaking the flask makes the balls roll to free spaces so that the density of packing is increased and the total volume of the system becomes smaller. As a result the level of balls in the container goes down. Consequently, the center of mass of the system gets lower and the potential energy is decreased.

Now, at last, we can work out clearly enough what happens to the wet sand. Incessant surf agitates it until a dense packing of grains is formed. By stepping onto the sand with your foot, you disturb the arrangement of granules and augment intergranular pores. Water from the upper sand layer soaks down to fill the pores. This looks like the sand is drying. Taking the foot away restores the dense packing and the depression left by the foot gets filled with water expelled from narrowing voids. However sometimes after strong compression the dense packing cannot be recovered. Then the footprint will become wet only as water rises from below and fills the widened pores.

It is interesting that these features of granular matter were well known to Indian fakirs. One of their tricks consisted of sticking a long thin dagger several times into a narrow-necked vase poured with rice. At some point the dagger gets stuck in the rice and it is possible to lift the vase holding it by the dagger handle.

Evidently the secret was that piercing randomly poured rice helped to "optimize" the packing of grains just like shaking would. One may imagine this like a sort of compression wave propagating in a loose medium. In the beginning the grain was packed compactly right around the blade but lay freely in the bulk and near the walls. The "front" of the wave (surely a rather smooth one) divided the dense core from the loose environment. The front advanced further with every stab and when at last it reached the walls of the vessel the rice throughout the volume became densely packed. In other words the possibilities to compress it further were exhausted. The properties of the substance changed drastically, it became "incompressible". And this was the moment when the dagger gets stuck, since the pressure of grains on the blade and consequently the friction are enough to hold it.

⚠ **Caution!** In case you have decided to surprise your party fellows, please, avoid using glass flasks and china vases. The result can be quite unexpected.

15.2 The long-range and short-range orders

Of course the atoms of which all bodies are built are not rigid balls. Yet simple geometrical arguments help us to understand the structure of matter.

The first to apply a geometrical approach was the German scientist Johann Kepler[c] who in 1611 put forward the idea that the hexagonal form of snowflakes is related to the dense packing of balls. Mikhail Lomonosov[d] in

---

[c]J. Kepler, (1571–1630), German astronomer, creator of celestial mechanics. The famous Kepler laws of planetary motion laid the base for Newton's discovery of the law of gravitation. Kepler's interest in polyhedra was a tribute to the idea of the world ruled by mathematical harmony. According to Kepler ratios of the radii of planetary orbits in the Solar system could be related to properties of regular and uniform polyhedra.

[d]M. V. Lomonosov, (1711–1765), the first scientist of world importance in Russian history. Successfully worked in natural sciences including physics, chemistry, material science, as well as in literature, poetry, and painting. Founded the Moscow University.

1760 was the first to delineate the most compact cubic packing of balls and used that to explain the forms of crystal polyhedra. The French abbot R.-J. Haüy[e] noticed in 1783 that all crystals may be constructed of a plethora of repeated parts, Figure 15.6. He explained the regular form of crystals by suggesting that they are built of identical little "bricks". Finally in 1824 the German scientist A. L. Seeber proposed the model of crystal composed of regularly set little spheres interacting like atoms. The dense packing of the spheres corresponded to the minimum of potential energy.

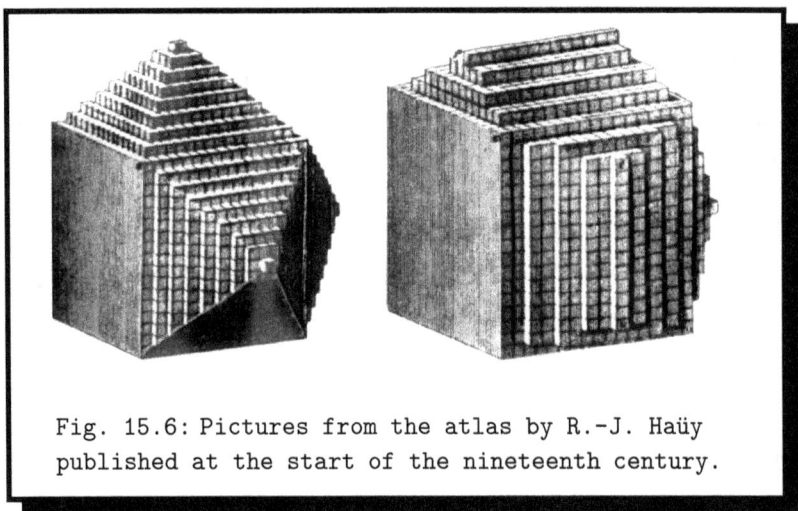

Fig. 15.6: Pictures from the atlas by R.-J. Haüy published at the start of the nineteenth century.

   Structures of crystals are the subject matter of a special science called *crystallography*. Presently, the periodic arrangement of atoms in crystals is a well-established fact. Electronic and tunneling microscopes offer the chance to see this. The tendency to close packing does exist in the atomic world. About 35 chemical elements crystallize so that their atoms are situated like the balls depicted in Figure 15.3. The centers of atoms (or, to be precise, atomic nuclei) make up in space a so-called *crystal lattice* that consists of periodically repeated units. The elementary lattices that may be constructed by periodically shifting a single atom are called *Bravais*

---

[e]R.-J. Haüy, (1743–1822), French crystallographer and mineralogist.

lattices (after the French naval officer Auguste Bravais[f] who was the first to develop the theory of spatial lattices).

There are not many Bravais lattices — only 14 of them. The reason is that most symmetrical elements do not survive in periodic lattices. For instance, a regular pentagon may be turned around the axis passing through the center and it will coincide with the original five times per revolution. One says that it has a fivefold symmetry axis. However a Bravais lattice cannot have a fivefold axis. If such a lattice existed its nodes would be vertices of regular pentagons and those, in turn, would cover the entire plane without breaks. But it is well known that there is no tiling of the plane by pentagons, Figure 15.7!

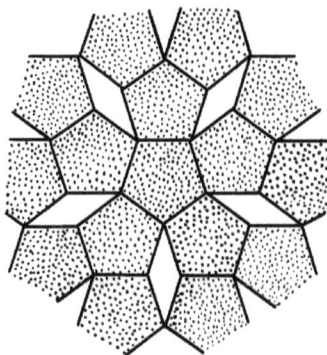

Fig. 15.7: It is impossible to tile the plane by regular pentagons.

So all crystals can be composed of repeated units. This property is called *translational symmetry*. One may also say that there is a *long-range order* in crystals. This is the main property that distinguishes crystals from other bodies.

There is, though, an important class of substances which are deprived of the long-range order. These are *amorphous* substances. Liquids are an example of the amorphous state. But solid matter also may be amorphous. The structure of glass is portrayed in Figure 15.8 together with that of quartz which has the same chemical composition. But quartz is a crystalline substance in distinction from the amorphous glass. The absence of a long-range order does not mean that atoms in glass are situated chaotically. You

---

[f]A. Bravais, (1811–1863), French crystallographer.

may see in the picture that a certain ordering of the nearest neighbors is preserved even in glass. This is called a *short-range order*.

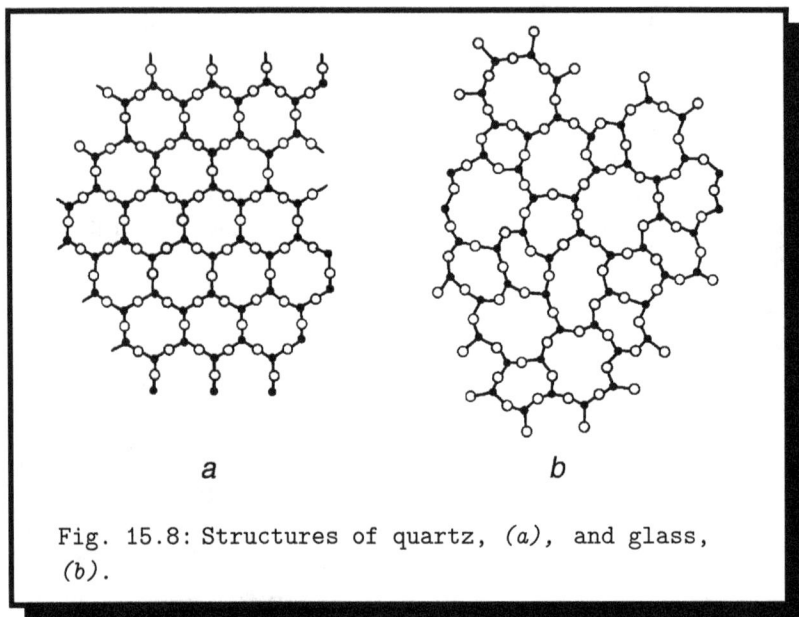

Fig. 15.8: Structures of quartz, *(a)*, and glass, *(b)*.

Recently amorphous materials have found important technical applications. For example, amorphous metal alloys *(metallic glasses)* have unique properties. It turned out that they possess enhanced hardness, high corrosion resistance, and exhibit an optimal compromise of electric and magnetic characteristics. Metallic glasses are obtained by means of an extremely fast cooling of liquid metal: the rate of which must be of the order of several thousands degrees per second. This may be realized by sputtering tiny droplets of metal onto the surface of a rapidly spinning cold disk. The droplets get squashed against the disk forming a film of several microns thickness and the instantaneous removal of heat simply leaves no time for the atoms to arrange properly when cooling.[g]

Interesting studies that shed light on the structure of amorphous solids

---

[g]In the method of melt spinning, a jet of molten metal is propelled against the moving surface of a cold, rotating copper drum. A solid film of metallic glass is spun off as a continuous ribbon at a speed that can exceed a kilometer per minute.

were carried out in 1959 by the English scientist J. Bernal.[h] Equally sized balls of plasticine were randomly put together and compressed into a big lump. The polyhedra obtained after disjoining them turned out to have mainly pentagonal faces. The experiment was repeated with lead pellets. If the pellets had been laid densely and regularly then the compression remolded them into almost regular rhombododecahedrons.[i] On the other hand, if the pellets had been poured unintentionally they transformed into irregular 14-faced polyhedra. Among the faces of those were tetragons, pentagons and hexagons forms, but pentagons prevailed.

In modern technology it is often necessary to densely pack the elements by way of contrivance. For instance, Figure 15.9 shows a cross section of a superconducting cable made of a large number of superconducting wires enclosed in a copper envelope. At first the wires were cylindrical but after rolling they became hexagonal prisms. The more densely and accurately the wires were packed, the more regular hexagons were at the section. This is evidence of the high quality of the cable. If the density of packing is interrupted, pentagons appear at the section.

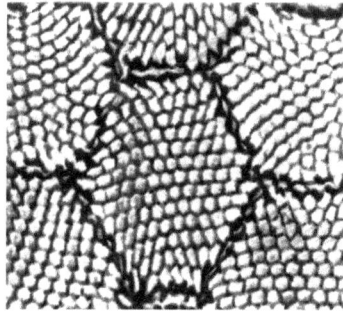

Fig. 15.9: Cross section of high-quality superconducting cable. After rolling cylindrical wires became hexagonal.

---

[h]J. Bernal, (1901–1971), English physicist, specialist in X-ray diffraction analysis, studied structures of metals, proteins, viruses, etc.

[i]Rhombododecahedron (or rhombic dodecahedron) is a polyhedron with 12 rhombic faces and 14 vertices. It may be obtained as a result of uniform squeezing of the hexagonal packing depicted in Figure 15.3, *a*. A way to construct it geometrically is to draw mutual tangential planes of a ball and all its neighbors. In the compact packing shown in Figure 15.3, *a* it plays the role of the unit cell which belongs to hexagons in the planar tiling, Figure 15.1, *b*. Each ball is inscribed into a rhombododecahedron and those fill the entire space.

Fivefold symmetry is widespread in nature. Figure 15.10 presents a photograph of a viral colony. What a striking similarity it has with the pentagonal packing of balls portrayed in Figure 15.5! Paleontologists even use the presence of fivefold axes in fossils as a proof of their biological (contrary to geological) origin. See how far away from the deserted beach the footprints have led.

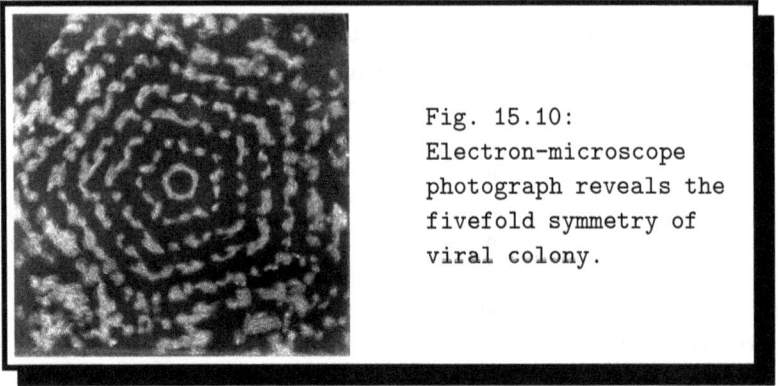

Fig. 15.10:
Electron-microscope photograph reveals the fivefold symmetry of viral colony.

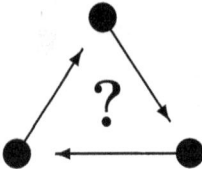

Was it important that Indian fakirs used metal vases with long slim necks for their tricks? What ratio of the volumes of the neck to the body of the vase must be used?

Chapter 16

How to prevent snowdrifts

Look how the snow drifts
flare on the Soracte's slopes
there, straining branches
barely sustain their white
load. Locked in ice, streams
buckle, send cracks
stuttering over the winters sharp still.

Horace, *"Snow Drifts."*

Sections of roads and railroad tracks passing through hollows are often covered with snow even if there have been no snowfalls recently. Why does this happen? Certainly, the answer is simple: the snow has been brought by wind. However it took a good deal of investigation to understand in detail the mechanism of the process.

In 1936 the English geologist Ralph Bagnold[a] studied wind transport of sand in an air tunnel. He discovered that unless the wind velocity is larger than some critical value v_1 the sand would not move. If the velocity of air flow is higher than v_1, but yet less than another value, v_2, the mass of sand may stay at rest. However, an occasional sand grain falling from above brings several rebound particles. These get caught by the wind and, when falling down, knock out more sand grains from the base layer. As a result the sand gets carried by the wind in a kind of leaping motion. In case that velocity exceeds v_2 the wind lifts and blows along substantial clouds of sand. The density of the clouds decreases with height. Trajectories of sand grains may be viewed in Figure 16.1.

---

[a]R. A. Bagnold, (1896–1990), English geologist, expert in the mechanics of sediment transport and eolian (wind-effect) processes.

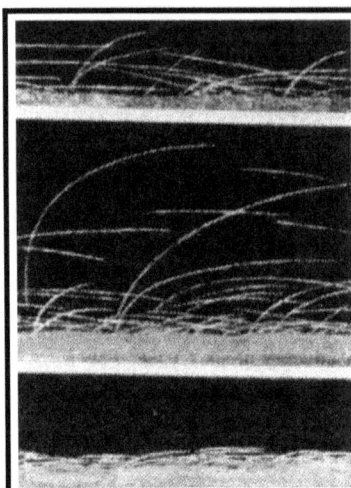

Fig. 16.1:
Trajectories of sand
particles in air flows
of different strength.

Now we can explain why the wind fills dips with snow. Look at the picture of flow lines in Figure 16.2. It is obvious that when traversing a hollow the air flow widens and its speed lessens. This disturbs the balance between deposited and lifted particles: more particles fall down than are taken away. So the depression gets gradually filled with snow.

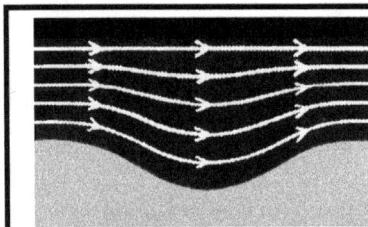

Fig. 16.2: Flow lines
rarefy when wind
crosses a hollow.

Analogous processes take place when the snow carried by wind encounters an obstacle, say, a tree. Meeting the trunk the incident flow turns up and an ascending current of air is formed. This current digs a deep hole on the windward side of the tree. At the same time in front of the hole and behind the tree the speed of wind is smaller and a mound is heaped up.

This phenomenon helps prevent snow-binding in low-lying road sections. A protecting fence is made of wooden planks at a certain distance windward of the road. Behind the fence a lee zone of steady light wind is established where all the snow precipitates.

The same mechanism explains the motion of sand dunes. A strong enough wind blowing against a dune picks up sand on the windward side. On the rear of the dune the air flow slows down and the sand falls back down. As a result, with time, dunes get moved along inch-by-inch with the wind and the dunes "wander".

Chapter 17

The incident on the train

In a station,
Beneath the rain,
Here I wait,
For the nine o'clock train.

Timothy Miller, *"The Nine O'clock Train."*

Not so long ago the authors of this book had to return from Venice to Naples by express train. The train ran fast (its speed was about $150\,km/h$) and landscapes that looked like paintings by Renaissance masters flitted by outside the window. In perfect accord with their canvases, the country was hilly, and we sometimes flew over a bridge or dove into a tunnel.

In one especially long tunnel between Bologna and Florence, we suddenly felt a dull pain in our ears, as happens to air passengers when taking off or landing. It was clear that the same sensation visited all our fellow travelers, who swayed their heads trying to get rid of the unpleasant feeling. But when the train finally burst out from the narrow tunnel the discomfort passed. However one of us, who was not used to such surprises on the railways, got interested in the origin of this phenomenon. Since it was evidently connected with a pressure jump, we started a lively discussion of the possible physical causes.

At first glance it seemed to us that the air pressure in the gap between the tunnel walls and the train had increased in comparison with the atmospheric one, but this assumption became less and less obvious as the discussion went on. In such matters mathematics is the best judge, so we attempted to approach the problem quantitatively. Soon we arrived at an explanation which follows.

Let us consider a train with a cross-sectional area S_t that moves at a speed v_t in a long tunnel with a cross-sectional area S_0. First of all, let us switch to the inertial coordinate frame associated with the train. We will take the air flow as stationary and laminar and ignore its viscosity. The motion of the tunnel walls relative to the train need not be taken into account in this case — because of the absence of viscosity it does not influence the air flow. We will also assume that the train is sufficiently long so that one could ignore end effects near the front and rear cars. We shall assume that the air pressure in the tunnel is constant and does not vary along the whole train.

As you can see, by gradually eliminating minor details, we have passed from the actual movement of the train to a simplified physical model that could be analyzed mathematically. Here it is. We have a long tube (the former tunnel) with air being blown through it and a cylinder with streamlined ends (the former train) coaxially fixed inside,[a] Figure 17.1. Far away from the train (at the cross section $A-A$) the air pressure equals the atmospheric one p_0. Velocity of the air flow at this section is equal in magnitude and opposite in direction to the velocity of the train $\vec{v}_t$ with respect to the ground. Let us examine some cross section $B-B$ (just in case, we may place $B-B$ far enough from the ends of the train so that our assumptions are justified). We denote the air pressure in this cross section by p and the air velocity by v. These values can be linked with v_t, and p_0 by means of the Bernoulli[b] equation:

$$p + \frac{\rho\,v^2}{2} = p_0 + \frac{\rho\,v_t^2}{2}, \qquad (17.1)$$

where ρ is the air density. Equation (17.1) contains two unknowns, p, and v; so in order to determine p we need one more relation. This is provided by the condition of conservation of the air flow. According to this the mass of air passing through any cross section of the tube in a unit of time is constant and equals:

$$\rho\,v_t\,S_0 = \rho\,v\,(S_0 - S_t). \qquad (17.2)$$

This equation expresses the fact that the air mass can neither appear nor disappear while it flows through the tube. It is usually called the condition of flow continuity.

---

[a] Note that actually we replaced the ordinary railway tunnel with an air tunnel like those where airplanes are tested.

[b] Daniel Bernoulli, (1700–1782), Swiss physicist and mathematician born in the Netherlands (son of the Swiss mathematician Johann Bernoulli); formulated the fundamentals of theoretical hydrodynamics.

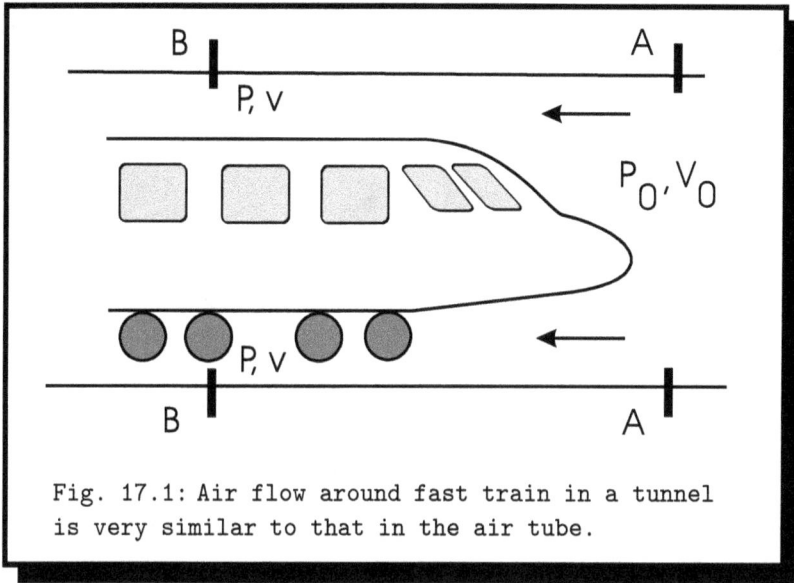

Fig. 17.1: Air flow around fast train in a tunnel is very similar to that in the air tube.

As you probably noticed, we took the air density in equations (17.1) and (17.2) to be constant. For this assumption to be valid, two conditions must hold. The first one requires that the pressure jump we are looking for, Δp, must be much less than the pressure itself: $\Delta p \ll p$. If the air temperature does not change along the tube, then, according to the ideal gas law (Mendeleev - Clapeyron equation) written far from the tunnel, the air density ρ can be expressed in terms of p_0:

$$\rho = \frac{m}{V} = \frac{p_0 \, \mu}{R \, T}. \tag{17.3}$$

Here $\mu = 29 \, g/mol$ is the molar mass of air, T is the absolute temperature and R is the gas constant per mole. We supposed above that Δp is small and neglected the density change $\Delta \rho = \rho \frac{\Delta p}{p} \ll \rho$. We shall see later that this is indeed the case. The second condition concerns flow speeds at different sections of the tunnel. In order that the density is uniform throughout the tube there must be enough time for air to come to equilibrium. This means that speed of flow must be much less than the root-mean-square speed of chaotic thermal motion of the molecules. It is just this speed that determines the characteristic time required to establish the constant equilibrium gas density on the macroscopic scale.

Eliminating the speed v from equation (17.1) by means of equation (17.2), we get:

$$p = p_0 - \frac{\rho \, v_t^2}{2} \left(\frac{S_0^2}{(S_0 - S_t)^2} - 1 \right). \tag{17.4}$$

Substituting Eq. (17.3) into (17.4), we get:

$$p = p_0 \left[1 - \frac{\mu \, v_t^2}{2RT} \left(\frac{S_0^2}{(S_0 - S_t)^2} - 1 \right) \right]. \tag{17.5}$$

The combination $\frac{\mu \, v_t^2}{2RT}$ in the right-hand side, evidently, is dimensionless. So the expression $\sqrt{RT/\mu}$ must have the dimension of speed. Up to a coefficient it is easy to recognize in it the root-mean-square speed of thermal molecular motion. But in aerodynamical problems another physical characteristic of gas, the sound speed v_s, is more relevant. It is determined by the same combination of temperature and molecular mass as the root-mean-square speed of molecules, but the numerical value of v_s includes in addition the so-called adiabatic index γ. The latter is a number of the order of unity characteristic of a gas (for air, $\gamma = 1.41$):

$$v_s = \sqrt{\gamma \frac{RT}{\mu}} \, . \tag{17.6}$$

Under normal conditions, $v_s = 1\,200 \, Km/h$. With the help of equation (17.6) we can bring the expression (17.5) to the form that will be convenient for the further discussion:

$$p = p_0 \left[1 - \frac{\gamma}{2} \frac{v_t^2}{v_s^2} \left(\frac{S_0^2}{(S_0 - S_t)^2} - 1 \right) \right]. \tag{17.7}$$

Now it is time to stop and think a little. We calculated the pressure along the skin of the train inside the tunnel. But our ears ached not because of the pressure itself but because it had changed in comparison with the pressure p' that was there when moving in the open.[c] We can easily determine this pressure directly from equation (17.7), noticing that the open air can be considered as a tunnel of infinite cross section $S_0 \to \infty$. So we have:

$$p' = p_0.$$

---

[c] Here we should point out two circumstances. First, in biophysics there is the so-called Weber–Fechner law. According to this law any changes in the environment can be detected by organs of sense only if the relative change of parameters exceeds some threshold value. The second is that in a long tunnel our organism adapts to the new conditions and the discomfort disappears. Nevertheless, it comes back at the exit.

This result was sufficiently evident without calculation, though.

It is interesting to observe that the relative pressure difference is negative:

$$\frac{\Delta p}{p_0} = \frac{p - p_0}{p_0} = -\frac{\gamma}{2} \left(\frac{v_t}{v_s}\right)^2 \left(\frac{S_0^2}{(S_0 - S_t)^2} - 1\right). \qquad (17.8)$$

From this we can see that when a train enters a tunnel the pressure around it decreases, contrary to what we might expect at first. Now let us estimate the magnitude of the effect. As we have mentioned before, $v_t = 150\, Km/h$ and $v_s = 1\,200\, Km/h$. For narrow railroad tunnels the ratio $S_t\,/\,S_0$ is about $1/4$ (for there were two tracks in our tunnel). So we find that:

$$\frac{\Delta p}{p_0} = -\frac{1.41}{2} \left(\frac{1}{8}\right)^2 \left(\left(\frac{4}{3}\right)^2 - 1\right) \approx 1\,\%.$$

This value may seem pretty small, but if we take into account that $p_0 = 10^5\, N\,/\,m^2$ and take the area of the eardrum to be $\sigma \sim 1\,cm^2$, we get an excess force $\Delta F = \Delta p_0 \cdot \sigma \sim 0.1\,N$, which may turn out to be quite noticeable.

So it seemed that the effect was explained, and we could call it quits. But something still worried us about the final equation. Namely, from expression (17.8) it followed that even in the case of an ordinary train moving with normal velocity, so that[d] $\frac{v_t}{v_s} \ll 1$, the value of $|\Delta p|$ might reach and even exceed the normal pressure p_0 in sufficiently narrow tunnels! Clearly, within the framework of our assumptions we were getting the absurd result that the pressure between the walls of narrow tunnel and train became negative!

Wait a minute! We probably missed something that hampers the validity of our formula. Let us take a closer look at our findings. If $\Delta p \sim p$, then

$$\frac{v_t}{v_s} \left(\frac{S_0}{S_0 - S_t}\right) \sim 1,$$

and, consequently,

$$v_t\, S_0 \sim v_s\, (S_0 - S_t).$$

Comparing the last equation with the continuity equation (17.2), we begin to understand the situation. If Δp becomes of the order of p_0 the speed of air flowing in the gap between the train and the walls of this

---

[d]The omnipresent ratio of velocities $M = v\,/\,v_s$ in aerodynamics is called the Mach number after the Austrian physicist Ernest Mach, (1838–1916).

narrow tunnel turns out to be of the order of the speed of sound. That is, we cannot speak of laminar air flow any longer and the previously smooth flow becomes turbulent.

So the correct condition for the use of equation (17.8) is not merely $v_t < v_s$ but the more rigid one:

$$v_t \ll v_s \left(\frac{S_0 - S_t}{S_0} \right).$$

It is evident that for real trains and tunnels this condition is always met. Nevertheless, our investigation of the limits of applicability of equation (17.8) was not just an empty mathematical exercise. Physicists must always understand the limits of validity of their results. Besides there is a quite practical reason to take it seriously. In the last few decades fundamentally new forms of transportation, including high-speed trains, were discussed more frequently. One of the projects exploited a magnetic cushion produced by a powerful superconducting magnet.

The prototype *maglev* (an abbreviation for *magnetic levitation*) of such type of transportation have already been created in Japan and China. The most advanced pilot sample carries passengers along the $40\,km$ test track at a maximum speed of $581\,km/h$ – almost half the speed of sound. The vehicle hovered above metal rails supported by a strong magnetic field and resistance to its motion was determined solely by aerodynamic effects.

The next step in developing this transport was the idea of, believe it or not, enclosing the train in a hermetically sealed tube and reducing the aerodynamical factor by pumping the air out! You see how close this problem is to the one that has captivated us. But here physicists and engineers have encountered a much more complex case of $v_t \sim v_s$ and $S_0 - S_t \ll S_0$. Therefore the air flow is not laminar, and the temperature of the air changes considerably along the train.

Modern science does not have ready answers to the questions which appear when solving these problems. But even our simple estimate allows, in principle, an understanding of when new effects come into play and become important.

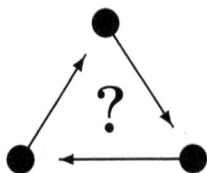

Why does the noise from a moving train increase considerably when the train enters a tunnel?
Why do express trains on close parallel tracks slow down when meeting?

PART 3

Physics in the kitchen

In this part of the book we will explain how the laws of physics govern preparation of different dishes and drinks, starting from everyday breakfast eggs and finishing with delicious wines and champagnes.

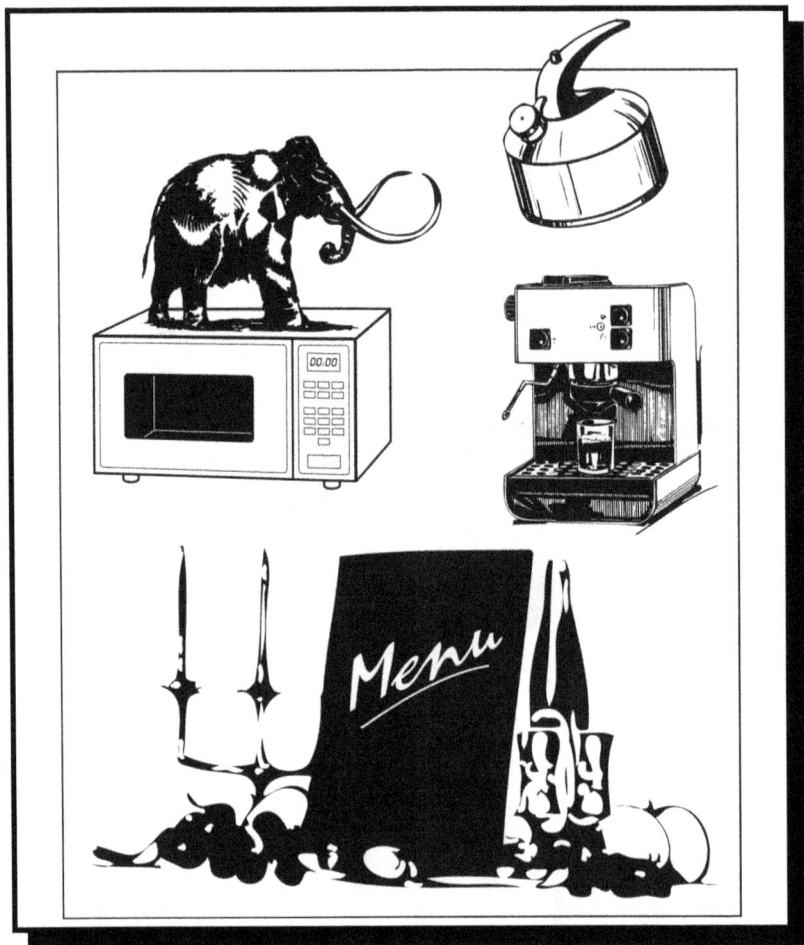

Chapter 18

Craving microwaved mammoth

I could not allow such a wonderful wildfowl to escape
from me and I loaded my gun with an ordinary ramrod.
Then I shot it straight through all of the partridges as
they always rise from the ground in a direct line before
each other. The rod had been made so hot with the
shot that the birds were completely roasted by the time
I picked them up.

E. Raspe, *"The adventures of Baron Munchausen."*

The day when pre-historic humans tamed fire opens *era humana.* Once
and for all HUMANITY broke from its ape ancestry. Fire opened the way
to smelting metals, manufacturing cars, flying to outer space, but above
all. . . it allowed humans to forget the taste of raw meat. A well-browned
beefsteak may symbolize civilization just as well as a model of the atom.

Civilization grew and methods of food preparation changed. The camp-
fire under a spitted mammoth[a] was succeeded by the hearth, followed by
wood, coal and gas stoves. Those were replaced by liquid fuel stoves, hot-
plates, electric stoves, grills, toasters, roasters, *etc.*

The interaction of fire with food changed in appearance but the physical
entity of the process stayed almost untouched: the heat required for cooking
was either transferred immediately (sometimes with the help of convection)
or by means of infrared radiation. The former mechanism works, for example, when making diet cutlets or "manty" in the *steamer,* where food is

---

[a]Think, is it really possible to grill a whole mammoth? Read about the preparation of
elephant legs by African tribes in the book "The plant hunters" by Mayne Reid.

cooked by hot steam rising from the water that boils below.[b] Cooking soup
uses two effects at once: direct heating at the bottom of the pot is followed
by convective mixing of upper and lower layers of the liquid. In the mean
time, electric grills or charcoal barbecues are examples of radiative heat
transport.

The evolution of the "fireplace" (let this denote the heat source) notably
effected kitchen technology and cooking recipes. The possibilities of making
new exquisite and delicious dishes arose. But justice calls us to confess
that along with the enhancement of the culinary arsenal, the development
of the fireplace drove many dishes away from the table. Some of those
were forgotten, others yielded to the ersatz. For example, the true *Pizza
Napoletana* can be baked in a few minutes by almost nothing but **only** a
special blazing wood-stove.[c]

However, we have strayed from the initial subject. Let us return, even
if not right away to the kitchen, then to the lecture on the history of metallurgy. There you will learn that furnaces at ironworks were changing
almost as often as culinary appliances. In particular, Faraday's[d] discovery
of the law of electromagnetic induction opened doors to the invention of
induction melting. Here is the basic idea: a piece of metal is placed in a
strong, rapidly-changing magnetic field. As metals are good conductors the
inductive electromotive force (EMF) that appears due to the variation,

$$\mathcal{E} = -\frac{d\Phi}{dt}, \qquad (18.1)$$

(Φ is the magnetic flux penetrating the specimen) gives rise to *eddy induction currents* or, otherwise, *Foucault currents*. The word "eddy" indicates
that the currents are closed because they follow closed lines of the induction electric field. Induction currents generate Joule[e] heat like any other
currents caused by an applied electric field. If the EMF of induction is large
enough (this requires a big amplitude and a high frequency of the magnetic
field) the evolved heat will suffice for melting the metal. Induction melting

---

[b] "Manty" is the speciality of Asian cuisine made of minced seasoned lamb wrapped in
thin dough like Italian tortellini. The physical aspects of steaming will be discussed in
Chapter 20.

[c] Physics of baking good pizza will be discussed in detail in Chapter 19.

[d] M. Faraday, (1791–1867), English physicist and chemist.

[e] J. P. Joule, (1818–89), English physicist, specialist in thermo- and electrodynamics.

is widely used in the production of high-alloy steel, aerospace metallurgy *etc.*

Even astronauts get hungry, so let us leave the orbital vacuum induction furnace and visit the kitchen module of the spaceship. Here we would find a fireplace of the sort that drops out of the long sequence listed previously. This oven rather resembles an induction melting facility than a conventional kitchen utensil. It heats the food with the help of ultrahigh frequency electromagnetic radiation.

By the sixties, astronauts who were spending more and more time in orbit got fed up with tubed food. However, for obvious reasons, taking a camp stove to space was absolutely out of the question. First, the flame would consume the priceless oxygen and, second, the ruthless weightlessness would have a frustrating effect on the earthly magic of old appetizing recipes. (Just try to imagine the cooking of a weightless soup. What other difficulties would you predict in space cooking?)

The solution was found in using a kitchen analogue of induction furnace. Remember that almost all human food contains a noticeable amount of water. Salty water[f] is an electrolyte and a conductor, although not a particularly good one. Hence, a changing magnetic field applied to a meat chop will induce Foucault currents as if it were a metal. The energy of the electromagnetic field will transform to Joule heat and, as a result, the meat will roast.

A well-known example of an electromagnetic field changing both in time and in space is an electromagnetic wave. But is hard to believe that a random electromagnetic wave will be able to fry a steak. One would go hungry bluntly irradiating it by flashlight. Therefore some criteria must be met. First of all, the field must be strong enough. For example, the field of a radar for watching aircraft would do. (Eyewitnesses tell that birds which accidentally cross the beam of a powerful radar fall dead not singed but boiled. Almost like in Baron Munchausen's tale.) Certainly for safety it would be nice to "confine" the field, localize and hold the wave.

Waves may be "stored" in resonators. For sound waves this may be a wooden box. A body of violin is a typical resonator. Standing sound waves can last in it comparatively long after having been excited by an external source. Naturally this role is played by the bow stick and strings.

Quite similarly it is possible to "store" an electromagnetic field in a metal box. The length of the box must be equal to a whole number of

---

[f]Remember the taste of a bleeding finger...

halves of the confined wave. Exciting (by means of some microradar) electromagnetic oscillations in the box turns it into a resonator with a standing electromagnetic wave. The nodes of the wave (these are the points where the amplitude is zero) are at the walls. A microwave oven is exactly such a resonator combined with a small microwave source. As the device is about a foot long we can readily estimate the maximal wavelength of the radiation inside. You can check this estimate by looking on the back of the oven, where you will find that the standard frequency is $\nu = 2450\ MHz$, the corresponding wavelength being $\lambda = c/\nu \approx 12.3\ cm$.

Let us continue exploring the microwave oven and make an experiment as it once has been performed by the author himself. Take a bulky cut of deep frozen meat, salt and pepper it, put onto a suitable dish (its time will come yet), place it in the microwave and switch it on high. At first sight nothing happens but to the muffled hum of the ventilator (the necessity of which will be explained later). But notice through the glass door that the meat will gradually become brown, looking absolutely done in thirty minutes. Take it out and cut in halves. It is quite probable that you will find inside a portion not simply raw, but even frozen. What is the explanation?

The simplest thing would be to blame a nonuniform field distribution in the oven. Indeed, the size of the compartment is about $25\ cm$ and the wavelength is ($\lambda \approx 12.3\ cm$). As the resonator is longer than four half-waves $\lambda/2 \approx 6.15\ cm$ the field inside must have at least three nodes where its intensity turns to zero. Spatial positions of the nodes of the standing wave are stationary and our piece of meat could have had bad luck. Still in modern microwave ovens this problem is solved by a slowly rotating table. This averages the effect of high-frequency field over the volume of the food. So, why not to take a merry-go-round for a turntable, load it with freshly extracted from eternal congelation mammoth and tug into a suitable microwave oven? What if the taste and the splendid size of the outcome are worth the effort?

Unfortunately an impassable obstacle thwarts the realization of the dream, called the *skin effect*. This is the well-known property of high frequency currents to localize near the surface of a conductor. As long as the frequency of the electromagnetic wave in the oven is very high this effect may be really important. The field will dampen in the bulk and the power will be not enough to heat the meat through.

In order to prove the last assumption let us dwell on the skin effect. We shall try to estimate the effective depth of penetration of the electromagnetic field and study its dependence on frequency and on the properties of the conductor. The questions can be easily replied by solving Maxwell's[g] partial differential equations for electromagnetic fields. Nevertheless this may scare some small minority of our readers and we shall recourse to qualitative estimates.

First, let us formulate the problem. Let an electromagnetic wave with frequency ω fall normally onto a flat surface of a conductor, Fig. 18.1. Inside the conductor the electric field of the wave drives electrons into motion. The currents lead to Joule losses and the wave dies out. It can be shown that the damping obeys an exponential law so common in nature (remember for example radioactive decay) which can be stated as:

$$E(x) = E(0) \, e^{-x/\delta(\omega)}. \tag{18.2}$$

Here $e = 2.71\ldots$ is the base of the natural logarithms; $E(0)$ is the amplitude of the electric field at the surface of the conductor and $E(x)$ is that of the depth x; $\delta(\omega)$ is the effective depth of the field penetration (this means that at the depth $\delta(\omega)$ the field decreases e times).

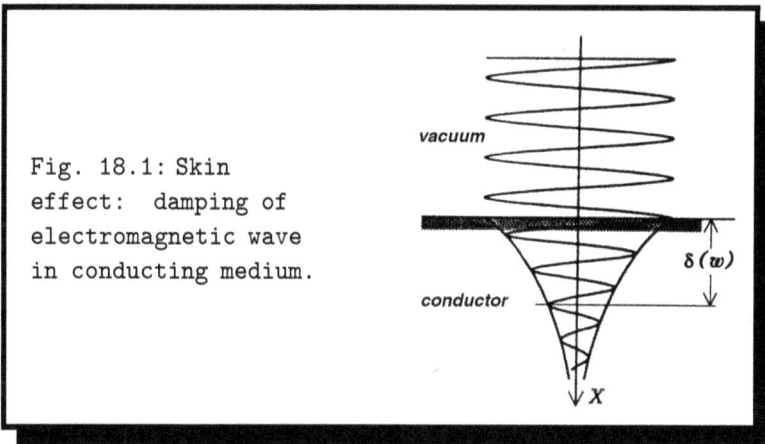

Fig. 18.1: Skin effect: damping of electromagnetic wave in conducting medium.

We shall find the value of $\delta(\omega)$ with the help of the dimensional analysis.

---

[g]J. C. Maxwell, (1831–1879), Scottish physicist, specialist in statistical physics, electro- and thermodynamics, optics *etc.*

Clearly the penetration depth must depend on the frequency. Remember that direct current ($\omega = 0$) flows through the full cross-section of the conductor. So the skin effect is weak, $\delta \to \infty$, at low frequencies and becomes pronounced at higher ones. It is quite natural to suppose that the frequency dependence of penetration depth obeys a power law:

$$\delta(\omega) \propto \omega^\alpha,$$

and expect α to be negative.

It is none the less obvious that the penetration depth must depend on conducting properties of the sample. Those are characterized by the resistivity ρ of the material or, just the same, by the conductivity $\sigma = 1 / \rho$. The essence of the skin effect is that the energy of the electromagnetic wave is converted to thermal energy. The rate of Joule losses in a unit volume is:

$$p = j\,E = \frac{E^2}{\rho} = \sigma\,E^2,$$

where E is the strength of the electric field and j is the current density at the point. (Try to derive this formula yourself. Remember that the local form of the Ohm[h] law is: $j = \sigma\,E$.) The more effective energy dissipation results in the faster damping. Thus the penetration depth must depend on the conductivity of the medium:

$$\delta \propto \sigma^\beta,$$

and one may believe that β is negative like α.

Finally let us note that the equations of electromagnetism when written in the International System of Units (SI) contain the dimensional magnetic constant $\mu_0 = 4\pi \cdot 10^{-7}\,H/m$, called the vacuum magnetic permeability. This constant enters into the expression for magnetic induction around electric current just like the vacuum dielectric permeability ε_0 appears in the formula for the electric field of a point charge. Let us assume that the

---

[h]G. S. Ohm, (1787-1854), German physicist; worked on electricity, acoustics, crystal optics.

penetration depth is a combination of these three parameters:[i]

$$\delta \propto w^\alpha \, \sigma^\beta \, \mu_0^\gamma, \tag{18.3}$$

and find the exponents α, β and γ by comparing the dimensions of the left and right sides of the equation. Write down the dimensions of all quantities in (18.3):

$$[\delta] = m, \quad [w] = sec^{-1}, \quad [\sigma] = \Omega^{-1} \cdot m^{-1}, \quad [\mu_0] = H \cdot m^{-1}.$$

Note that as *Henry* is the unit of inductance the relation

$$\mathcal{E} = -L \frac{\Delta I}{\Delta t}$$

makes possible to represent it as follows:

$$H = V \cdot sec/A = \Omega \cdot sec,$$

and

$$[\mu_0] = \Omega \cdot sec \cdot m^{-1}.$$

The two sides of the equation (18.3) must have the same dimension:

$$m = (sec)^{-\alpha} \, (\Omega \cdot m)^{-\beta} \, (\Omega \cdot sec \cdot m^{-1})^\gamma,$$

or

$$m^1 = (\Omega)^{\gamma-\beta} \cdot sec^{\gamma-\alpha} \cdot m^{-\gamma-\beta}.$$

This is equivalent to the three equations:

$$\begin{cases} \gamma - \beta & = & 0, \\ \gamma - \alpha & = & 0, \\ -\gamma - \beta & = & 1. \end{cases}$$

Solving those we find: $\alpha = \beta = \gamma = -1/2$. Substitution of these values converts the equation (18.3) into

$$\delta \propto \sqrt{\frac{1}{\mu_0 \, w \, \sigma}}.$$

---

[i]This assumption implies that magnetic field is produced only by real currents. Neglecting the electric component of the wave (the so-called displacement currents) is well justified for frequencies in question. Otherwise the dimensional constant ε_0 would appear in the result.

The dependence of the penetration depth on ω and σ confirms the preliminary physical analysis, since both α and β are negative. The accurate computation based on Maxwell's equations gives the same expression up to the numerical factor $\sqrt{2}$:

$$\delta = \sqrt{\frac{2}{\mu_0 \, \omega \, \sigma}}. \tag{18.4}$$

Now we can estimate the penetration depth at the standard frequency $\nu = \omega \, / \, 2\pi = 2.15 \cdot 10^9 \, Hz$ used in microwave cooking. The conductivity of the bulk of the meat is practically equal to that of muscle which, according to biophysical handbooks, is $\sigma_m \approx 2.5 \, \Omega^{-1} \cdot m^{-1}$. It is interesting to compare it to the model conductor, copper, with $\sigma_c \approx 6 \cdot 10^7 \, \Omega^{-1} \cdot m^{-1}$. The formula (18.4) gives for these values:

$$\delta_m \approx 0,7 \; cm \qquad \text{and} \qquad \delta_c \approx 10^{-4} \; cm.$$

The effect is rather strong for good conductors and leads to the enhancement of the resistance of wires at high frequencies. The analysis proves that alternating currents are concentrated in the layer of thickness $\delta(\omega)$ near the surface. At high frequencies δ is small and the effective cross-section of wire falls down. However even for such a poor conductor as meat, the effect is quite noticeable. Our test (10 cm thick) beef cut turned out to be too large. The field strength and correspondingly the released heat diminish many times in the center. In result the only source of heat delivery there remains the thermal conductivity. And turning back to the mammoth, here the skin effect shows up mockingly literally. At best the thick hide of the giant would fry whereas the meat would remain untouched.

Well, one can survive that. Just do not put into your oven slices that are too thick and the meat will roast safely and thoroughly. A big but flat steak is not sensitive to the skin effect. Maybe you will even invent some delightful applications of the skin effect. For example, try to imagine preparing in a microwave such an exotic dessert as ice cream in a hot freshly baked pastry crust.

Like everything in the world the microwave oven has its drawbacks. On the one hand it opens prospectives of pioneer cooking dishes that nobody has ever heard of, it preserves vitamins and provides a means of making dietary products but on the other it is incapable to imitate a soft-boiled egg. Indeed, let us **imagine** an egg placed in the oven. After the power has been switched on, the Foucault currents emerge in the liquid contents. The

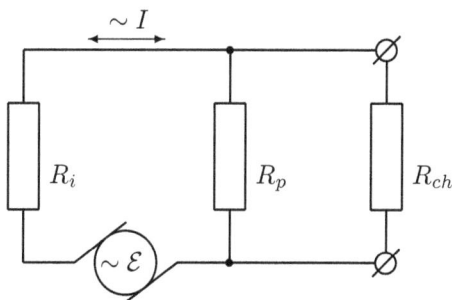

Fig. 18.2: Equivalent scheme of microwave oven.

fast heating causes the release of vapor which can not get out of the shell. The pressure grows precariously, till... BLAST! The egg blows up. The whole compartment has gotten splashed with your anticipated breakfast and you passionately disparage the lack of foresight while scrubbing it.

Now, after the relatively harmless explosion, it is the right time to talk about safety. Remember that the maximum power of microwave ovens is rather high, being about $1\,kW$. The power can be regulated but still there must always be something inside to dissipate the heat. Therefore it is strictly forbidden to operate an empty oven. Missing the object to act upon, the high frequency field "starts searching" where it could dump the energy and turns the firepower against itself. The heat will be released in the emitting elements and destroy them.

In order to illustrate what happens let us look for an analogy. Suppose that we placed into our microwave a seasoned chicken. The physics of the frying process are reflected by the following equivalent scheme, Fig. 18.2. A source of alternating voltage with the electromotive force $\mathcal{E}$ (that imitates the microwave generator) with the internal resistance R_i is closed by a parasite resistor R_p (in our case this role belongs to emitting elements and walls of the oven) and connected to some external load R_{ch} (probably you have already guessed that ch stands for the chicken). No doubt engineers do their best in order to maximize R_p and minimize the losses. But although $R_p \gg R_i$ it is impossible to make it infinite. Inasmuch as R_p and R_{ch} are connected in parallel the net resistance is $R_n = R_p \cdot R_{ch} / (R_p + R_{ch})$. The current in the circuit is $I = \mathcal{E} / (R_n + R_i)$ and the power spent by the source $P = \mathcal{E} I$ is divided between the three resistances and converted to heat. The optimal working regime corresponds to $R_{ch} \approx R_i \ll R_p$ and the heat released in the circuit is distributed as follows:

$$P_i^{\text{opt}} \approx P_{ch}^{\text{opt}} \approx \frac{P_0}{2} = \frac{\mathcal{E}^2}{4R_i}; \qquad P_p^{\text{opt}} \approx \frac{\mathcal{E}^2}{4R_p} = P_0 \frac{R_i}{2 R_p} \ll P_0;$$

here P_0 is the power consumed under the optimum conditions. Note that one half of the energy (up to $0.5\,kW$) is dissipated in the source itself. This explains why a ventilator is an indispensable feature of a microwave oven.

Now, what happens if bad boys have replaced the true chicken by a plastic copy with $R_{ch} \to \infty$? (This is almost the same as leaving the oven empty.) The total energy consumption will substantially fall down and the power will be shared as follows (∞ stands for the dielectric chicken, $R_{ch} = \infty$):

$$P_i^{\infty} \approx \frac{\mathcal{E}^2 R_i}{R_p^2} = 2P_0 \frac{R_i}{R_p} \ll P_i^{\text{opt}} \qquad \text{and} \qquad P_p^{\infty} \approx \frac{\mathcal{E}^2}{R_p} \approx 4\,P_p^{\text{opt}}.$$

Note that the parasite heat production has increased four times and the greater part of it falls on the emitters. Getting them fried is hardly desirable.

Operating an empty microwave oven is fatal for the emitting elements.

But a much more serious danger is concealed in choosing wrong utensils. Of course some prefer to buy special glass-ware *"Pyrex"* and use it. But we recommend you to think over the physical requirements of kitchenware and take an old earthen pot and save the money. The main demand is that the pot must be transparent for the microwave radiation. It must remain dielectric even at high frequencies. Electric and magnetic properties of the material can markedly depend on the frequency of electromagnetic field. So, clearly, not every dish serves the purpose. And under no pretext should metal pots, foil-wrapped foods, or even gold-rimmed china be put into the oven. In the twinkling of an eye the sympathetic kitchen accessory will turn into its fire-spouting relative from the blacksmith's shop and wreak havoc in and outside the microwave.

Let us turn to the equivalent scheme in Fig. 18.2 once again. Suppose that grandma stuffed a chicken with prunes and chestnuts, sprinkled it with ground cloves and cinnamon, then arranged all on her trusty cast iron pan and put that into a microwave. Because of the pan, the effective resistance of such a masterpiece is zero, $R_{ch} = 0$, and it will short the current source. The current will bypass the parasite resistance and take the short-circuit value, $I_0 = \mathcal{E} / R_i$. The frying pan will get no heat, $P_{ch}^0 = I_0^2 R_{ch} = 0$ and the power will dissipate only in the microwave source:

$$ P_i^0 = \frac{\mathcal{E}^2}{R_i} = 4P_i^{\text{opt}}. $$

You see that now the heat released in the electric part exceeds the standard value four times and reaches $2\,P_0 \approx 2\,kW$! Good if the device alone will be fused.

Metal in the chamber of a microwave oven short-circuits the high-frequency generator.

Nevertheless a lot of earthen pots and ceramic plates will do. To make sure whether a utensil fits microwave cooking you may put it into the chamber along with a glass of water (what for?) and turn the switch on. If two minutes later the object of testing remained cool leave your worries, it has passed the exam.

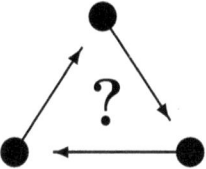

Try to estimate the depth of penetration into the brain of the radio waves emitted by a cellular phone ($\nu = 1800\ MHz$).

Chapter 19

The physics of baking good pizza

At first, pizza seems like a simple dish;
after examination, it is a compound dish.
The pizza is in oil, the pizza is in bacon,
the pizza is in lard, the pizza is in cheese,
the pizza is in tomatoes, the pizza is in anchovies.
It is the gastronomic thermometer of the market:
it rises or falls in price, according to the price of
the above-mentioned ingredients, according to
the abundance or the scarcity of the year.

Alexandre Dumas, *Sketches of Naples, 1843*

The history of pizza is crowded with tales, legends and anecdotes. Notwithstanding that, Italians are trusted as the inventors of this humble, delicious, and "universal" flatbread. The precursors of pizza are the Neolithic unleavened flatbreads baked on fire-heated rocks and made from coarse grained flour (spelt, barley, emmer), developed autonomously in several areas from China to the Americas. The Italian word "pizza" first appeared on a Latin parchment (Codex Diplomaticus Cajtanus) reporting a list of donations due by a domain tenant to the bishop of Gaeta (Naples). The document dated 997 AD fixes a supply of *duodecim pizze* ("twelve pizzas") every Christmas Day and Easter Sunday. Etymologists debate on different origins of the word "pizza": from the Byzantine Greek $\pi\iota\tau\alpha$- bread, cake, pie, pitta; from Greek $\pi\eta\kappa\tau\eta'$ -congealed; from Latin *picta* - painted, decorated, or *pinsere* -flatten, enlarge. Since the 8th century BC, the Greek colonies in southern Italy (Napoli itself was founded around 600 B.C. as a Greek town), produced the *Plakuntos*, a flat baked grain-based sourdough covered with oil, garlic, onion, herbs, occasionally minced meat or small fishes, almost similar to actual Turkish pide. Plato mentions "cakes" made from barley flour, kneaded and cooked with olives and cheese.

Greeks familiarized the whole Mediterranean with two Egyptian procedures: the leavening and kneading of the dough to get a more digestible bread, and the use of cupola-roof-ovens instead of open fires. Greek bakers who became popular in Rome since the 4th century BC converted flavored and garnished *Plakuntos* into the roman *placenta* - flat bread.

No doubt that more than twenty centuries ago Greeks and Romans created the prototypes of pizza, but it was the Neapolitans, "inventors" of this inexpensive food that could be consumed quickly, who were responsible for the addition of the ingredients universally associated with pizza today — tomato and mozzarella cheese (Figure 19.1). Neapolitans become familiar with the "exotic" tomato plant after Columbus but the fruit was believed to be poisonous. Barely and hazardously eaten by peasants, tomatoes first appeared in the recipe book "Il Cuoco Galante" (The Gallant Cook) written in 1819 by chef Vincenzo Corrado. The first mentioning of "mozzarella" can be found in the recipe-book Opera (1570) by Bartolomeo Scappi. Up to the present time, genuine mozzarella is still produced from fat buffalo milk in the vicinity of Naples (Figure 19.2). It is a very delicate product, not only in taste, but also to store; mozzarella does not handle low temperatures well. It becomes "rubbery" If kept in a refrigerator. It should only be stored in its own whey at room temperature for a few days.

Fig. 19.1: Mozzarella di buffala and cherry tomatoes.

Documents demonstrate that until the 18th century, the Neapolitan pizza was a simple dish of pasta baked or fried, flavored with lard, pecorino-cheese, olives, salt or small fishes called *cecinielli*. During the 19th century up to two hundred pizzaioli (pizza-makers) crowded the streets of Naples selling for a nickel baked or fried pizzas dressed with tomato sauce and basil leaves. In 1889, a few years after the unification of Italy, the pizzaiolo Raffaele Esposito decided to pay homage to the Italian Queen by adding mozzarella to the traditional tomato and basil pie. The combination of red, white, and green symbolized the colors of the new Italian flag and the tri-color pizza, still known as *Pizza Margherita*, transformed the pie to a success that the modest pizzaiolo could never have imagined.

Fig. 19.2: Buffalo ranch near Capaccio(Campagna region), Italy.

In Italy today, pizza is made in different regional styles. Neapolitans are famous for their round-shaped pizzas with a high crusty edge or *cornicione*. Beyond the mentioned *Margherita* the standard is *Pizza Napoletana* (protected by the European Union as a Guaranteed Traditional Speciality), dressed with tomato, mozzarella and anchovies. The simplest version is called *Marinara*, simply topped with tomato, garlic, oregano and oil. Elaborate and opulent is the *Calzone* a circle of pizza dough folded into a half-moon stuffed with ricotta, mozzarella, salami or prosciutto. Romans, devoted to crunchy flatbreads, add more water to the flour (up to 70%) and add olive-oil or lard to the dough, so their pizzas can be stretched to the thickness of canvas without losing its toothsome chewiness; condiments are the same for the *Napoletana* but the pizza served at the table of a roman pizzeria is without *cornicione*. Roman bakeries and groceries usually sell *pizza bianca* by weight, a "white", rectangular, crusty pizza topped only with oil and salt. *Sardenaira* is a flatbread with tomatoes, olives and anchovies typical of Liguria but originated in the nearby Provence where it is called *pissaladire* (from the anchovy paste called *pissalat* — salt fish).

Typical of the Abruzzis tradition is the *Pizza di Sfrigoli* made by deeply kneading lard, flour, and salt, then incorporating little pig bits (sfrigole) before baking, while in Apulia the beloved recipe is *Pizza Pugliese*, thin and covered with tomato sauce and a lot of stewed onions (anchovies and olives are optional). Another Apulian specialty is the *Panzerotto*, a little pizza pocket served to mark the beginning of Carnival. Panzerotto differs from the Neapolitan calzone in both size and method of cooking (it is deep-fried, not baked). The classic filling is tomato sauce and fresh mozzarella, but many variations exist. Calabrians — devoted to sharp and spicy flavors — enrich their pizzas with hot salami (*soppressata*) or *nduja*, a spreadable and fiery blend of lard and chili peppers probably introduced into Calabria by the Spanish. The Sicilian version of pizza is called *sfinciuni* (from the Latin spongia — sponge), a rectangular thick and cushioned flatbread generously dressed with olive oil, onions, sheep's cheese and sun-dried tomatoes. *Scaccia* — a specialty of the Ragusa province in Sicily is a thinly rolled dough spread with tomato sauce and cheese and then folded up on itself to resemble a strudel; the long, rectangular pizzas are then sliced, revealing the layers of crust, sauce, and cheese. Sardinian *panada* is a nourishing pizza shield filled with eggplant, lamb and tomatoes or in a seafood version, stuffed with fish or buttery local eel.

Essentially an open sandwich, Neapolitan pizza disembarked to the USA in the late 19th and early 20th centuries when Italian immigrants, along with millions of other Europeans, were coming to New York, Trenton, New Haven, Boston, Chicago and St. Louis. Flavors and aromas of humble pizzas sold on the streets by Neapolitan pizzaioli to the country fellows began to intrigue the Americans. Neapolitan immigrant Antonio Pero began making pizza for Lombardis grocery store, which still exists in NYs Little Italy and in 1905 Mr. Gennaro, Lombardis owner, was licensed to open the first Pizzeria in Manhattan on Spring Steet. Due to the wide influence of Italian immigrants in American culture, the U.S. has developed regional forms of pizza, some bearing only a casual resemblance to the Italian original. Chicago has its own style of a deep-dish pizza. Detroit also has its unique twice-baked style, with cheese all the way to the edge of the crust, and New York City's thin crust pizzas are well-known. St. Louis, Missouri uses thin crusts and rectangular slices in its local pizzas, while New Haven-style pizza is a thin crust variety that does not include cheese unless the customer asks for it as an additional topping.

Being curious, the authors began looking into the secrets of making pizza. Rule number one, as Italians told them, was to always look for a pizzeria with a wood burning oven (not with an electrical one). Good pizzerias are proud of their "forno" ("oven" in Italian), in which you can see with your own eyes the whole process of baking. The pizzaiolo forms a dough disc, covers it with topping, places the fresh pizza on top of a wooden or aluminum spade, and finally transfers it into the oven. A couple of minutes later it is sitting in front of you, covered with mouth-watering bubbles of cheese, encouraging you to consume it and wash it down with a pitcher of good beer. The authors received useful advice from a friendly pizzaiolo who was working in a local Roman pizzeria, frequently visited by them when they lived in that neighborhood: "Always come for a pizza either before 8 p.m. or after 10 p.m., when the pizzeria is half-empty." The advice was also confirmed by one of the pizzerias frequent visitors: a big grey cat. When the pizzeria was full, the cat would leave, and did not show any interest in what was on the patrons plates.

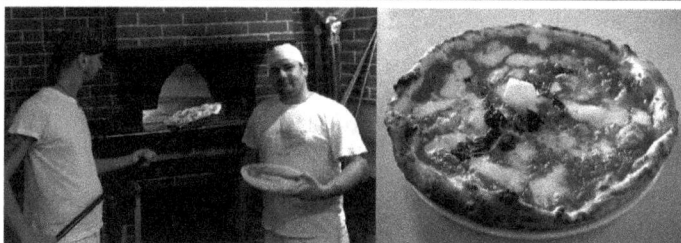

Fig. 19.3: Two modern pizzaiolos in Roman pizzeria ''Due Leoni'' in front of a brick pizza oven and their pizza ''Margherita''.

The reason for this advice was very simple: oven capacity. As the pizzaiolo explained, $325 - 330°C$[a] is the optimal temperature for Roman

---

[a]Obviously, the temperature depends on the way the dough was prepared and stored. The pizzaiolo Antonio prepares the dough well in advance: 24 hours before baking the pizza. After mixing all ingredients and kneading it well, he leaves the dough to rest for several hours, and then cuts it into pieces and forms ball-shaped portions. For Naples-style pizza the portion is 180-250 grams, for Roman-style less. These portions are used for a single pizza. Then he puts the dough balls into wooden boxes, where the dough

pizza baked in a wood burning oven with a fire-brick bottom. In this case, a thin Roman pizza will be done in 2 minutes. Thus, even putting two pizzas into the oven, the pizzaiolo can serve 50-60 clients within an hour. During peak hours, about one hundred customers frequent the pizzeria and at least ten clients are waiting for a take-out pizza. To meet the demand, the pizzaiolo increases the temperature in the oven up to $390°C$, and pizzas "fly out" of the oven every 50 seconds (hence, each one requires a "baking time" of around 1.5 minutes). However, their quality is not the same: the bottom and the crust are a little "overdone" (slightly black), and the tomatoes are a little undercooked.

Since it is not always easy to find a pizzeria with a brick oven, let us take a look at what advantages it has compared to an electric oven and whether there is a way to improve the latter to produce a decent pizza.

To illustrate the physical principles involved in baking pizzas, let us consider a common example of how heat is transferred. Imagine when you were a child and had a fever, but no thermometer at hand. Your mother would put her hand on your forehead and quickly say: "you have a high temperature, no school for you tomorrow". To investigate this process scientifically, we start by simplifying the problem. Let us imagine that your mom is touching your forehead with her own forehead rather than her hand. In that case, if the temperature of your forehead would have been $38°C$, and your mothers $36°C$, it is clear by the symmetry of the problem that the temperature at the interface (T_0) between the two foreheads will be $37°C$, and that your mother would feel the flow of heat coming from your forehead (the actual temperature distribution in time is shown in Figure 19.4).

Now let us assume that your mothers head is made of steel, and her temperature is the same $36°C$. Intuitively, it is clear the temperature at the interface will decrease, let us say, to $36.3°C$. This is related to the fact that the steel will draw off the heat from the interface region effectively to its bulk, since its heat conductivity is large. It is also clear that this removal becomes more efficient, when a smaller amount of heat needs to be drawn away from the interface region (i.e., the heat removal increases when the specific heat capacity of the "mothers head" material decreases, an illustration is shown in Figure 19.5).

---

ferments for 4-6 hours. After that it is ready to be baked or placed in a refrigerator for later use.

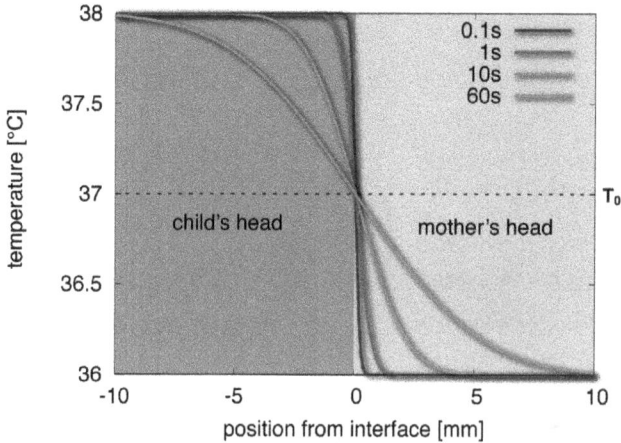

Fig. 19.4: Temperature profile within mother's and child's head -- 0.1 s, 1 s, 10 s and 60 s after they made contact.

Let us now analyze the process of pizza baking more scientifically. We start by reminding the reader of the main concepts of heat transfer. When we speak about "heat",we usually have in mind the energy of a system (like the mothers head, the oven, or the pizza itself) associated with the chaotic motion of atoms, molecules and other particles it is composed of. We inherited this concept of heat from the physics of a past era. Physicists say that heat is not a function of the state of a system, its amount depends on the way the system achieved this state. Like work, heat is not a type of energy, but rather a value convenient to use to describe energy transfer. The amount of heat, necessary to raise the temperature of a mass unit of the material by one degree, is called a specific heat capacity of the material:

$$c = \frac{\Delta Q}{M \Delta T}. \tag{19.1}$$

Here M is the mass of the body and ΔQ is the quantity of heat required for heating it by a temperature ΔT. From this expression it is clear that the heat capacity is measured in $J \cdot kg^{-1} \cdot K^{-1}$ in SI units.

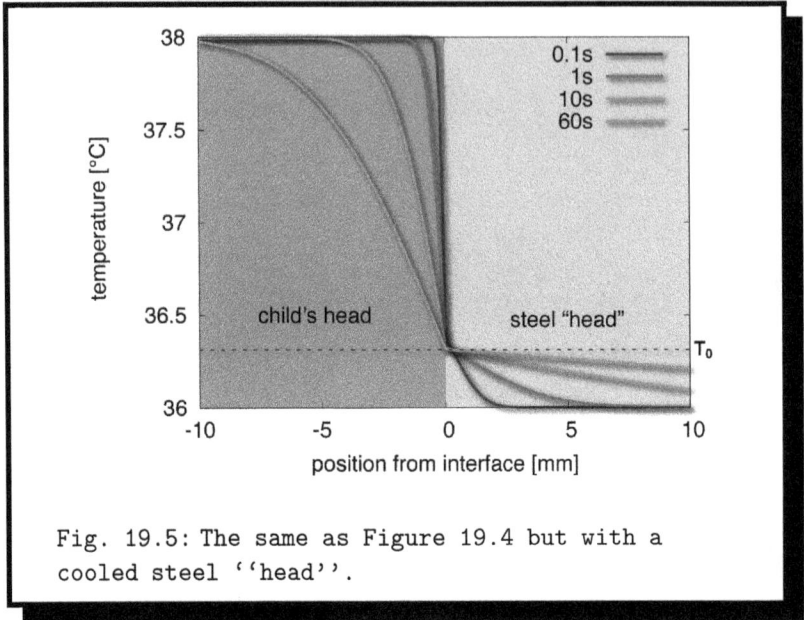

Fig. 19.5: The same as Figure 19.4 but with a cooled steel ''head''.

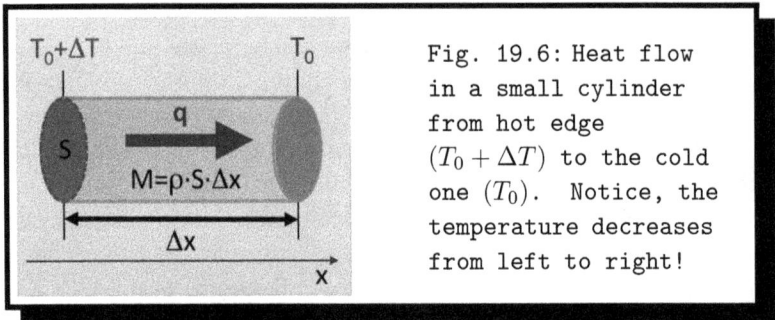

Fig. 19.6: Heat flow in a small cylinder from hot edge $(T_0 + \Delta T)$ to the cold one (T_0). Notice, the temperature decreases from left to right!

In the case of a thermal contact between the two systems with different temperatures, the heat will go from the warmer part of the system to the cooler one. The heat flux density is the amount of heat that flows through a unit area per unit time in the direction of temperature change:

$$q = \frac{\Delta Q}{S \Delta t}. \tag{19.2}$$

Here S is the area of the cylinder cross-section. In the simplest case of a homogeneous non-uniformly heated system, using Eq. (19.1), one finds

$$q = \frac{cM\Delta T}{S\Delta t} = c\rho\frac{(\Delta x)^2}{\Delta t}\left(\frac{\Delta T}{\Delta x}\right) = -\kappa\frac{dT}{dx}, \tag{19.3}$$

where ρ is the mass density. Assuming that Δx is small, we identified the value in parentheses as the derivative of the temperature by the coordinate x and took into account the fact that the temperature decreases in the x-direction (see Fig. 19.6). In the general case, $\mathbf{q}$ is a vector and the derivative in Eq. (19.3) is replaced by the gradient ∇T, which describes the rate of temperature change in space. The coefficient κ in Eq. (19.3) is the thermal conductivity, which describes the ability of a material to transfer heat when a temperature gradient is applied.[b] Eq. (19.3) expresses mathematically the so-called Fourier's law, which is valid when the temperature variation is small.

Next, let us analyze how a temperature front penetrates a medium from its surface, when a heat flow is supplied to it (see Fig. 19.6). Assume that during time t the temperature in the small cylinder of the height $L(t)$ and cross-section S has changed by ΔT. Let us get back to Eq. (19.3) and rewrite it by replacing Δx by $L(t)$:

$$\frac{c\rho L(t)\Delta T}{T} = \kappa\frac{\Delta T}{L(t)}. \tag{19.4}$$

Solving it with respect to the length one finds:

$$L(t) \sim \sqrt{\frac{\kappa t}{c\rho}} = \sqrt{\chi t}, \tag{19.5}$$

i.e., the temperature front enters the medium by the square-root law of time. The time after which the temperature at depth L will reach a value close to the one of the interface depends on the values κ, c, and ρ. The parameter $\chi = \kappa/(c\rho)$ is called the thermal diffusivity or coefficient of temperature conductivity. The heating time of the whole volume can be expressed in its terms: $\tau \sim L^2/\chi$.

Of course, our consideration of the heat penetration problem into a medium is just a simple evaluation of the value $L(t)$. A more precise

---

[b]The definition of the thermal conductivity $\kappa = c\rho\frac{(\Delta x)^2}{\Delta t}$, used in Eq. (19.3), requires a clarification: while our simplified derivation suggests a geometry dependence, we emphasize that in reality it is determined only by microscopic properties of the material.

approach requires solution of differential equations. Yet, the final result confirms our conclusion (19.5), just corrected by a numerical factor:

$$L(t) = \sqrt{\pi \chi t}. \tag{19.6}$$

Now that we know how heat transfer works, let us get back to the problem of calculating the temperature at the interface between two semi-spaces: on the left with parameters κ_1, c_1, ρ_1 and temperature T_1 at $-\infty$, and on the right with parameters κ_2, c_2, ρ_2 and temperature T_2 at ∞. Let us denote the temperature at the boundary layer as T_0. The equation of the energy balance, *i.e.*, the requirement of the equality of the heat flowing from the warm, right semi-space through the interface to the cold, left semi-space, can be written in the form

$$q = \kappa_1 \frac{T_1 - T_0}{\sqrt{\pi \chi_1 t}} = \kappa_2 \frac{T_0 - T_2}{\sqrt{\pi \chi_2 t}}. \tag{19.7}$$

Here we simplified the problem assuming that all temperature changes happen at the corresponding temperature dependent length (19.6). Solving this equation with respect to T_0 one finds that

$$T_0 = \frac{T_1 + \nu_{21} T_2}{1 + \nu_{21}}, \tag{19.8}$$

where

$$\nu_{21} = \frac{\kappa_2}{\kappa_1} \sqrt{\frac{\chi_1}{\chi_2}} = \sqrt{\frac{\kappa_2 c_2 \rho_2}{\kappa_1 c_1 \rho_1}}. \tag{19.9}$$

One notices, that time does not enter in expression (19.8) (*i.e.*, the interface temperature remains constant in the process of the heat transfer, see Figures 19.4, 19.5 and 19.7). In the case of identical media with different temperatures one can easily find: $T_0 = \frac{T_1 + T_2}{2}$. This is the quantitative proof of the intuitive response we provided in the beginning of the article for the temperature $37°C$ of the interface between the mother's hand and the child's forehead. If the mother's hand was made of steel, her temperature would remain almost unchanged after contact with the hot forehead, meaning that she would not be able to notice the child's fever.

Finally, we are ready to discuss the advantages of the brick oven. Let us start from the calculation of the temperature at the interface between the pizza placed into the brick oven and its heated baking surface. All necessary parameters are shown in the table below:

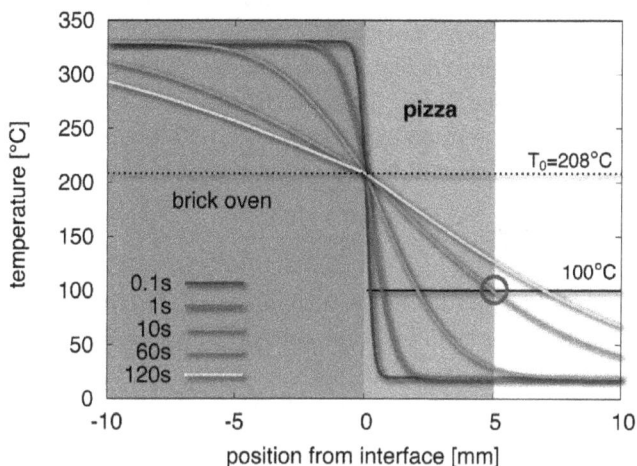

Fig. 19.7: Temperature profile in a brick oven with pizza at different times. At 60 s the top surface of the pizza reaches $100°C$ (circle). Here we only take thermal diffusion into account. Evaporation and radiation are neglected.

Table 19.1

Material	Heat capacity $c\ [\frac{J}{kg \cdot K}]$	Thermal conductivity $\kappa\ [\frac{W}{m \cdot K}]$	Mass density $\rho\ 10^3 [\frac{kg}{m^3}]$	Temperature conductivity $\chi\ 10^{-7} [\frac{m^2}{s}]$	ν_{21}
dough (2)	$2-2,5 \cdot 10^3$	0,5	$0,6-0,8$	$2,5-4,2$	1
steel (1)	$4,96 \cdot 10^2$	18	$7,9$	45	0,1
fire brick (1)	$8,8 \cdot 10^2$	0,86	$2,5$	$4,0$	0,65

Assuming the initial temperature of the pizza dough as $20°C$, the temperature inside the oven, as our pizzaiolo claimed, being about $330°C$, and the ratio $\nu_{d,fb} = 0,65$ we find for the temperature at the boundary layer between the oven surface and pizza bottom

$$T_0^{bp} = \frac{330°C + 0,65 \cdot 20°C}{1,65} \approx 208°C.$$

As we know from the words of the same pizzaiolo, a pizza is perfectly baked in two minutes under these conditions.

Let us now repeat our calculations for the electric oven with its baking surface made of steel. For an electric oven the ratio $\nu_{d,s} = 0,1$, and if heated to the same temperature of $330°C$, $T_{1*}^{eo} = T_1^{wo} = 330°C$, the temperature at the bottom of the pizza will be equal to

$$T_{0*}^{eo} = \frac{330°C + 0,1 \cdot 20°C}{1,1} \approx 300°C.$$

That is too much! The pizza will just turn into coal! This interface temperature is even much higher than in Naples pizzerias, where oven temperatures between $400 - 500°C$ are common.

Well, let us formulate the problem differently. Let us assume that generations of pizza makers, who were using wooden spade to transfer pizzas into the oven, are right: the temperature at the (Roman) pizzas bottom should be about $210°C$. What would be the necessary temperature for an electric oven with steel surface?

The answer follows from the Eq. (19.8) with the coefficient $\nu_{d,s} = 0,1$ and solved with respect to T_1^{eo} when the temperature at the bottom of the pizza is the same as in the wood oven: $T_0^{eo} = T_0^{wo}$. The result of this exercise shows that the electric oven should be much colder than the brick one: $T_1^{eo} = 230°C$.

It seems that if one is able to forget the flavor of burning wood, sizzling dry air in the brick oven and the other natural features, the problem would be solved let us set electric stove controls to $230°C$ and in a couple of minutes we can take an excellent pizza out of the oven. But: Is it that easy?

In order to answer this question, we first need to consider the second important mechanism of heat transfer: *thermal radiation*. Its intensity, the amount of radiation energy arriving each second to $1\,m^2$ of surface in the oven, is determined by the Stefan-Boltzmann law:

$$I = \sigma T^4, \tag{19.10}$$

where $\sigma = 5{,}67 \cdot 10^{-8}\,W/(m^2 \cdot K^4)$ is the so-called Stefan-Boltzmann constant.

A typical brick oven has a double-crown vault filled with sand, which is kept at almost constant temperature. Its walls, as well as the bottom part, are also heated meaning that the complete volume of the oven is "filled" by infrared radiation. With a temperature that high, this radiation becomes significant: the pizza here is continuously "irradiated" from both sides by a "flow" of infrared radiation of the intensity

$$I^{bo} = \sigma(T_1^{bo})^4 = 5{,}67 \cdot 10^{-8}(603)^4 = 7{,}5\,kW/m^2,$$

i.e., each second an amount of energy close to $0{,}75\,J$ arrives at $1\,cm^2$ of pizza.

Here one should notice, that, in its turn, the pizza also irradiates out a "flow" of the intensity. Since the major part of the baking time is required for the evaporation of water contained in the dough and toppings, we can assume $T_{pizza} = T_b = 100°C = 373\,K$, which results in a radiation intensity of $I_{pizza} = \sigma(T_b)^4 = 1{,}1\,kW/m^2$, *i.e.*, 15% of the obtained radiation, the pizza returns to the oven.

For the much less heated electric oven, the corresponding amount of energy, incident to $1\,cm^2$ of pizza surface, is more than twice less:

$$I^{eo} = \sigma(T_1^{eo})^4 = 5{,}67 \cdot 10^{-8}(503)^4 = 3{,}6\,kW/m^2,$$

while the returned radiation is the same: $1{,}1\,kW/m^2$.

Now it is a time to evaluate what amount of heat $1\,cm^2$ of pizza receives per second through its bottom. By definition it is determined by the heat flow (19.3) and in order to get its numeric value we will evaluate the temperature gradient at the oven surface in the same way as was already done in Eq. (19.7):

$$q(t) = \kappa \frac{T_1^o - T_0}{\sqrt{\pi \chi t}},$$

where T_1^o is the temperature of the oven. One can see, that, contrary to the Stefan-Boltzmann radiation, the heat flux arriving into the pizza by means of a heat conductance depends on time. Correspondingly, the amount of heat transferred to $1\,cm^2$ of pizza in this way from the oven during time is determined by

$$Q(\tau) = \int_0^\tau q(t)\,dt = 2\kappa(T_1^o - T_0)\sqrt{\frac{\tau}{\pi \chi}}.$$

Therefore, the total amount of heat, arriving at $1\,cm^2$ of pizza during time τ, is

$$Q_{tot}(\tau) = \sigma[(T_1^o)^4 - (T_b)^4]\tau + 2\kappa(T_1^o - T_0)\sqrt{\frac{\tau}{\pi\chi}}. \qquad (19.11)$$

This value is used to heat $1\,cm^2$ of pizza from the dough temperature $20°C$ to the boiling temperature of water $100°C$:

$$Q_h = c^d \rho^d d_p (T^b - T_0^d).$$

Here d_p is the thickness of the pizza, which we assume to be $d_p = 0.5cm$.

Yet, this is not all. During the process of baking the perfect pizza we apparently evaporate water from the dough, tomatoes, cheese, and other ingredients. We need to take the required energy for this into account as well. If one assumes that the water mass fraction evaporates from the dough and all topping one gets:

$$Q_{ev} = \alpha \mathcal{L} \rho^w d_p.$$

Here $\mathcal{L} = 2264.76J \cdot g^{-1}$ is the latent heat of evaporation for water. Collecting both these contributions in one, we can write

$$Q_{tot} = Q_h + Q_{ev} = c^d \rho^d (T^b - T_0^d)d_p + \alpha \mathcal{L} \rho^w d_p. \qquad (19.12)$$

Equating (19.11) and (19.12) one finds the final equation determining the baking time of pizza:

$$\sigma[(T_1^o)^4 - (T_b)^4]\tau + 2\kappa(T_1^o - T_0) = c^d \rho^d (T^b - T_0^d)d_p + \alpha \mathcal{L} \rho^w d_p. \quad (19.13)$$

In order to obtain a realistic answer for the baking time, it is important to know the amount of water, which is evaporated during the baking process. A typical recipe for pizza Margarita calls for $240\,g$ of dough and $90\,g$ of toppings (consisting of tomatoes and mozzarella). The dough contains about one-third of water and the toppings 80% (the rest is mostly fat from the cheese). Together with a weight loss of $30\,g$, a good assumption is a 20% loss of water, i.e. $\alpha = 0.2$. Using this with the values of specific heat capacity and density for dough from the table above, one finds that $Q_{tot} = (70 + 226)\,J/cm^2$, which gives for the baking time in the wood oven $\tau_{wo} \approx 125\,s$. For the electric oven an analogous calculation results in an almost 50% longer time: $\tau_{eo} \approx 170\,s$. We see that we have succeeded to reproduce the value disclosed to us by our pizzaiolo: 2 minutes for baking of the Roman pizza in a wood oven. The result of an attempt to bake a pizza in the electric oven will be the mentioned unbalanced product.

Using Eq. (19.8) one can easily find that the temperature at the interface between the pizza and oven surface reaches $240°C$, when the temperature in wood fired brick oven increases to $390°C$. Replacing correspondingly in Eq. (19.13) one can find the baking time under these extreme conditions to be approximately 82 seconds, hence the productivity of the oven increases by almost 50%!

A final "trick" disclosed to us is the importance for pizzas with water-rich toppings (eggplants, tomatoes slices, or other vegetables). In this case, the expert first bakes the pizza in the regular way on the oven surface, but when the pizza's bottom is ready he lifts it with the wooden/aluminum spade and holds it elevated from the baking surface for another half minute or more in order to expose the pizza to just heat irradiation. In this way they avoid burning the dough and get well cooked toppings. Certainly, as is routinely done in physics, in order to get to the core of the phenomenon, we examined only the simplest model here (in particular, we ignored the third mechanism of the heat transfer: convection, which we can assume to have only a small effect), which captures the essential physical processes.

As a final note, we remark that it is difficult to build a classic brick oven, and many customers do not appreciate the difference between an excellent and decent pizza. These are the reasons why engineers invent all sorts of innovations: for example, inserting a ceramic bottom made of special ceramics to imitate the bottom of brick ovens in a modern professional electric oven. To bake a pizza evenly, rotating baking surfaces are available, convection ovens emulate the gas flows in brick ovens, and many other things. But, the dry heat and the smell of wood in traditional firebrick ovens remains the ideal way to bake the perfect pizza.

Chapter 20
Boiling, steaming or rinsing?

The art of Chinese cooking is to make the meat taste
like vegetables and the vegetables taste like meat, without either the meat or vegetables losing their original
texture.

Richard Hughes, *"Foreign Devil."*

Globalization, rapidly taking place in the world, is vividly manifested
by the ubiquitous availability of dishes of various cuisines from all over the
world. Of course, as a rule, this is just some semblance of true masterpieces
of culinary art: in addition to the skill of the cook, the creation of the latter
requires the corresponding products. As our familiar Italian gastronomic
critic, Sergio Grasso, says, "food does not go to a person, this person should
travel to food."

Chinese cuisine is one of the richest and most interesting in the world.
There everything, or, well, almost everything, can be eaten. The ways of
food cooking in Chinese cuisine are very different, being interesting not only
due to the exotic tastes of the final product but also due to the unusual
physical processes lying in their base.

Speaking about the Chinese cuisine, the first that comes in mind, probably, are the dumplings, the most worldwide popular Chinese dish, which
you can eat in Chicago, Canberra and Moscow. Dumplings are honored all
over China, especially in Jiangnan (close to Yangtzi Delta) region of China.
Three types of dumplings are commonly seen, especially in Shanghai and
Suzhou, namely, Siaulon Pau, Santsie Moedeu and Wonton. Made of a
thin dough, filled with pork, fat of which melts into soup when cooked,
and served with Chinkiang vinegar, they are pretty much the same. The

most significant difference is the method of their heat treatment. Santsie Moedeu (large size dumpings) are pan-fried, this process is called in Shanghai "Santsie". Wonton dumplings are boiled in water at $100°C$, whereas Siaulon dumplings are processed also at $100°C$, but in the atmosphere of saturated steam in small bamboo steaming baskets which are called "Siaulon".

Fig. 20.1: The three kinds of dumplings in Jiangnan, China. The left one is Santsie Moedeu, the middle one is Siaulon Pau and the right one is Wonton.

Hotpot is also a popular type of Chinese cuisine. Originated in Mongolia more than a thousand years ago and gaining its popularity in times of the Qing Dynasty[a] all over the country, hotpot boasts a profound history. During its spread, hotpot has been diversified into many variations.

Beijing hotpot lays particular emphasis on the soup base and sauces, Chongqing hotpot boasts a stimulating and refreshing "Ma La" ("numb and spicy") flavor, and Chaoshan hotpot is famous for its deliberately-prepared thin-cut mutton, named "Shuan Yangrou" in Chinese.

Fig. 20.2: Cooking in the hotpot.

---

[a]The Qing Dynasty was the last imperial dynasty of China, established in 1636 and ruling China from 1644 to 1912 with a brief, abortive restoration in 1917. It was preceded by the Ming Dynasty and succeeded by the Republic of China.

When enjoying hotpot, one puts ingredients such as beef balls, fish balls, crab meat, or vegetable slices into the elaborately prepared soup base and wait for it to be done. After picking it up and dipping it in the sauce, delicious food is ready to eat. The whole process is called "Zhu" ("to boil") in Chinese, and takes 5-10 minutes or so, being applied to meat balls and vegetables. Remarkably, in the time between, another process can be used to cook a different kind of food in the same hotpot, but much faster. It is called in Chinese "Shuan" ("rinse" or "instant-boil"). It consists of soaking the thin-cut sliced beef or mutton in the boiling soup. Surprisingly that in only 10 seconds the sliced beef changes its color from pink to white or gray, indicating the slice is ready to eat. The beef slice becomes ready even while remaining between the chopsticks.

Today, cooking has become not only a giant industry, not just art, but also a vast field of science. Here, biology, chemistry, physics, economics, ethics and much more intersect. The tasks of this science are infinite. All the time, the new methods of cooking appear. We do not even try to list them - neither the frying of meat, nor baking turkey, nor about the preparation of a BBQ on charcoal will be discussed here. Let us talk about the physical processes underlying cooking on the example of the dishes described above - about the physics of boiling, steaming, and "rinsing in hotpot".

20.1 Boiling

What is the essence of the process of meat boiling? In the everyday language, raw meat should become cooked. And what does this mean "scientifically"?

Meat, basically, consists of complex organic macromolecules called proteins (the type of protein varies from the type of meat). In raw meat, protein molecules are in a state of entangled long chains (see Figure 20.3). In the course of heat treatment, the temperature rises and these chains straighten, and when the temperature reaches the specific for each type of meat value T_d, they are compactified into some kind of "carpet". This process is called protein denaturation. It occurs at relatively low temperatures: for meat it is $55 - 80°C$, for fish the temperature is even lower. In any case, anyone who has ever eaten a chicken soup can be sure that boiling at $100°C$, turns out to be sufficient for complete compacting of proteins in the meat.

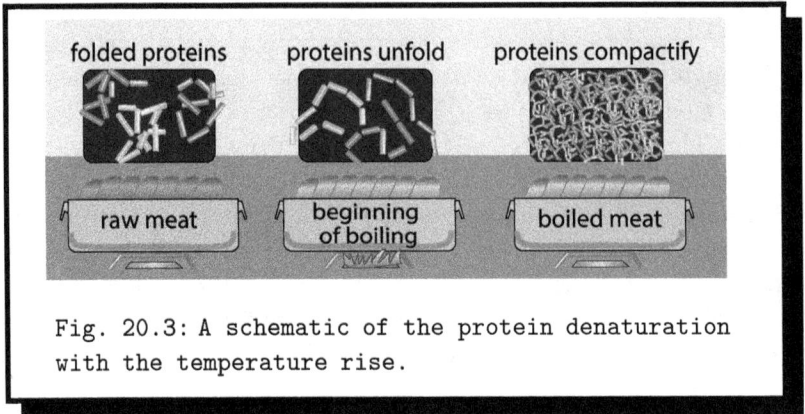

Fig. 20.3: A schematic of the protein denaturation
with the temperature rise.

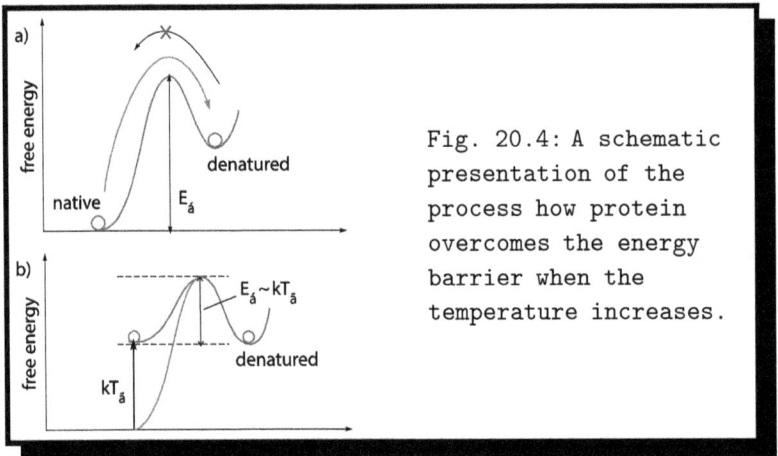

Fig. 20.4: A schematic
presentation of the
process how protein
overcomes the energy
barrier when the
temperature increases.

From the point of view of physics, the states of proteins in raw meat and
boiled meat differ in their energy. To turn the protein from its native state
to the denatured one, an energy barrier must be overcome (see Figure
20.4). At room temperature, this barrier is high. In the process of cooking,
temperature raises. Correspondingly this changes the energy of the protein,
as is shown in Figure 20.5. Having reached the top of the "hill", the protein
falls down to the new state a denatured protein the meat is cooked! This is

what happens in a pot of boiling soup.[b] So, the first task of the cook when boiling meat in terms of physics is to increase the temperature throughout the volume of the piece to at least the temperature of denaturation.

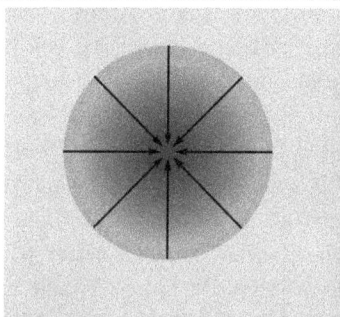

Fig. 20.5: Penetration of heat in the spherically symmetrical piece of meat.

In the light of the above, we formulate the simplest model of the process of meat cooking. Let a spherically symmetric homogeneous piece of meat (radius R) with an initial temperature and a thermal conductivity coefficient be placed in an environment with a fixed temperature maintained. How much time does it take for the temperature of the meat in the center of the ball to reach the same temperature as the environment?

In mathematical physics, the process of heat transfer inside a sphere is described by a complicated differential equation

$$\frac{\partial T(r,t)}{\partial t} = \frac{\kappa}{\rho c} \frac{\partial}{\partial r} \left(r^2 \frac{\partial T(r,t)}{\partial r} \right), \tag{20.1}$$

where $T(r,t)$ is the temperature at a point r at time t, κ is the thermal conductivity of the meat, ρ is its density, and c is the specific heat. Since the water is boiling in a saucepan, the temperature at the surface of the sphere at any instant of time remains constant and equal to $100°C$

$$T(r = R, \forall t) = 100°C. \tag{20.2}$$

---

[b]It should be noted that in recent years it became fashionable to cook meat at relatively low temperatures, the so-called "sous-vide" method. The meat is placed in a thermostat with a temperature somewhat lower than that of denaturation. Each separate macromolecule lacks energy alone to jump over the barrier. However, it can occasionally "borrow" it from the environment. So, gradually (it takes a long time - many hours, maybe even a day), all bulk of the meat transits into the specific denaturated state.

We took the meat from the refrigerator, so at the time when it was dropped into the water, the temperature was $4°C$ throughout its volume:

$$T(\forall r, t = 0) = 4°C. \tag{20.3}$$

Eqs. (20.1)-(20.3) determine so-called problem of solution of differential equation with the boundary conditions. How to deal with them is well known for mathematicians and knowing the numerical values of the thermal conductivity of meat, its density and specific heat they will be able to accurately write a recipe for cooking broth. Nevertheless, let us try to figure it out by ourselves using the dimensional analysis method. The temperature of denaturation of meat coincides by an order of magnitude with the boiling point of water (differs from it by $20 - 25\%$). Therefore, we assume that the time of "delivery" of the necessary temperature to the center of the solid sphere depends only on its material parameters and size: the thermal conductivity of the meat, its density, specific heat and radius. Therefore, we seek the dependence of the required time on the size of the sphere in the form:

$$\tau = \kappa^\alpha \rho^\beta c^\gamma R^\delta. \tag{20.4}$$

By comparing dimensions, we write:

$$[\tau] = [\kappa]^\alpha [\rho]^\beta [c]^\gamma [R]^\delta. \tag{20.5}$$

The dimension of the thermal conductivity $[\kappa] = \frac{kg \cdot m}{s^3 \cdot K}$. Substituting it, side by side with the dimensions of density ($[\rho] = \frac{kg}{m^3}$), specific heat ($[c] = \frac{J}{kg \cdot K}$) and radius ($[R] = m$), into Eq. (20.5) and then comparing them in the right and left hand sides, one finds: $\alpha = -1$, $\beta = \gamma = 1$, $\delta = 2$. Thus, we conclude that

$$\tau = C_0 R^2 / \chi, \tag{20.6}$$

where $\chi = \frac{\kappa}{\rho c}$ and C_0 is an unknown constant of the order of unity. Substituting the quantities $\kappa = 0,45 \frac{W}{m \cdot K}$, $\rho = 1,1 \cdot 10^3 \frac{kg}{m^3}$, $c = 2,8 \cdot 10^3 \frac{kJ}{kg K}$ we find that for the meat $\chi = \kappa / (\rho c) = 1,4 \cdot 10^{-7} \frac{m^2}{s}$. The latter value is called the coefficient of temperature conductivity.

Consequently, a half kilogram piece of meat should be cooked for about an hour and a half. The estimate is in some way exaggerated, since we do not distinguish in its process the temperature of denaturation from the boiling point, but the order of magnitude is correct. Returning to the dumpling whose diameter is about 2 cm, we find that it should be cooked for several minutes, which corresponds to our life experience.

20.2 Steaming

Now let us discuss the physical aspects of "Siaulon" dumplings preparation. Here the meat ball of radius R (our model of the dumpling) is placed into the atmosphere of saturated steam at $100°C$. The pressure here is the same as atmospheric pressure, i.e., is equal to $1 \, atm$. Formally the siaulon dumpling here can be considered under the same boundary conditions as the wonton dumpling in the boiling water. Indeed, it is taken from the same refrigerator and is placed into the environment with a temperature of $100°C$. Therefore, from the point of view of a mathematician, the propagation of heat in the siaulon dumpling is described by the same Eq. (20.1) with the boundary conditions (20.2) and (20.3). Therefore, if condition (20.2) is satisfied, then the temperature distribution inside it will be the same as for the wonton dumpling of the same size, and its preparation should take about the same time. However, a physicist is obliged to ask: how how does one ensure a temperature of $100°C$ on the surface of a siaulon dumpling?

In the case of a wonton this was easy: although immediately after its placement in the pan the boiling around it temporarily terminates, however, due to the high heat capacity of the water, its good enough heat conduction and convection, and the constantly supplied heat to the pan, the water very quickly will boil again providing the condition (20.2) and hence, the required heat flow into the dumpling.

In the case of the siaulon, cooked in the atmosphere of the saturated steam, the mechanism of the heat transfer into the dumpling is not so evident, starting from ensuring that the boundary condition (20.2) was satisfied. Here the heat transfer has a completely different character from the one discussed in the previous section. At the first moment, the vapor molecules near the still cold surface of the siaulon locally are in the state of a strongly supersaturated vapor. They begin to condense on the surface rapidly increasing its temperature up to the temperature of the ambient, $100°C$. Assuming that the temperature jump occurs in a very narrow region close to the surface, we return to the same equation (20.1) with boundary conditions (20.2) and (20.3). That is, the temperature distribution inside the siaulon should change with time in the same way as in the case of wonton boiled in the water. Consequently, the heat flux (see Eq. (19.2))

$$q(R,t) = -\kappa \left[\frac{\partial T(r,t)}{\partial r}\right]_{r=R} \tag{20.7}$$

at its surface should be the same. Yet, now this flow is provided not by the thermal conductivity of the water, but by the molecules of the vapor "landing" at $1\,cm^2$ of the surface during 1 second:

$$q(R,t) = H_v \cdot m(t) = H_v \frac{\mu_{H_2O}}{N_A} N(t).$$

Here H_v is the specific heat of vaporization, $N(t)$ is the number of molecules condensed per second, $m(t)$ is their mass, N_A is the Avogadro number, μ_{H_2O} is the molecular mass of water. Thus, the number of molecules "landing" in the steam atmosphere at a square centimeter of "siaulon" per one second

$$N(t) = N_A \frac{\kappa}{H_v \cdot \mu_{H_2O}} \left[\frac{\partial T(r,t)}{\partial r}\right]_{r=R}.$$

Perfect! Mathematicians find this number by solving a complex equation, the siaulon itself "feels" what heat flux it needs to keep the temperature at $100°C$ on the surface. It remains only to understand where from the molecules of the vapor "learn" how many of them should condense in a given second at a square centimeter of the siaulon surface.

Let us suppose that at some time instead of $N(t)$ molecules condense $N(t) + \delta N$ of them. The first of them are hospitably absorbed by the siaulon: in fact, they are necessary to keep in harmony a centigrade surface and a still cold inner part. The remaining δN molecules are persons *non-grate* they were not expected here, the temperature conductivity of the siaulon does not allow their heat released to penetrate into the dumpling. What do they have to do? To take off back? Too troublesome, so they stay on the surface locally increasing its temperature (see Figure 20.6). As a consequence, the point at the phase diagram (see Figure 20.6) that represents the local balance between vapor and water moves up along their coexistence line. Let us notice that the pressure in the system remains the same, equal to $1\,atm$. Therefore, above the selected square centimeter, the vapor locally ceases to be saturated. As a result, the next moment here will land somewhat less number of molecules with respect to the required amount. Consequently, the surface temperature will go down.

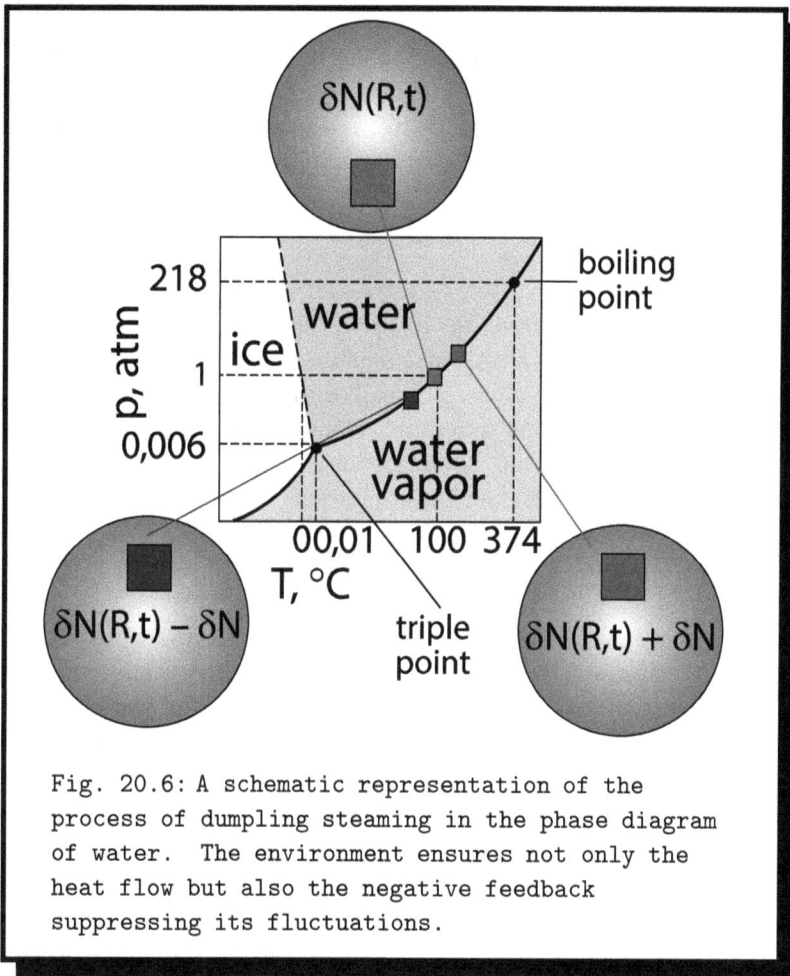

Fig. 20.6: A schematic representation of the process of dumpling steaming in the phase diagram of water. The environment ensures not only the heat flow but also the negative feedback suppressing its fluctuations.

Exactly the same mechanism works also in the case of the accidental lack of condensing of the required number of molecules per second for returning of the temperature to $100°C$ (see Figure 20.6). Such a mechanism of the self-regulation is called "the negative feedback".

Let us note the important culinary difference between boiling and steaming of dumplings. In the former case water penetrates into a dumpling due to the diffusion process. Interacting with its filling it creates a tasty juice. Due to the same diffusion, but in the opposite direction, this juice partially flows out from dumpling to the ambient, transforming the water in diluted broth. In the case of dumpling steaming the ambient is the saturated steam. Condensed at the surface of a dumpling water also diffuses into its bulk, but there is no inverse process. Hence, the juice of steamed dumpling is richer than that in the case of boiled dumpling.

20.3   Rinsing

At the beginning of the article we discussed the "hot pot" and the different ways of cooking meat in it. Yet, there remains the mystery of why the same amount of meat in the form of a meatball or thin slice differs in its cooking time tens of times. Why oil soup base "Hongyou" in Chongqing pot boils much earlier than water soup "Qingtang" used in its Chaoshan and Beijing versions? We will try to answer these questions below.

20.3.1   *On cooking times*

"Shuan" and "Zhu" is actually the same thermal process which was discussed in the previous sections. Here heat conductivity and convection provide a constant temperature $100°C$ on the surface of the object and heat enters inside it, raising the internal temperature. So, what causes this striking difference in time between rinsing the thin-cut beef and boiling the meatball? The answer requires an inspection of the thermal process of boiling.

What concerns the time of "zhu" everything is clear: the way of heat propagation in the ball, which can serve as a meatball model, we already discussed (cooking of wonton dumpling), and we can use Eq. (20.6) to estimate the corresponding time.

The cooking time of the "shuan" process requires separate consideration. We consider a body occupying the semi-space $x > 0$ (see Fig. 20.7). We assume its temperature to be T_0 at the initial moment $t = 0$ and rapidly increase it at the surface $x = 0$ to T_1. Let us fix $T = T_1$ at the surface $x = 0$ creating the corresponding heat flow. In Figure 20.7 is shown how the temperature evolves with time in the bulk of the body. One can see that heat penetrates into the medium little by little. We already discussed this process in Chapter 19 and demonstrated that the characteristic length $L(t)$ describing propagation of the temperature front with time is determined by Eq. (19.6).

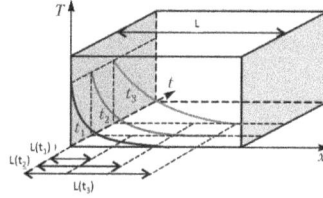

Fig. 20.7: schematic of temperature penetration into a semi-space.

Now we can return to comparison between the rinsing time of a thin-cut beef versus the boiling time of a meatball. We consider an $a x b$ rectangular slice of thickness d and a meatball of radius R of the same volume of meat. During the boiling process, temperature should penetrate into the specimen up to the farthest from the surface. Thus, the required length for thin-cut slices and meat balls are respectively half its thickness and its radius. The times for them to be done are

$$\tau_{\text{slice}} = \frac{(d/2)^2}{\pi \chi} \qquad \tau_{\text{ball}} = \frac{R^2}{\pi \chi}. \qquad (20.8)$$

The ratio between their cooking times is

$$\frac{\tau_{\text{ball}}}{\tau_{\text{slice}}} = \left(\frac{2R}{d}\right)^2.$$

The size of a thin-cut beef slice in China typically is $a \cdot b = 3\,cm \cdot 15\,cm$ and its thickness is $0,1\,cm$, which is marvelously thin. Out of the same amount of beef, according to the equality in volume

$$a \cdot b \cdot d = 4\pi R^3/3$$

we can make a beef ball whose radius is $R = 1,0\,cm$. The ratio $2R/d = 20$ and the estimated difference in cooking times

$$\left(\frac{\tau_{\text{ball}}}{\tau_{\text{slice}}}\right)_{\text{est}} \sim 400. \qquad (20.9)$$

Stop! As we have mentioned above the common times for the meat ball (dumpling) and the meat slice to be cooked are 5 minutes and 10 seconds, i.e. the realistic value for the ratio is

$$\left(\frac{\tau_{\text{ball}}}{\tau_{\text{slice}}}\right)_{\text{real}} \sim 30. \tag{20.10}$$

This number differs by one order of magnitude with Eq. (20.9), it is too much.

The reason for this discrepancy is in our assumption that all the cooking time is spent on "delivery" of the necessary temperature through the whole volume of the meat. However, we ignored the time required for carrying out of the denaturation process itself. In most of cases, this time, τ_{denat}, is so short that it can be neglected with respect to the "delivery" time. Yet, the "shuan" process is so quick that the denaturation time cannot be neglected and has to be added so that:

$$\tau_{\text{slice}} = \tau_{\text{denat}} + \frac{(d/2)^2}{\pi\chi}, \tag{20.11}$$

where the temperature conductivity of the beef, as it was already mentioned above, is $\chi_{\text{beef}} = 1,4 \cdot 10^{-7}\,m^2 \cdot s^{-1}$. The value of the second term in Eq. (20.11) is only $\frac{(d/2)^2}{\pi\chi} \approx 0,5\,s$, while one rinses it in the hotpot around $10\,s$. It is why we conclude that all this time is required for the chemical reaction of denaturation i.e., $\tau_{\text{chem}} \approx 10\,s$. Of course, this value is negligibly small compared with required one to increase the temperature throughout the entire meatball volume above the denaturation point, but it turns out to be dominant when rinsing a slice of meat in a hot pot. The new estimation

$$\tau_{\text{slice}} \approx 10s \qquad \tau_{\text{ball}} \approx 200s$$

agrees well with Chinese experience.

Table 20.1. Comparison among three ways of cooking mentioned above boiling, steaming, and rinsing.

Process	Boiling	Steaming	Rinsing
Size	Radius R	Radius R	thickness $d \ll a, b$
Cooking time	$\dfrac{R^2}{\pi\chi}$	$\dfrac{R^2}{\pi\chi}$	$\dfrac{(d/2)^2}{\pi\chi}$
The way of heat transfer	Thermal conduction and natural convection	Latent heat of the condensing steam	Thermal conduction and forced convection
Environment	Boiling water/ diluted broth	Saturated steam	Boiling soup-base

20.3.2 *On soup base*

Besides the process of boiling sliced beef itself, the soup-base also contains some physics in it. First of all, it differs strongly from the diluted aquatic broth formed during dumplings boiling (see above). Soup-base for the hot pot is prepared in advance, mixing water with oil and other ingredients. Below we discuss some of its specific properties having also interesting physical grounds.

Qingtang soup base commonly is used in the Beijing and Chaoshan hotpots. It consists of the mix of water with a small amount of oil, salt and other condiments. In contrast, for cooking in Chongqing hotpot the Hongyou (spicy oil) soup base serves. A large fraction of Hongyou soup is oil stewed with chili powder, Chinese prickly ash, etc. The remaining part is a small portion of water. If one starts to heat both soups simultaneously, the oil soup base will start to boil much earlier than the water soup. It seems to be rather strange, since the boiling point of oil (if any) is much higher than that of water. Actually, what boils here is not the oil, but the small portion of water in the soup. Hence, the Hong You soup starts to boil at the boiling point of water instead of that of oil.

There are two reasons which explain why the Hongyou soup needs a shorter time to boil. The first is that the specific heat capacities of oil and water are respectively $c_{\text{oil}} = 2 \cdot 10^3 \frac{J}{kg \cdot K}$ and $c_{\text{water}} = 4,2 \cdot 10^3 \frac{J}{kg \cdot K}$, so that half less heat is needed to raise the temperature of the oil soup base from room temperature to $100°C$.

The second reason concerns heat dissipation. The process of the heat dissipation is due to both heat conduction convection processes. In the latter energy is transferred by the movement of a heated substance as a result of differences in densities of lower and higher layers. The heat flux at the surface is loosely described by Newtons law of cooling, which claims that the latter is proportional to the temperature difference:

$$q = h\Delta T.$$

The coefficient h is called the heat transfer coefficient and it describes how violent the convection is. The typical values of the heat transfer coefficient for boiling water and oil are respectively $(2.5 - 25) \cdot 10^3 \frac{W}{m^2 K}$ and $(0.05 - 1.5) \cdot 10^3 \frac{W}{m^2 K}$. The significant difference in the values of h may be explained by the viscosity and the poor heat conductivity of oil. As a result, the heat dissipation at the interface between the oil soup and air is two orders of magnitude less significant than that in the case of water soup.

Manman Chi![c]

---

[c] "Eat slowly", Chinese version of "enjoy your meal".

Chapter 21

The mystery of the Christmas turkey

> It arrived upon Christmas morning, in company with a
> good fat goose, which is, I have no doubt, roasting at
> this moment in front of Peterson's fire.

Sir Arthur Conan Doyle, *The Adventure of the Blue Carbuncle*.

Different nations developed a variety of traditions regarding the festive menu of a Christmas dinner. The best of the local gifts of nature are usually represented on the table. Thus, in Naples, which is located on the sea coast, it should be a fried eel, in Rome - a capon, and in England - a specially fed goose. Nevertheless, in the time of globalization the food is getting more standardized: baked turkey is frequently becoming a central treat in Christmas menus of different countries. This dish is also a symbol of Thanksgiving - the annual holiday celebrated in the United States as a day of giving thanks for the blessings of the harvest of the preceding year.

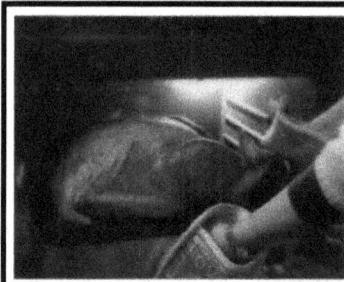

Fig. 21.1: Christmas
turkey.

Baking turkey for Thanksgiving[a] is not a simple process. Its mass sometimes can reach about 13-14 kilograms, and all this meat should be baked well without burning the external layers. Baking time recommendations are usually posted on the packaging with frozen or fresh turkey. But, if you accidentally discharged of the packaging, you can click on the Internet and find the following table of cooking time for different turkey sizes:

Table 21.1. Cooking time for different turkey sizes

Mass M (pounds)	Baking time τ (*hours*)
8	3, 5
12	4, 5
16	5, 5
20	6, 5
24	7, 35

As you can see in the table above, the dependence of turkey baking time τ on its mass M is essentially nonlinear. That is why the hostess of a family dinner who was used to cooking a 3 kg (or 6 lb) turkey for a festive meal, may face problems when cooking an enormous bird weighing 13 kg (or 26 lb) for a big party gathered for Christmas celebration (Fig. 21.2). Let us try to help her by deriving a formula for baking time τ depending on turkey mass M.

Just above, in the Chapter 20 we have discussed what meat boiling means in scientific terms. Yet, fried meat differs vastly in taste from boiled meat, which demonstrated that the culinary art cannot be explained by protein denaturation only. The fact of the matter is that meat frying or baking requires a much higher temperature than boiling. It is reached during frying by adding oil, which allows the raising of the temperature in the frying pan up to $160 - 200°C$; when baking or grilling this effect is achieved by heating meat with infrared radiation and hot air. At a temperature T_M of about $140°C$ (this value naturally depending on the sort

---

[a]Thanksgiving is celebrated on the fourth Thursday of November. The most typical festive dish on Thanksgiving table is turkey (that is why sometimes Thanksgiving is called a Turkey day). The tradition of eating turkey as a main course of Thanksgiving feast dates back to president Lincoln, who in 1863 proclaimed it a national holiday. Hundreds of thousands turkeys were sent to the Army, and proved to be more nourishing for troops and more cost-effective for American Treasury. According to the data of US Department of Agriculture, in 2003 there were 45 million turkeys on the Thanksgiving tables in the United States.

of meat) the so-called Maillard reaction takes place. During this reaction glycogen (animal starch) is getting caramelized, which adds a specific taste to fried or baked meat. This reaction was discovered by Louis Maillard[b] in 1910. Indeed, in order to cook a Christmas turkey the recommended temperature inside the oven should be $T_0 = 160 - 170°C$. The question is: how much time will it take for this temperature to set on inside the whole turkey, if we will heat the bird through its surface only? Also, how long will it take for Maillard reaction to take place, if the temperature at a given point of turkey has already reached the required level?

Formally, this problem is similar to the one we already solved above, in Chapter 20, discussing the dumplings cooking. Indeed, in both cases a constant temperature T_0 is set in the oven or in the boiling soup and heat propagates from the inside to the internal part of the specimen. The temperature on the surface of the turkey reaches the oven one, $T_0 = 160 - 170°C$, in a short time after the turkey was placed in the oven due to the processes of heat transfer, infra-red radiation and convection of hot air (see Chapter 19). Nevertheless, the turkey remains cold inside — so the flow of heat will propagate from the surface to the center. The turkey will be completely baked when the Maillard reaction takes place in all its extent, i.e., the temperature will reach the value of T_M (which is somewhat lower, but close to T_0), even in the center of the turkey (which we, like the dumpling above, consider as a solid sphere). Discussing the cooking time of dumplings we already found that the time needed to deliver necessary heat up to the center of the dumpling is determined by the coefficient of temperature-conductivity of meat and the square of its radius. In the case of dumplings boiling we spoke about the proteins' denaturation process but it does not matter, both T_{denat} and T_M are supposed to be close enough to the temperature of the ambient. In the case of a turkey, the square of its radius in Eq. (20.6) is more conveniently expressed in terms of its mass:

$$\tau = C_1 \frac{c \cdot \rho^{1/3}}{\kappa} M^{2/3}, \qquad (21.1)$$

where C_1 is numerical coefficient of the order of 1.

Let us get back to the mysterious digits in Table 21.1 and present them in the form of graphic dependence of τ as the function of $M^{2/3}$.

---

[b]Louis Maillard (Louis-Camille Maillard, 1878-1936) — French chemist and physicist.

Table 21.2. Dependence of τ as the function of $M^{2/3}$

Baking time τ (hours)	$M^{2/3}$	$\tau/M^{2/3}$
3, 5	4	0, 875
4, 5	5, 24	0, 86
5, 5	6, 35	0, 87
6, 5	7, 37	0, 88
7, 35	8, 32	0, 88

As seen in Fig. 21.2, within the chosen coordinates this dependence with fair precision represents a direct line, and thus, our theory brilliantly confirms the experience accumulated by generations of cooks. The fact of the direct line $\tau \sim M^{2/3}$ passes almost through zero indicates that most of the baking time is required for delivery of the necessary heat to the central part of the turkey, while the Maillard reaction occurs in the scale of hours almost instantly.

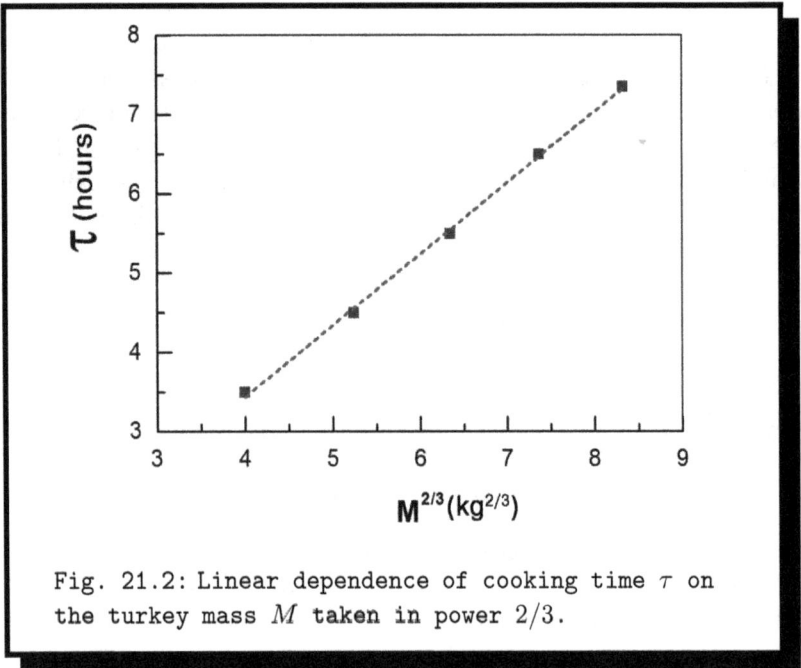

Fig. 21.2: Linear dependence of cooking time τ on the turkey mass M taken in power $2/3$.

Chapter 22

Macaroni, spaghetti, and physics

If Narcissus was turned into a flower,
I want to be metamorphosed into Macaroni.

Fellippo Sgruttendio.[a]

Each and everyone knows a lot about spaghetti, and many of you have cooked it at home. But have you ever thought about the physical processes taking place in the pot where pasta is cooking, and producing a correctly boiled "pasta" (in accordance to Italian standards) as a result? Have you ever asked yourself what happens inside spaghetti when it is floating in the boiling water? Why should we always follow the cooking time instructions specified on spaghetti packaging? Why is the cooking time different for different types of pasta? How does cooking time depend on the shape of pasta (classical spaghetti with different diameters, rigatoni or bucatini, etc.)? Does the cooking time depend on the location: are you making pasta at the seaside or in the highlands? When bent, why does an uncooked spaghetti strand ("spaghetto") almost never break into two pieces but into three or four? Why do spaghetti strands never tie into knots during cooking? How do you select the type of pasta, if you already have a sauce, so that the dish will be hot and tasty?

Below we will try to gain insight into the physics of the spaghetti cooking process and find the answers to some of the questions that may arise while the interested audience waits for a pot of pasta to boil in anticipation of a hearty dinner.

---

[a]Fellippo Sgruttendio, Neapolitan poet of the eighteenth century

22.1   A glimpse into a history of pasta and its manufacturing

Contrary to popular opinion, pasta was not brought to the West by Marco Polo after his trip to China (1295). In fact, its history began much earlier on the Mediterranean coast, at the time when the prehistoric man was leaving the nomadic lifestyle behind and started settling down and growing grain for food. The first flat cakes baked on top of hot stones were mentioned in the Old Testament (*Genesis* and *Kings*). In the first millennium B.C., Greeks were already making a thin layered pasta; they named it *"laganon"*. This word came to ancient Romans in the form of *"laganum"*, and may well have became the source for modern term *"lasagna"*.

Etruscans were also preserving thin layers of pasta. With the growth of the Roman Empire, pasta started spreading throughout Western Europe. Reliance on pasta as a way to preserve grain products emerged with the need to transfer food supplies during the periods of tribal migration. In Sicily, pasta was introduced by Arabs when they conquered the island in the tenth century. Sicilian pasta named *"trie"* may be considered the ancestor of spaghetti. It was shaped in thin strands, and the name originated from the Arabic word *"itryah"* (flat cake cut into strands). People living in Palermo started making pasta in the beginning of the second millennium. Based on a detailed will, probated by a Genoa notary public Ugolino Scarpa, we can positively argue that by 1280 macaroni products were already consumed in Liguria. It is known from the History of Italian Literature that pasta has attracted the attention of such writers as Jacopone da Todi, Cecco Angiolieri, and Felippo Sgruttendio. Ultimately, in Boccaccio's *Decameron* macaroni became a symbol of sophisticated gourmet food.

The first guilds of pasta makers (*"Pastai"*) with their own charters were created in the sixteenth century in Italy, where they received political and public recognition. At that time macaroni was regarded as a food for the rich, especially in the provinces which did not grow their own durum wheat (e.g., Naples). The invention of the mechanical press resulted in lowered production costs and hence the price of the product. As a result, by the seventeenth century pasta turned into one of the staple foods consumed by all social classes; now it is widespread in all countries of the Mediterranean basin. Naples has become a major center of pasta manufacturing and export. There, pasta with basil tomato sauce or sprinkled with grated cheese is sold on every street corner. In Northern Italy pasta became popular at the end of the eighteenth century mainly due to Pietro Barilla, who opened a small factory in Parma and later became a major producer in the Italian food industry.

Modern methods of pasta production are primarily based on the process of extrusion (pushing out through holes) and dragging. Extrusion was invented and used for the first time in the manufacture of long metal components with specified cross sections (see Figure 22.1). Extrusion processes are based on the property of fluidity and on subsequent pushing of material through a rigid die by means of compression. It can work both in cold and hot conditions. Dragging is a process similar to extrusion, the only difference being that in the case of dragging the material is pressed through the die located at the vessel's outlet and thus it becomes a process of stretching rather than compression. This method is used by the metal-processing industry for cylinder, wire, and pipe manufacturing. It allows reduction of the diameter of metal wire to 0.025 *mm*. Other materials that can be processed by extrusion are: polymeric compounds, ceramics, and food. Dies used to produce spaghetti are shown in Figure 22.2.

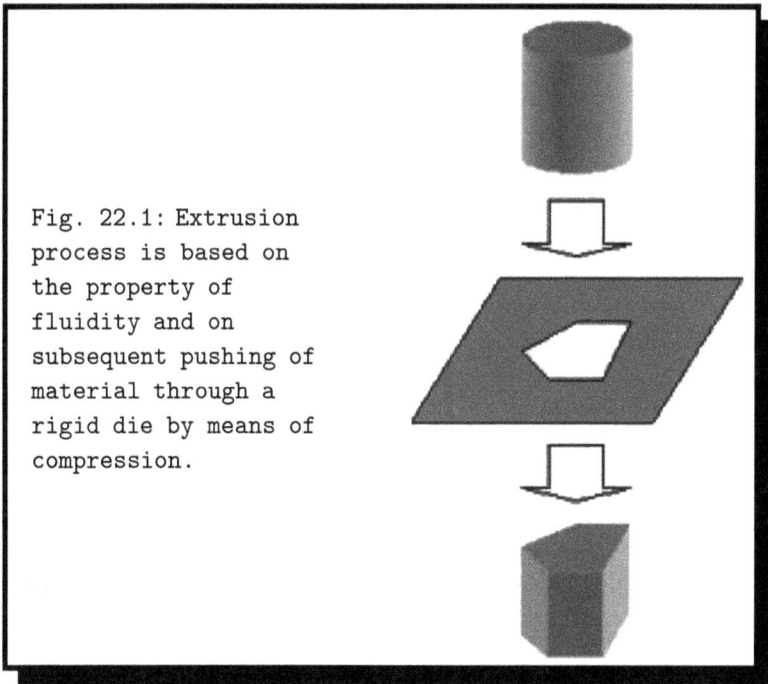

Fig. 22.1: Extrusion process is based on the property of fluidity and on subsequent pushing of material through a rigid die by means of compression.

Fig. 22.2: Spaghetti
die.

22.2 A scientific way of spaghetti cooking

Above, when discussing the process of a meat boiling, we got the expression
for the cooking time of a dumpling. It turns out that the same Eq. (20.6)-
(20.11) (but, naturally, with the other coefficients) can also be used to
determine the cooking times of many types of pasta, in particular, of the
pasta of cylindrical shape.

Before we begin, let us try to understand what is involved in the process
of pasta cooking. In the flour, molecules of starch are grouped into granules with diameter of 10–30 microns, which, in turn, are surrounded with
different proteins. In the process of pasta fabrication two of them, gliadin
and glutenin, combine with water, unite, and form a continuous net, called
gluten, which is strong and has low permeability for water molecules. This
net covers the starch granules. Cooking time is directly related to the capacity of the starch molecules (surrounded by gluten in the process of pasta
drying) to absorb water, which begins penetrating through the gluten network and diffuses to the inward of pasta as soon as it is placed in the pan
of boiling water. At a temperature of about $T_g = 70°C$, starch molecules
begin forming a gel-like compound, which hinders water absorption. Pasta
is considered "al dente" (to the tooth) the moment when the gel-like starch
absorbs the minimum amount of water necessary to make it sufficiently
soft. Hence in order to cook pasta it is necessary to deliver hot water inside
the initially dry spaghetto.

Thus, one of the necessary conditions for the preparation of pasta is the
same as when cooking meat: the temperature inside it should be increased
to a certain specific value (in the case of pasta it is $70°C$). The second

condition, in comparison to the boiling of meat is new one: the water here should penetrate inside the originally dry pasta. Both of these conditions have to be fulfilled during the cooking process. Indeed, neither the heating of dry spaghetti in the stove, nor their prolonged holding in cold water will lead to the appearance on the table of a plate of mouth-watering pasta.

The Eq. (20.6) was obtained as a result of the dimensional analysis of the heat transfer equation. Although we analyzed it in the simplest case of a spherically symmetric specimen, the result remains valid for a cylindrically symmetric "spaghetto". Moreover, fortunately, the process of water diffusion from the outside to the center of the "spaghetto" is described by the equation of the same type (the diffusion equation) as that one for the heat transfer, but the latter being written for the concentration of water instead of the local value of temperature. As a result, the spaghetto's cooking time is related to its diameter by the relation of the same type as Eq. (20.11):

$$\tau_{sp} = ad^2 + b. \tag{22.1}$$

where d is the spaghetto's diameter. The coefficient a is determined by the physical properties of the pasta (its temperature conductivity and, what is new, the diffusion coefficient), while the coefficient b characterizes ... the nationality of the eater. In fact, if the first term in the Eq. (22.1) determines the delivery time of water and heat to the center of the spaghetto, then the second tells us how long these factors affect the central part. That is why for Italians, who prefer to eat the pasta in a degree of readiness "al dente" ("on tooth"), the process of the gel formation of starch occurs not in the whole volume: the latter remains relatively firm at the central part of spaghetto. As a result, as we will see below from the analysis of experimental data, the coefficient b turns to be negative for them. In other countries, spaghetti lovers believe that the pasta should be well-boiled and the time that they cook pasta can significantly exceed the one indicated on the packaging by the Italian manufacturer.

The starch gelification temperature, T_g, is constant while the boiling temperature of water, T_b, depends on the height with respect to the sea level, H. Consequently, the coefficient $a = a(H)$ depends on H and hence, cooking time $\tau_{sp} = \tau_{sp}(H)$. The recommended cooking time on the packaging corresponds to the sea level, where $T_b = 100°C$. At high altitudes, where water boils at lower temperatures, cooking time should be extended. In the extreme case of Everest (its height is 8,848 m) $T_b = 73°C$, which is very close to T_g and pasta can be hardly cooked well at all.

Let us now go to a supermarket and buy every kind of cylindrically shaped pasta: capellini, spaghettini, spaghetti, vermicelli, bucatini. Read and collect in Table 22.1 the recommended cooking times (the column "Experimental cooking time". Then take a slide gauge, measure the corresponding diameters and fill out the same table (the column "diameter, external/internal"):

Table 22.1

Type of pasta	Diameter, external/internal (mm)	Experimental cooking time (min)
capellini no.1	1.15/-	3
spaghettini no.3	1.45/-	5
spaghetti no.5	1.75/-	8
vermicelli no.7	1.90/-	11
vermicelli no.8	2.10/-	13
bucatini	2.70/1	8

In order to find the numerical values of the coefficients a and b it is enough to write the equation (22.1) using the data from two rows of our Table 22.1:

$$t_1 = ad_1^2 + b$$
$$t_2 = ad_2^2 + b \tag{22.2}$$

and to solve the system of these two simple equations. As the reference case, we choose the data for spaghettini no. 3 and vermicelli no. 8. This gives:

$$a = \frac{t_2 - t_1}{d_2^2 - d_1^2} = 3.4\,\text{min}\,/mm^2$$

$$b = \frac{d_2^2 t_1 - d_1^2 t_2}{d_2^2 - d_1^2} = -2.3\,\text{min}\,.$$

We bought the Italian pasta, and the recommended cooking times on the packaging are given to obtain the "pasta al dente". Consequently, one can see that the coefficient b for Italians is indeed negative.

Having the numerical values of the coefficients a and b we can check how our formula works for other types of cylinder-shaped pasta. The results of our calculations can be found in Table 22.2 and show very good agreement with the experimental data for all rows except the two extremes: capellini and bucatini.

Table 22.2

Type of pasta	Experimental cooking time (min)	Theoretical cooking time (min)
capellini no.1	3	2.2
spaghettini no.3	5	5.0
spaghetti no.5	8	8.1
vermicelli no.7	11	10.0
vermicelli no.8	13	13.0
bucatini	8	22.5 ?!

This is a typical situation in theoretical physics: predictions of a theory have a range of validity corresponding to the assumed simplifications made for this theory. For instance, let us take a look at the last row of the table: for bucatini the difference between the theoretical and experimental values is striking: 22.5 min versus 8 min. This contradiction reflects an important fact: the range of possible thicknesses for all varieties of a whole cylinder-shaped pasta (capellini, spaghetti, vermicelli) is very narrow: 1 *mm* only. Indeed, our calculation of the "al dente" cooking time for bucatini in the approximation of the uniform cylinder gave 22.5 min, which would completely boil the periphery of such a thick "spaghettone" into mush.

The way to make such a thick spaghetti edible was found empirically: a hole should be made in the spaghetti strand along its axis. In the process of cooking, water enters through this hole, and there is no longer a need to deliver it to the core from the outside. One can try to modify the formula (22.1) deducting the internal diameter from the external one, and immediately the theoretical result comes close to the reality:

$$t = a \left(d_{\text{ext}} - d_{\text{int}}\right)^2 + b \approx 7.5 \, \text{min}.$$

Nevertheless, one should keep in mind that the hole in the pasta cannot be less than ~ 1 mm in diameter, otherwise, due to capillary pressure,

$$P_{\text{cap}} = \frac{4\sigma \left(T = 100°C\right)}{d_{\text{int}}} \sim 200 \, Pa,$$

water will not be able to enter into it. The value $200 \, Pa$ corresponds to the pressure a couple of centimeters of water above the cooking pasta.

Another deviation of theory from reality occurs for very thin pasta. The reason for this error is obvious: speaking about cooking pasta "al dente" we have chosen the parameter b as negative: $b = -2.3 \, \text{min}$. Formally, this means there should exist a pasta thin enough to not even need cooking at all to be eaten "al dente". The corresponding critical diameter d_{cr} can be determined from the relation:

$$\tau_{\text{cr}} = a d_{\text{cr}}^2 + b = 0,$$

which gives

$$d_{\text{cr}} = \sqrt{|b|/a} \approx 0.82 \, mm.$$

One can see that the real diameter of capellini (1.15 mm) is not far from this critical value, so the underestimated capellini cooking time in Table 22.2 is the result of this limitation of our model.

22.3   Spaghetti knotting

Cooked spaghetti strands entangled with each other present a complex tangle in hot water, however, the authors of this book have never seen them knot themselves. The reasons why this does not happen can be learned from a new field in statistical mechanics: the statistics of polymers.

The probability of a long polymer chain self-knotting is determined by the expression[b]

$$w = 1 - \exp\left(-\frac{L}{\gamma\xi}\right),$$

where L is the full length of a polymer, ξ is the characteristic length at which the polymer can change its direction by $\pi/2$ and $\gamma \approx 300$ is a large factor, obtained as a result of numerical and theoretical modeling. Applying

---

[b]A.V. is grateful to A.Y. Grosberg for introduction to the theory of knots.

this formula to spaghetti, where $\xi \approx 3$ *cm*, one can find the length L_{min} when the probability of self-knotting becomes noticeable ($w \sim 0.1$) :

$$\exp\left(-\frac{L_{min}}{\gamma\xi}\right) \approx 0.9$$

which gives

$$L_{min} \approx \gamma\xi \ln 1.1 \approx 30\xi \approx 1 \ m.$$

The length of a standard spaghetto is 23 *cm* and this is not long enough to form knots.

22.4 The secrets of mixing of pasta with sauce

The rules of good etiquette prescribe to start eating at the same time everyone is sitting at the table. However, at the Italian table, an exception is made for the first dish, pasta: it is eaten immediately, as the plate appears on the table. The pasta should be hot. However, it gets to the table not directly from a pot of boiling water: first it is discarded in a colander and mixed with the sauce. It is clear that this process takes some time and can be long. Then the paste will cool down and the pleasure will be destroyed.

In order to understand what time is required to mix the sauce and pasta, we start with a simple model: let the viscous liquid flow through the cylinder under the action of gravity (this is the simple model of the pasta in which the sauce flows). The stationary flow ($Q = \Delta M/\Delta t$) of a liquid in a pipe of diameter D under the effect of pressure difference ΔP is determined by the Poiseuille formula

$$Q = \frac{\pi\rho\Delta P}{2^7\eta l}D^4,$$

where η is the viscosity of the fluid, ρ is its density, and l is the pipe length. On the other hand, the magnitude of the flow is determined by the mass of fluid flowing through the cross section of the pipe per unit time:

$$Q = \Delta M/\Delta t = \frac{1}{4}\pi\rho D^2 \Delta l/\Delta t.$$

Comparing these two expressions and assigning the pressure difference to the effect of the gravitational force $\Delta P = \rho g \Delta l$ we find

$$\frac{\Delta t}{\Delta l} = \frac{32}{\rho g}\left(\frac{\eta}{D^2}\right).$$

We see that the rate at which a viscous liquid fills the tube is proportional to the ratio η/D^2.

This formula is obtained in the model of fluid flow in a gravitational field through a vertically arranged tube, so the acceleration of free fall g is included in the answer. Yet, it is clear that the nature of the acceleration is insignificant: with the same success, g in this formula can be replaced by the acceleration which pasta acquires from being stirred by a ladle.

Above we have considered liquid flowing inside the pipe. Nevertheless, it is clear that if several tubes are tightly arranged side by side, the liquid between the tubes will flow more or less at the same speed as inside them. Thus, we arrive at the conclusion that the characteristic time of pasta stirring is

$$\tau_{\text{mix}} \sim \left(\frac{\eta}{D^2}\right).$$

The greater the viscosity of the sauce, the larger the diameter of the pasta should be. It would be hard to mix well the finest capellini with a viscous "pesto genovese": the latter naturally combines with short "trofie". And *vice versa*, liquid tomato sauce just drains to the bottom of the plate being mixed with huge "paccheri". So it is better to leave the cherry tomatoes in it, and also dress the sauce with pieces of zucchini and shrimps. The

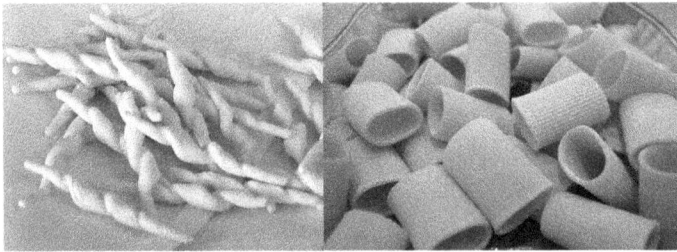

Fig. 22.3: Different types of Italian pasta: trofie, paccheri.

obtained formula also helps us to understand the empirical rules of Italian cuisine. Usually, pasta and sauce are prepared at the same time, the pasta is cooked in a saucepan, and the sauce in a frying pan. In the hands of a good chef, they arrive to the state of readiness almost at the same time. Discarding the pasta in a colander and decanting the water, he sends pasta to the pan, where it mixes with the sauce. Viscosity drops noticeably with increasing temperature, so the mixing time of the pasta with the sauce boiling in the frying pan will be significantly shorter than if the sauce was taken out of the refrigerator. In addition, the pasta does not lose heat during its mixing with the sauce. Another subtlety. The pasta should be boiled in a saucepan for slightly less time in comparison with the time recommended on the package. Being stirred with sauce in a pan it continues to be cooked and "gathers additionally" the missing minutes in order to achieve the "al dente" condition.

22.5   Breaking a spaghetti strand

In the beginning of this chapter, we mentioned another interesting property of spaghetti related to its mechanical fracture. Take both ends of a spaghetto and bend it into an arch, gradually increasing its curvature. One can suppose that sooner or later it will break into two parts somewhere near the middle. It turns out, though, that in this case our intuition is not correct: nearly always it will break into three or more pieces.

Such unusual behavior of a spaghetti strand attracted the attention of numerous scientists, Richard Feynman among them. But only a short time ago in 2005, owing to the research studies conducted by the two French physicists Audoly and Neukirch, was a quantitative description of this phenomenon obtained.

The scientists studied the behavior of a thin, elastic rod under the effect of flexural deformation. They wrote the differential equation which describes the distribution of the tension (so-called Kirchhoff equation) in a curved elastic rod, first with both ends fixed. Then they studied what will happen to the tension distribution along the rod after the instantaneous release of one of its ends. Only a numeric solution was obtained, but it provided an understanding of the essence of the process. Qualitatively, the explanation is as follows.

Let us suppose that due to the applied mechanical stress, the first fracture occurs at some (weakest) point of the strand. It would seem that after being broken, both parts of the rod should return to their equilibrium positions. This is true, however the transition to the equilibrium state occurs in a very nontrivial way. The first fracture generates flexural waves in both fragments of the rod which start to propagate along each of the fragments. Evidently, flexural waves of this kind (generated by the first fracture) will dampen out with time, but, at certain ratios between the rod length and its elastic modulus, the wave propagation can lead to a subsequent rod fragmentation. Indeed, propagation of such a wave means periodic growth and decrease of the local flexural stress along the rod. It is important to note that these flexural waves propagate on the background of the already existing initial homogeneous bend of the rods, which relaxes much more slowly than the flexural wave period. As a result of the summation of these two quasistatic and dynamic stresses, further rod fractures may occur at other points, where such sums exceed a critical value. It is also noteworthy that after intricate computations, the researchers confirmed their theoretical conclusions by filming the experimental spaghetti break studies with a high-speed camcorder (Figure 22.4).

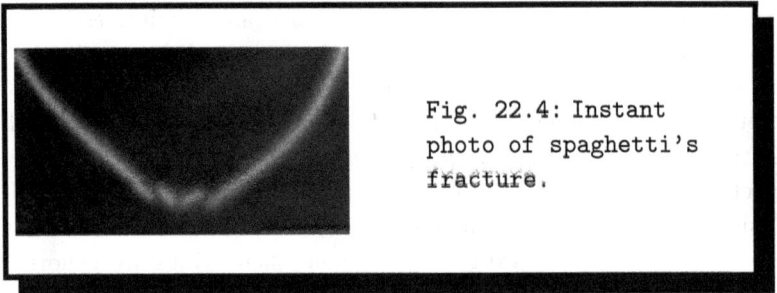

Fig. 22.4: Instant photo of spaghetti's fracture.

Very recently, Ronald Heisser and Vishal Patil, two students of the Massachusett Institute of Technology (USA), working on a scientific project proposed to them by their tutor Jörn Dunkel, managed to overcome the magic reluctance of the spaghetto to break into two parts. For this, as it turned out, it is enough to twist spaghetto, say 360 degrees, along its axis. In this case, after the first break, only part of its released energy is spent on the flexural wave excitation, the rest into spinning the spaghetto (excitation of the "twisting wave"). As a result, the amount of released energy is no longer sufficient for the second break...

While the spaghetti is cooking, you can entertain yourself with several remaining dry spaghetti by experimentally verifying the findings of Audoly and Neukirch.

Chapter 23

Ab(out) ovo

For the Words are these:
That all true Believers shall break their Eggs at the convenient End:
and which is the convenient End, seems, in my humble Opinion, to be left to every Man's conscience, or at least in the power of the Chief Magistrate to determine.

Jonathan Swift, *"Gulliver's Travels."*

Literally, the Latin expression *ab ovo* means "from the egg". Figuratively, it indicates the earliest possible stage, the very origin of something. The ancients used it in order to emphasize the fundamental importance of an event. Thus, without noticing, they resolved the famous paradox "which came first, the chicken or the egg?" in favor of the egg. But we shall leave this perennial dispute aside[a] and look instead at various physical phenomena around this seemingly unexciting object, the chicken's egg.

As everyone remembers from the school reading of *Gulliver's Travels*, the two great empires of Blefuscu and Lilliput were engaged in a most obstinate war that started after the Emperor of Lilliput published an edict commanding all his subjects, upon great penalties, to break the smaller end of their eggs. Gulliver believed (and was in agreement with the doctrine of the great Lilliputian prophet Lustrog) that it was really a private matter which particular end of an egg to break. The authors respect their opinion but, for the sake of curiosity, it would be interesting to find out which end

---

[a]For those interested in paradoxes, we would recommend the excellent collection *Paradoxicon* by Nicholas Falletta.

of the egg shell is it *easier* to break. The solution would give one a clear winning strategy in the fierce battles of "hard-boiled eggs" that often break out at the Easter table.

What is the right tactic: to attack the opponent or to grant him the first strike? Which egg to take: a big or a little one? To hit the small end or the big one? These are the main strategic questions of such a fight. The common opinion is that the attacking side gains the advantage. However, if both eggs move uniformly it makes no difference which is moving and which is static. One may see that without breaking the eggs. Recall Galileo's relativity principle and look at the problem from the moving reference frame where the attacker is at rest. In this frame he automatically changes from the "aggressor" to the suffering side.

Now, let us look closer at the actual collision of the two eggs. Let the eggs be exactly alike, i.e., suppose that their size, their shape, and the strength of their shells (the breaking stress σ_b) are the same. The eggs collide along the common axis, one with its small end and the other with the big end at the point of collision (Figure 23.1).

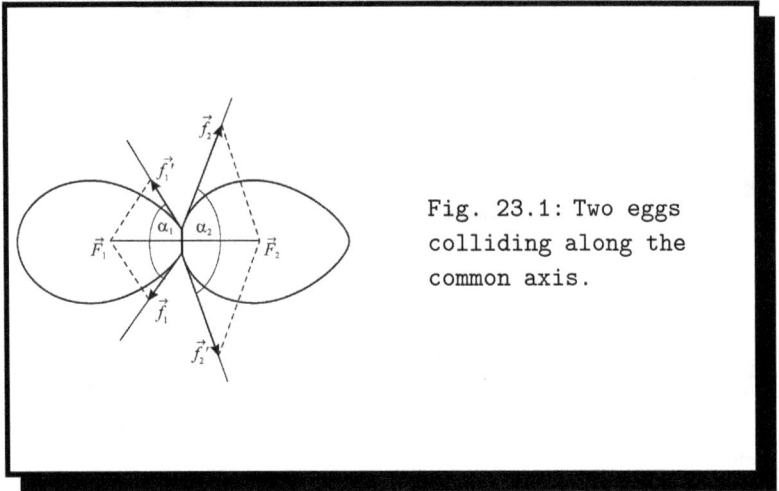

Fig. 23.1: Two eggs colliding along the common axis.

According to Newton's third law, the forces $\mathbf{F}_1$ and $\mathbf{F}_2$ acting on the colliding eggs are equal and oppositely directed. Let us represent $\mathbf{F}_1$ and $\mathbf{F}_2$ as sums of distributed forces $\mathbf{f}_1$, $\mathbf{f}_1'$ and $\mathbf{f}_2$, $\mathbf{f}_2'$, directed along the corresponding tangents to the colliding surfaces.[b] The magnitude of stress

---

[b]Note that due to the elasticity of the shells, the colliding eggs touch each other at a tiny but finite spot.

f depends on the angle between the tangents: $f = \frac{F}{2\cos\alpha/2}$. Therefore, the greater the angle, the greater the "cracking" forces acting on the shell during the collision. The angle is defined by the radius of curvature of the surface. The picture clearly proves that it is indeed advantageous to strike the opponent's egg with the small end of your egg. Another reason in favor of this strategy is that the air cell in the large end weakens the wide side of the egg even more (try to explain this).

Note that the previous analysis suggests a practical ruse: even if the experienced opponent turns his egg with its small end forward, you can heighten your chances by attacking him and hitting his egg slightly to the side of the tip, rather than straight on. There the curvature of the shell is less and therefore the stress produced by deformation increases.

In toy and souvenir stores one can find a funny toy called a *"tippee top"*. It is shaped like a truncated ball with a cylindrical stem sticking out from the center of the flat side. When spun fast enough by twisting the stem between the finger and thumb, the top performs a peculiar trick: after a while it turns upside down and continues the rotation on the tail. Obviously, the potential energy of the top is rising due to such a somersault! Long ago this eccentric behavior was explained by W. Thomson[c] and since then the toy is sometimes called a Thomson's spinning top.

It turns out that an ordinary hard-boiled egg also can behave like a Thomson's top. Take an egg and twirl it as fast as possible on a flat, unpolished, hard surface, say, on smooth oilskin. After some revolutions the egg will stand up onto the end and spin about the vertical axis! Only after being noticeably slowed by friction the egg will start swaying from side to side and finally lie down onto the side. Note that for this trick you need a really *hard*-boiled egg. The soft-boiled ones will not do. The reason is that the viscous friction between layers of liquid yolk and white and between the liquid white and the shell will slow down the rotation and the egg will lose the angular momentum. This distinction in behavior offers an easy way to find out whether an egg has been boiled hard or soft. When spun, a soft-boiled egg will stop after a couple of turns, whereas a hard-boiled one will keep on spinning much longer.

Now, after this extensive discourse upon boiled eggs, let us look closer at what happens to eggs during thermal treatment. What should we do, for instance, in order to prevent an egg cracking or bursting (like in a microwave oven)? When should we stop in order to cook eggs soft and not hard? Why does a knowledgeable cook always boil eggs in salted water? The answers

---

[c]See Chapter 7, Footnote a.

to these questions are not easy to find even in hefty cookbooks.

In essence, the process of egg-boiling and many other ways of heat treatment of food (such as frying, baking *etc.*) are based on the fundamental process of protein denaturation. At high temperature these complex organic molecules break into fragments and change their shape and spatial structure. Denaturation may be caused by many factors including chemicals, enzymes *etc.* In contrast to physics of nonorganic matter, where at fixed pressure one can operate with precise values of temperatures of such phase transitions as melting and boiling, this is not so simple with such complex organic compounds as proteins. Nevertheless one can judge about the state of protein for instance by studying its turbidity.[d] Choosing the heat flow rate more or less corresponding to the conditions of the egg boiling one can determine the denaturing temperature of the egg's white as $68°C$ (this makes $154°$ Farenheit), whereas the yolk begins to denature at lower temperatures, already at $63°–65°\,C$ (or $145°–150°\,F$). It makes no sense to specify these temperatures precisely since proteins denature not abruptly but gradually. Besides, the temperatures depend on such factors as the salt content of the egg, the age and feed of the layer *etc.*

So how one can explain the existence of soft-boiled eggs, which many of us eat at breakfast? Placing the egg into boiling water evidently leads to heat front propagation over its volume. Yet, why does the less resistant yolk remain liquid while the denaturing at higher temperature egg's white hardens?

Physically speaking, as soon as the egg gets into hot water a heat flow from the shell to the center appears. Problems of heat transfer in the most general form were in the scope of mathematical physics for a long time. Hence, in order to determine how long will it take to cook a particular egg we need to know only the thermal conductivities, heat capacities, densities of the egg's white and yolk, and the geometrical size of the egg.

In order to resolve the above mentioned paradox let us estimate how fast the temperature front-line "penetrates" in the cold medium. As we already demonstrated above the corresponding dependence of "penetration depth" versus time is determined by Eq. (19.6). The necessary values for calculation of the thermal diffusivity of the egg's white are: $\kappa = 0.56 W/(m \cdot K)$; $c = 3.14 kJ/(kg \cdot K)$; $\rho = 1040 kg/m^3$, what gives $\chi = 1.8 \cdot 10^{-7} m^2/s$. Hence, during a cooking time of the order of 5 minutes the front of temperature $100°\,C$ will penetrate into the egg volume for a

---

[d]See Z. Akkouche, L. Aissat, K. Madani "Effect of Heat on Egg White Proteins", Proceedings of International Conference on Applied Life Sciences (ICALS2012) Turkey, September 10-12, 2012, p. 407.

distance of the order of $L \approx 0.7cm$. i.e., the white of the egg will warm up in its volume to the temperature exceeding the denaturing one $(68°C)$, while the yolk's volume, being a remote from the shell, remains at lower temperature below the corresponding denaturing value.

The exact calculation of the cooking time for an ellipsoidal egg was carried out by English physicist Peter Barham. In his book "The science of cooking" he derives a formula relating the cooking time t (min) to the boiling-point of water (T_b), the egg's minor diameter d (cm), (Fig. 23.2) and the initial (T_0) and intended final (T_f) temperatures of the yolk:

$$t = 0.15d^2 \log 2 \frac{(T_b - T_0)}{(T_b - T_f)}.$$

Fig. 23.2: Measuring
the egg's minor
diameter

Here log stands for the natural logarithm. For minor diameter measured in inches the overall coefficient should be substituted by 0.023.

Under the standard conditions (usually that means at normal atmospheric pressure at the sea level) the boiling-point of water is $T_b = 100°C$ $(212°F)$, see Table 24.1. Therefore, according to Barham's formula, the time required to boil soft $(T_f = 63°C \sim 145°F)$ a typical egg with $d = 4\ cm$, just from the fridge $(T_0 = 5°C \sim 40°F)$ must be $t_1 = 3\ min\ 56\ s$. For a bigger egg with $(d = 6\ cm)$ the time is twice as long: $t_2 = 8\ min\ 50\ s$.[e]

Note that the formula also indicates a necessity to enhance cooking times at higher altitudes, as the boiling-point of water notably drops with altitude (see page 214). Therefore, the times recommended in books written at sea level should be appropriately extended for Alpine cooking. For instance,

---

[e]Of course, cooking is not an *exact* science and times may vary from these estimates.

at an altitude of 5000 m (16.5 thousand feet) water boils at 88°C (190°F). Thus, making a soft-boiled egg will take 4 min 32 s instead of the 3 min 56 s that would suffice at sea level.

Let us note that the insignificant at first glance difference in the denaturing temperatures of the white and yolk makes it possible for a Japanese chef to prepare a dish unusual for Europeans called "Onsen Tamago". The Onsen Tomago is the egg boiled very slowly at temperature around 65°C. The resultant food has a solid yolk at the center and liquid egg white around contrary to the soft-boiled egg. The difference resides in the temperature regimes of cooking: one is low temperature and slow, the other is high temperature and rapid. "Onsen Tamago" eggs are very famous in the hot SPA area all over in Japan. Japanese people have conserved their patience and control, which is not currently observed in our own daily life.

Now, let us discuss salting the water for boiling eggs. The density of a fresh egg is greater than that of fresh water. So in unsalted water the egg lies on the bottom of the pan. Turbulent streams of boiling water toss the egg to and fro, knocking it against the bottom and walls, and may crack the shell. The white will leak through the crack and thicken into dangling flakes. The meal is ruined! Yet if a prudent cook has added to the water just a half spoonful of salt, the "catastrophe" is normally avoided. Salt stimulates the white's protein denaturation. The coagulated white patches the leak like caulk.

Finally, the egg is cooked. Spoon it out of the water and try to touch while it is still wet. Certainly it is pretty hot but still possible to hold in your hands. It gets harder though, as the shell dries. Soon the egg dries completely and becomes too hot to hold. Why?

After having answered this question, try to remove the shell from the egg. You will discover that the shell has stuck in places and takes with it chunks of white. One could have prevented that by putting the freshly cooked egg into cold water. The shell and the white of the egg have different coefficients of thermal expansion; the white would have shrunk more quickly and came loose from the shell.

So far we have witnessed how the laws of mechanics, hydrodynamics, and molecular physics reveal themselves in the behavior of an ordinary egg. But could it help us to observe electric phenomena? Consider how an egg heated in a microwave oven will explode; this is due to the rapid heating of the egg by the electromagnetic field. There is another, less hazardous electrical phenomenon one can observe with the help of our "protagonist". The egg shell is a dielectric. Let us follow the famous Michael Faraday who used this property in his demonstrations of static electricity.

Take a fresh egg and make two small holes in the ends: one a little bigger than the other. Then prick the yolk through the hole with a long needle and stir the contents until liquid. If you now blow into the smaller hole, the liquid will flow out and leave you a hollow almost intact shell. Wash it and let dry. Now it is ready!

Try to approach it with a charged object, such as a comb which you have just used on dry hair or an ebonite stick electrified by rubbing against dry wool. Electrostatic forces will draw the empty shell and it will follow the comb like a devoted dog follows his master.

When the game is over, you may convert the empty shell into a small jet engine. Seal the bigger hole with a piece of chewing gum (everything works for a true experimentalist) and half-fill the shell with water.[f] Fix this reservoir on a light carriage with freely turning wheels, as shown in Figure 23.3. Finally, put under the shell a tablet of solid alcohol and light it up. In a short while the water will come to boil and the vapor will escape through the smaller hole. According to the principles of jet propulsion this will cause our vehicle to run in the opposite direction.

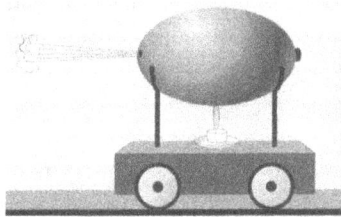

Fig. 23.3: The egg-propelled jet carriage.

We have already discussed how to recognize hard-boiled eggs from uncooked ones. But is it possible to find out without breaking whether an egg is fresh or if it has spoiled? Yes, of course. Just put it into a glass of water, Figure 23.4.

A fresh egg, laid a few days before, sinks right away to the bottom. An egg laid a week ago stays under water vertically but does not rise to the surface. A bad one (say, three and more weeks old) will float by the surface and even stick out a little. The thing is that as an egg "ages" its white and yolk start to degrade and dry up. As a consequence, a gas (hydrogen sulfide) is produced. Some of the gas, along with a bit of water vapor,

---

[f]Change the order if necessary.

Fig. 23.4: The older the egg, the better it
floats.

escapes through the tiny pores in the shell, but the rest stays inside. These
processes decrease the mass of the egg, while its volume remains the same.
As a result the egg becomes buoyant.

A practical difficulty of this method is that it is rather tiresome to take
a bucket of water to the store every time you are going to buy some eggs.
A simpler way is just to look through the examined egg at a bright source
of light. If the light passes through the egg then it is most likely fresh.
Otherwise it must be spoiled. It is the same hydrogen sulfide that makes
a spoiled egg opaque. Long ago in grocery stores there were even special
devices called *ovoscopes* for testing eggs. They consisted of a shelf with
oval holes for tested eggs and a lamp that was lit beneath to see whether
the eggs were transparent.

In conclusion, you could try to perform this elegant trick. Put two egg-
holders next to each other and place a hard-boiled egg into one of them.
Now blow as hard as you can into the gap between the egg and the inner
wall of the holder. The egg will jump from its holder into the next one.
When after some attempts you succeed, try to explain why!

Chapter 24

Waiting for the kettle to boil

A bright idea came into Alice's head.
"Is that the reason so many tea-things are put out here?" — she asked.
"Yes, that's it," — said the Hatter with a sigh:
"it's always tea-time, and we have no time to wash the things between whiles."

Lewis Carroll, *"Alice's Adventures in Wonderland."*

There are thick Eastern manuscripts as well as long, detailed chapters in special books devoted entirely to the tea drinking ritual. Yet when taking another, unconventional, peek at the process, one surely finds interesting and edifying physical phenomena galore, which are not described even in the most reverent culinary "oracles".

To limber up a little, let us perform the following experiment. We will take two identical tea-kettles, each with an equal amount of cold water (same initial temperature in both), and put them on the burners with the same heating power. One of the subject kettles will be covered with the lid, and the other one will remain bareheaded. Which will boil first? Any housewife (no offence to the intelligence of housewives is intended) will give you the correct answer right away. If she wants to have hot water faster, she will put the lid on and reply that the water will boil first in the covered kettle. Well, not taking this statement for granted, let us check it as we are supposed to, experimentally, and wait until the water does indeed start to boil, then have a discussion on the resulting observation afterwards.

In the meantime, while our two kettles are getting hot, we put one more, identical kettle on a third burner. The volume of water and its initial temperature again are the same as for the two previous kettles, as well as the power of the burner. Now, we aim to get the water in this last tea-kettle boiling somehow faster than in the other two. How could we possibly raise the water temperature in this kettle faster? A trivial way would be to stick an extra heating coil in it. Let us say we do not have any available. Perhaps we need to just add some hotter water to the kettle to reach the boiling point faster? It turns out that on the contrary, it will slow it down. To prove this, let us assume that the original amount of water of mass m_1 at a temperature T_1 did not mix with the added water (m_2 at T_2) and did not exchange any heat with this new mass of water either. The amount of heat required for the original volume of water in the kettle to boil would be $Q_1 = c\,m_1\,(T_b - T_1)$ where c is the specific heat of water. But now, besides this energy, the mass m_2 has to be heated to the same boiling temperature, so that the total amount of heat needed is:

$$Q_1 = c\,m_1\,(T_b - T_1) + c\,m_2\,(T_b - T_2).$$

Even if we pour boiling water straight into the tea-kettle, the added water will cool down somewhat during the filling due to heat transfer, lowering its temperature below T_b. Obviously our quite naive conjecture that the two portions of water in the same pot remain unmixed after the addition did not affect in any way the law of conservation of energy in the system, yet permitted us to handle the evaluation faster and more simply.

By the time we have finally given up the idea of adding water into the third kettle, the first of our two "original" kettles begins hissing. And what is the physical mechanism behind this familiar sibilant sound, and what is its characteristic frequency? Next, we shall try to answer these questions.

As the first candidate to generate this whistling discord, one could suggest oscillations excited in the liquid when the steam bubbles take off from the walls and bottom of the reservoir (the kettle in this case). These bubbles always start developing on all sorts of microcracks and other defects, ever present on any real surface. The typical size of such bubbles, before the water starts to boil, is about $1\,mm$ (after that they can reach as much as $1\,cm$). To estimate the frequency of sound produced by the simmering liquid, we need to know how long it takes for the bubbles to release from the bottom. This time actually measures the length of the push that the liquid experiences when each bubble takes off, and therefore, the period of the vibrations they excite. Within our assumptions, the sought-after frequency is determined by the reciprocal of this time: $\nu \sim \tau^{-1}$.

When the nascent bubbles are resting on the bottom, there are two forces acting upon them:[a] the Archimedean buoyancy force, $F_A = \rho_w\, g\, V_b$ (here V_b is the volume of the bubble and ρ_w is the density of water) pushes it upward, and the surface tension that keeps it attached to the surface, $F_S = \sigma\, l$ (where l is the length of the contact area between the bubble and the surface). As the bubble grows (V_b increases), the Archimedean force is rising too, and at a certain moment, it exceeds the retaining force of surface tension. The bubble takes off, starting on its journey upward, Figure 24.1. Hence, the resulting force acting on the bubble during its "departure" stage should be of the order of F_A. Whereas, the acceleration of a bubble in liquid is defined, of course, not by its own negligible mass (consisting mostly of the mass of air trapped in it), but by the mass of liquid involved in the motion. For a spherical bubble this so-called *associated mass* equals $m^* = \frac{2}{3}\pi\,\rho_w\,r_0^3 = \frac{1}{2}\rho_w\,V_b$ (r_0 being the bubble's radius).

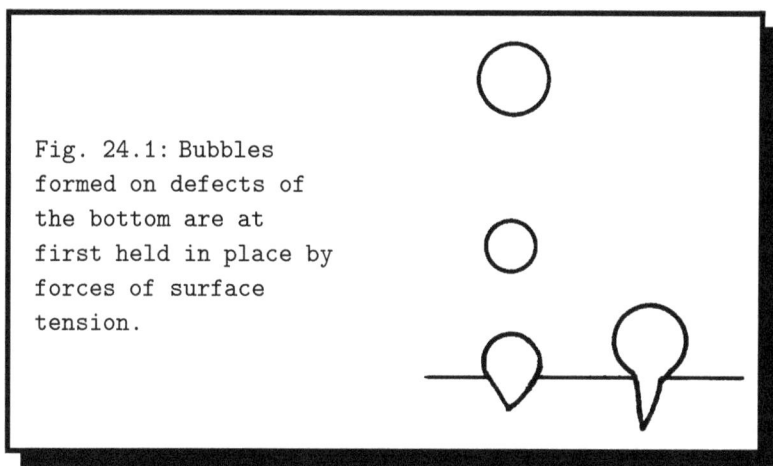

Fig. 24.1: Bubbles formed on defects of the bottom are at first held in place by forces of surface tension.

Thus, for the acceleration of the bubble during the initial stage, one finds:

$$a \sim \frac{F_A}{m^*} = 2g.$$

Now we can evaluate the bubble release time, considering (again, for

---

[a]Here we disregard the tiny weight of the bubble.

simplicity's sake) the motion to be uniformly accelerated. Our subject bubble will climb up to the height comparable to, say, its size within the time of

$$\tau_1 \sim \sqrt{\frac{2r_0}{a}} \sim 10^{-2}\, s. \qquad (24.1)$$

Then, the corresponding characteristic frequency of the sound generated at the bubble's take off should be equal to $\nu_1 \sim \tau_1^{-1} \sim 100\, Hz$. This seems like maybe an order of magnitude less than the sibilant tone one hears when a tea-kettle is being heated on a stove (long before the water starts actually boiling).[b]

So it turns out that there must be another cause for the tea-kettle's hissing when it is warming up. To establish this second reason, one would have to closely follow the bubble's fate after it leaves its parental surface. Having taken off from the hot wall (or bottom) where the vapor pressure in the bubble was around atmospheric (otherwise it could not expand enough to start moving upward), our protagonist hurries up in the higher, and, naturally, still colder, layers of water. So, the saturated water vapor, filling the bubble, cools down too, causing the inside bubble pressure to drop and no longer be able to compensate for the external pressure of liquid exerted on the poor bubble. As result, the squashed bubble flops or gets squeezed into a tiny one (the latter happens if, besides water vapor, it also had a little bit of air inside), generating a sound pulse in the liquid (see Figure 24.2). This process of the massive death or maiming of numerous steam bubbles on their way to the water surface is indeed what we hear as the hissing noise. And now, we will try to estimate its frequency, of course.

From Newton's second law, for the mass m of water rushing into the bubble when it collapses, we can write:

$$m\, a_r = F_p = S\, \Delta P.$$

Here $S = 4\pi r^2$ is the surface area of the bubble, F_p is the total pressure squeezing the bubble, ΔP is the pressure difference across the bubble's envelope, and a_r is the inward acceleration of the envelope. It is pretty clear that the mass involved in the process of "squeezing" should be of

---

[b]Note that the surface tension did not make it into (24.1) indicating, in a way, that the bubbles produce sound not only when leaving the surface but also during their accelerated motion upwards. This lasts until the buoyancy force gets compensated by the viscous friction, which is proportional to velocity.

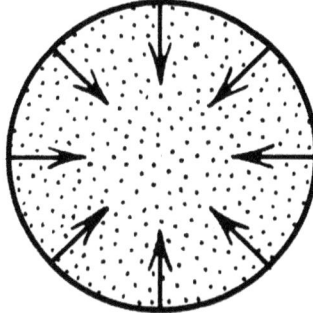

Fig. 24.2: Before the active boiling starts hundreds of tiny bubbles collapse making the tea-kettle sing.

the same order as the product of water density times the bubble's volume: $m \sim \rho_w r^3$. So, we could rearrange Newton's equation in the following manner:

$$\rho_w r^3 a_r \sim r^2 \Delta P.$$

Further, neglecting the fraction of pressure arising from the bubble's surface curvature, as well as the smidgen of air possibly ensnared inside the bubble, we will consider ΔP to be constant (more precisely, depending only on the temperature difference between the bottom and surface layers of water in our tea-kettle). Now, evaluating the acceleration $a_r = r'' \sim r_0 / \tau_2^2$ where τ_2 is the "flop" time we are looking for, we find that

$$\rho_w \frac{r_0^2}{\tau_2^2} \sim \Delta P,$$

which gives

$$\tau_2 \sim r_0 \sqrt{\frac{\rho_w}{\Delta P}}. \tag{24.2}$$

Compare this result with Eq. (9.4).

Near $T_b = 100° C$, the saturated water vapor pressure drops by about $3 \cdot 10^3 \, Pa$ per degree Celsius of the falling temperature (see Table 24.1). Hence, we could assume $\Delta P \sim 10^3 \, Pa$, and finally write for time $\tau_2 \sim 10^{-3} \, s$, leading to the noise frequency of $\nu_2 \sim \tau_2^{-1} \sim 10^3 \, Hz$. This answer is already much closer to the value perceived by our ears.

Table 24.1: Temperature dependence of pressure of saturated water vapor.

Temperature, $°C$	96.18	99.1	99.6	99.9	100	101	110.8
Pressure, kPa	88.26	98.07	100	101	101.3	105	147

One more fact supporting our conclusion that the proposed mechanism is indeed responsible for the tea-kettle noise is that, according to equation (24.2), its characteristic (high) frequency drops as the temperature grows. Right before the boiling begins, the bubbles cease flopping even in the upper layers of water. Then the only remaining sound becomes that produced by the bubbles taking off from the bottom. The frequency of the "tune" drops noticeably when the water in the kettle is about to start boiling. After it finally happens though, the kettle's "voice" may change again, especially if one opens the lid: the gurgling sound we hear is generated now by the bubbles rupturing right on the water's surface. This pitch depends also on the water level, as well as on the kettle's shape.[c]

Thus, we have established that the kettle's noise before the water starts boiling is related to the hundreds of steam bubbles produced on the hot bottom departing to the surface, and then perishing in the cooler upper layers of water. All these processes become especially apparent if one is heating water in a glass pot with transparent walls. Let us not hurry though to congratulate ourselves that we were the first to sort things out on this interesting question of singing boiling water. Back in the eighteenth century, the Scottish physicist Joseph Black[d] was studying the phenomenon and had established that the sound was produced by a duet consisting of the steam bubbles ascending to the surface and the vibration of the vessel's walls.

By now the water in the first of our subject tea-kettles (the one covered with a lid, remember?), the predicted favorite winner, has started to boil. The moment is quite resolutely announced by the stream of steam jetting out of the kettle's spout. And what could be the speed of the stream, by the way? One can solve this (honestly, not the most challenging) problem

---

[c] One more argument in favor of the proposed mechanism is that bubbles in fizzy drinks do not make the sound we have been talking about. The difference with boiling water is that the carbon-dioxide filled bubbles cannot collapse.

[d] J. Black, (1728–1799), Scottish physicist and chemist, he was the first to point out the difference between heat and temperature. He also introduced the idea of heat capacity.

by noticing that during the steady boil, all the energy provided to the kettle is being spent on vaporizing the water. Let us assume that in this case the only way steam is escaping is through the spout. Suppose further that a mass ΔM of water is vaporized in time Δt, at the expense of the supplied heat. Then, one could write the following balance equation:

$$H_v\, \Delta M = \mathcal{P}\, \Delta t,$$

where H_v is the specific heat of vaporization (heat per unit mass), and $\mathcal{P}$ is the power of the heater. During the same time Δt, the same mass ΔM is supposed to leave the kettle through its spout since otherwise the vapor would accumulate under the lid. If the area of the outlet (pretty much the perpendicular cross section of the spout) equals s, the steam density is $\rho_s(T_b)$, and v is the velocity in question, then the following relation holds:

$$\Delta M = \rho_s(T_b)\, s\, v\, \Delta t.$$

From the previous table, we pick $\rho_s(T_b) = 0.6\,kg/m^3$ as the density of saturated water vapor at $T_b = 373°\,C$. If one does not happen to have a suitable table handy, one can use the Clapeyron–Mendeleev gas law:[e]

$$\rho_s(T_b) = \frac{P_s(T_b)\, \mu_{H_2 O}}{R\,T_b} \approx 0.6\,kg/m^3. \tag{24.3}$$

Thus, for the velocity of steam escaping the spout, one finds:

$$v = \frac{\mathcal{P}\,R\,T_b}{H_v\,P_s(T_b)\,\mu_{H_2 O}\,s}.$$

After the required substitutions: $\mathcal{P} = 2000\,W$, $s = 2\,cm^2$, $H_v = 2.26 \cdot 10^6\,J/Kg$, $P_s(T_b) = 10^6\,Pa$, and $R = 8.31\,J/K \cdot mole$, we find the velocity to be $v \sim 1\,m/s$.

Now, finally, the water is boiling in the second (open) tea-kettle. It has noticeably lagged behind the capped victor. One should be very careful taking this one off the stove; if you just grab it by the handle, you can very easily scald yourself (we are sure, though, that our reader understands

---

[e]The expression for gas density ρ can be obtained from the equation of state of the ideal gas $PV = \frac{m}{\mu}RT$, where P, T, and m are the pressure, temperature, and mass of gas enclosed in the volume V respectively. This can be easily brought to the form

$$\rho = \frac{m}{V} = \frac{P\mu}{RT},$$

where μ is the molecular mass of the gas and R is the gas constant per mole.

that safety is first in any kind of experiment). Anyway, our next question is a safety-related one: which would scald worse, the steam or the boiling water? If we add some required physical constraints to this question, we could paraphrase it in the following way: which would scald worse, a certain mass of steam or the identical mass of boiling water?

Suppose there was $V_1 = 1\,l$ of saturated, $100°\,C$, steam under the lid of the tea-kettle. Imagine further that after the lid is opened, one tenth of this steam condenses on an unlucky hand. We know already that the density of water vapor at $T_b = 100°\,C$ equals $0.6\,Kg/m^3$. Hence, the mass of water condensed on the hand is going to be $m_s \approx 0.06\,g$. The amount of heat produced during the condensation and subsequent cooling of the condensed water from $100°\,C$ down to the room temperature $T_0 \approx 20°\,C$ then will be $\Delta Q = r\,m_s + c\,m_s(T_b - T_0)$. And, therefore, as $c = 4.19\,KJ/(Kg \cdot K)$ and $r \approx 2.26 \cdot 10^3\,KJ/Kg$, it takes about seven times more of the boiling water than the hot steam to get the same thermal effect! Besides, the area affected by the steam (due to the much higher mobility of the vapor molecules) is always significantly larger than that of hot water poured on a surface. Thus, the answer to our question is that the steam would be unequivocally much more dangerous than water of the same (boiling) temperature.

However, performing all these dangerous evaluations, we have swerved quite a bit from our original test with the two tea-kettles. Why did it take water in the open one so much longer to start boiling in the first place? Let us look at the phenomenon more closely. The answer seems to be almost trivial: during the heating, the nimblest molecules of water (those with higher velocity) can easily escape from the open kettle, pilfering some energy from the water remaining inside and, by doing so, effectively cooling it (this process is nothing but evaporation). Hence the heater in this case is supposed to provide not only energy to heat the water to boiling temperature, but also the heat required to evaporate some of the water. Thus, it is clear that it is going to take more energy (and, therefore more time because the heating power is fixed) than just heating the water in the capped kettle, where those swift fugitives have no other options but to congregate under the lid, building up the saturated water vapor, and eventually return to water, bringing along the stolen energy surplus.

Yet, there are two more effects occurring at the same time, although opposite to the one above. First, during the evaporation, the mass of water which needs to be warmed to T_b drops somewhat; secondly, in the open vessel, the pressure on the water is atmospheric and, hence, the boiling process starts at exactly $100° C$. On the other hand, in the covered kettle, if it is filled so that the steam cannot make it out through the spout, the pressure on the water surface will rise because of the intensive vaporizing. Note that now it is the sum of the partial pressures of a small amount of air under the lid as well as of the steam itself. With increasing external pressure, the boiling temperature should rise too, for it is determined by the equality of the saturated vapor pressure inside the bubbles to the external pressure. So, which of these factors should we prefer as the decisive one?

Each time such an uncertainty comes, one should retreat to a precise calculation or, at least, evaluate the magnitudes of the effects involved. So, first, let us estimate the amount of water which escapes from the open kettle before it starts boiling.

Molecules in liquids strongly interact with each other. In crystals, the potential energy of molecules is much greater than their kinetic energy, in gases, the kinetic part dominates. In liquid, the potential and kinetic energies are of the same order. So, molecules in the liquid state most of the time fluctuate around some "ascribed" equilibrium positions, yet once in a while manage to hop to a different neighboring equilibrium location. "Once in a while" just means a far longer duration compared to the period of oscillations about equilibrium points. However, in our conventional time scale, such hops occur quite frequently: in just one second, a jittery molecule of liquid can change its equilibrium position billions of times!

However, not every single molecule which happens to be wandering by the surface of the liquid can actually escape from it. To finally free themselves, such molecules ought to spend some energy performing work against the interaction forces. One may say that the potential energy of a molecule of water is less than that of a molecule of steam by the amount of the heat of evaporation, normalized per single molecule. Then, if H_v is the specific heat of evaporation, the molar vaporization heat is μH_v, and the "molecular" heat of vaporization will be $U_0 = \mu H_v / N_A$ (N_A being Avogadro's number). This work is performed at the expense of the kinetic energy of the molecule's thermal motion E_k. The corresponding average value $\bar{E}_k \approx kT$ ($k = 1.38 \cdot 10^{-23} J / K$ is the Boltzmann[f] constant) turns out to be far less than U_0. Nevertheless, according to the laws of molecular physics there will

---

[f]L. Boltzmann, (1844–1906), Austrian physicist; one of the founders of classical statistical physics.

always be some number of molecules with kinetic energies high enough to overcome the attraction forces and flee away. The concentration of these extremely swift molecules is given by the following expression:

$$n_{E_k > U_0} = n_0\, e^{-\frac{U_0}{kT}}, \tag{24.4}$$

where n_0 is the total density of molecules, and $e = 2.7182\ldots$ is the base of natural logarithms.

Now for the time being, we shall forget about the hops of molecules in liquid, and consider these high-energy molecules as a gas. A molecule of such a gas can reach the surface from inside in a short instant Δt provided its speed v is directed outwards and it has started less than $v\,\Delta t$ away from the surface. For a surface area S, these are the molecules from the cylinder of height $v\,\Delta t$ with the base S. Let us assume for simplicity that $\sim \frac{1}{6}$ of all the molecules in the cylinder (that is $\Delta N \sim \frac{1}{6} n\,S\,v\,\Delta t$) move towards the surface. Taking the density of the molecules with energies greater than U_0 from equation (24.4) we obtain for the evaporation rate (that is the number of molecules that escape the liquid in unit time):

$$\frac{\Delta N}{\Delta t} \sim \frac{n\,S\,v\,\Delta t}{6\,\Delta t} \sim S\,n_0 \sqrt{\frac{U_0}{m_0}}\, e^{-\frac{U_0}{kT}},$$

where we have taken the speed $v \sim \sqrt{U_0/m_0}$. So, the mass carried away from the liquid per unit time is equal to:

$$\frac{\Delta m}{\Delta t} \sim m_0 \frac{\Delta N}{\Delta t} \sim m_0\, S\, n_0 \sqrt{\frac{U_0}{m_0}}\, e^{-\frac{U_0}{kT}}. \tag{24.5}$$

It becomes more useful to recalculate this mass normalizing it to the $1\,K$ temperature increase while the kettle is being heated. To do so, we will use the energy conservation law: our kettle, in a time Δt, receives the amount of heat from the burner $\Delta Q = \mathcal{P}\,\Delta t$ ($\mathcal{P}$ is the burner's power), and, consequently, the water temperature rises by ΔT, entailing:

$$\mathcal{P}\,\Delta t = c\,M\,\Delta T,$$

where M is the mass of water in the kettle (here we disregard the heat capacity of the kettle itself). After plugging $\Delta t = c\,M\,\Delta T/\mathcal{P}$ into the equation for evaporation rate (24.5) we have:

$$\frac{\Delta m}{\Delta T} = \frac{\rho\,c\,S\,M}{\mathcal{P}} \sqrt{\frac{H_v\,\mu_{H_2 O}}{N_A\,m_0}}\, e^{-\frac{r\,\mu_{H_2 O}}{N_A kT}} = \frac{\rho\,c\,S\,M\,\sqrt{H_v}}{\mathcal{P}}\, e^{-\frac{H_v\,\mu_{H_2 O}}{N_A kT}}.$$

As the tea-kettle is being heated, the temperature increases from room temperature to the boiling point, $373\,K$. However, we conjecture at this point (and rightfully) that the majority of the mass is being lost while the temperature of water is already pretty high (close to its boiling value) so that we could put, say, $\bar{T} = 350\,K$ as the average temperature into our exponential expression above. For the rest of the terms we assume: $\Delta T = T_b - T_0 = 80\,K$, $S \sim 10^{-3}\,m^2$, $\rho \sim 10^3\,Kg/m^3$, $\mu_{H_2 O} = 0.018\,Kg/mol$, and $c = 4.19 \cdot 10^3\,J/Kg$. After putting all these values into the formula, we finally find:

$$\frac{\Delta m}{M} \approx \frac{\rho\, c\, S}{\mathcal{P}} \sqrt{H_v}\, e^{-\frac{H_v\, \mu_{H_2 O}}{R\bar{T}}} (T_b - T_0) \approx 1\,\%.$$

Thus, while being warmed up to the boiling temperature, just a small percentage of the total mass of water actually leaves the tea-kettle. The evaporation of such a mass takes extra energy from the heater, and naturally protracts the heating before it reaches the boiling point. To understand by how much, one could calculate and find out that the vaporization of this volume of water requires an amount of energy from the heater equivalent to heating from room temperature to boiling point about one fourth of the total mass of water in the tea-kettle.

Now let us turn to the second (or maybe the first – we do not remember already) covered kettle, and look more closely at the effects inhibiting the boiling. The first of them (potential change of mass of water during the heating) should have dropped off right away for, as we have just shown, the evaporation of about 1 percent of the water is energetically equivalent to the heating of about 25 percent of the water and, therefore, the heat required to bring this extra 1 percent of the water mass to the boil in the closed tea-kettle can be disregarded.

It turns out that the second of the effects (the increase of pressure over the water in the covered vessel) cannot play any noticeable competitive role either. Indeed, if the tea-kettle is completely filled with water (steam cannot escape from the spout), then the additional (to the atmospheric) pressure obviously cannot exceed the lid's weight divided by its area, for otherwise the lid would start to bounce, releasing the steam. Assuming $m_{lid} = 0.3\,Kg$ and $S_{lid} \sim 10^2\,cm^2$, we can limit this extra pressure by:

$$\Delta P \leq \frac{m_{lid}\, g}{S_{lid}} \approx \frac{3\,N}{10^{-2}\,m^2} = 3 \cdot 10^2\,Pa.$$

Having checked once more in Table 24.1, one finds that such an increase shifts the boiling temperature by no more than just $\delta T_b \sim 0.5°\,C$. Hence, to make the water boil it would take an extra heating energy of $\delta Q = c\,M\,\delta T_b$. Comparing this to $r\,\Delta M$, we see that the inequality $r\,\Delta M \gg c\,M\,\delta T_b$ holds with the safe ratio of at least as much as 30:1 allowing us to conclude that the increase of the boiling point of the water in the fully filled tea-kettle covered with the lid cannot seriously compete (regarded energywise) with the evaporation of water from the open water surface in the " bareheaded" vessel.

Digressing a bit, we would like to mention here that the described phenomenon of the increase of pressure during the heating of liquid in a closed volume has been successfully employed in the design of a utensil called the pressure cooker (familiar maybe to those who still do real cooking, at least once in a while). Instead of the spout though, it has a safety relief valve which opens only if the pressure inside goes over a certain limit; the rest of the time, the vessel remains entirely sealed. As liquid inside gets vaporized and all the steam builds in the cooker, the internal pressure rises to about $1.4 \cdot 10^5\,Pa$, before the relief valve opens, so that the boiling point (going back to our useful Table 24.1) moves up to $T_b^* = 108°\,C$. This allows one to cook food much faster than in a regular pot. However, one should be extremely cautious while opening the pressure cooker after taking it off the stove because when unsealed, the inside pressure drops and the liquid becomes significantly *superheated*. Therefore a certain mass δm, such that $r\,\delta m = c\,M\,(T_b^* - T_b)$, will instantly evaporate and can cause severe scalding. (In this situation, the liquid starts boiling explosively in the whole volume of the pot at once).

By the way, at high elevations (in addition to the typical beautiful scenery), the atmospheric pressure is lower and it becomes quite an undertaking to cook, say, a piece of meat, as water starts boiling at $70°\,C$ (the ambient pressure at the elevation of Everest, for example, is about $3.5 \cdot 10^4\,Pa$). So the pressure cooker is usually a quite welcome part of the climber's equipment. Due to its ability to reach acceptable cooking temperatures, it saves fuel as well. This is another weighty justification to have this massive utility in one's backpack.

But let us go back to our kettles, boiling now at full tilt, still on the stove. It is time to take them off. Take notice that the one with the lid does not stop boiling right away, after having been taken from the stove: the steam continues to puff out for some time. What fraction of water does actually evaporate during this ("postheating") boil?

To answer this one, we have to look at the chart in Figure 24.3, representing the dependence of the water temperature on height while the water is boiling (the heat, of course, is being supplied through the bottom). From the picture, we see that the slim bottom layer of about $\Delta H = 0.5\ cm$ thick is quite hot, with the temperature drop across it from $T_{bot} = 110°\ C$ (T_{bot} is the bottom temperature) to $T_i = 100.5°\ C$. The rest of the water stays, according to the graph, at about $100.5°\ C$, yet undergoes another step down of $\Delta T = 0.4°\ C$ when approaching the free surface of the liquid (the graph corresponds to the water level in the kettle being $H = 10\ cm$). Thus, for the amount of heat (beyond the equilibrium) stored in the vessel after its heating has stopped, we can write the following formula:

$$\Delta Q = c\,\rho\,S\,\Delta H \left(\frac{T_{bot} - T_i}{2}\right) + c\,\rho\,S\,H\,\Delta T,$$

where S is the kettle's bottom area (the kettle is considered to have a cylindrical shape). Sure enough, the tea-kettle's base is overheated somehow, but because of the much higher specific heat of water, we can safely disregard the contribution of this effect.

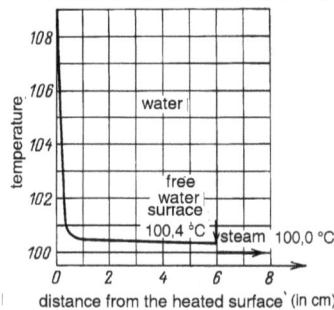

Fig. 24.3: Temperature near the bottom of a boiling kettle is much higher than that of the mass of water.

The ΔQ amount of extra heat is then spent evaporating the layer of liquid of thickness δH. The mass δm of such a slice of water can be found from the equation of heat balance:

$$H_v\,\delta m = \rho\,S\,\delta H\,r = c\,\rho\,S \left[\Delta H \left(\frac{T_{bot} - T_i}{2}\right) + H\,\Delta T\right]$$

consequently, giving us for δH:

$$\frac{\delta H}{H} = \frac{c}{H_v}\left[\frac{\Delta H}{H}\left(\frac{T_{bot} - T_i}{2}\right) + \Delta T\right] \approx 2 \cdot 10^{-3}.$$

So it shows that after we take the kettle from the stove, it still loses, due to continuing boiling, about 0.2 percent of its water content.

The typical time it would take to boil out all water of, say, mass $M = 1\,Kg$ from a kettle provided with heat of $\mathcal{P} = 2000\ W$, can be calculated as:

$$\tau = \frac{H_v\,M}{\mathcal{P}} \approx 1,25 \cdot 10^3\ s.$$

Therefore, the 0.2 percent of its mass will be vaporized within about 10 s (supposing that the evaporation rate does not change with respect to the stationary regime).

And now, after all this discussion, the time arrives to serve our long-awaited tea. By the way, in the Eastern countries, it is customary to drink tea using tiny tea bowls rather than tea cups or tea glasses. The former were most likely first introduced by the nomadic tribes of Asia – the small bowls are much easier to pack and they are far less fragile, which are obvious conveniences when one has an itinerant lifestyle. Besides, they have one more serious advantage over ordinary glasses: the bowl shape, which is wider at the top, lets water cool down faster in the upper layers precluding a possible burning by the scorching liquid, while the tea in the lower part of the utensil remains hot.

However in Azerbaijan you may meet another type of tea-drinking vessel, called an Armudi (see Figure 24.4). Here the wide top helps the safe and pleasant cooling of the beverage whereas the spherical part below has a minimum surface area and therefore keeps the tea hot for a longer time. Thus you may enjoy sipping the warm tea as the long and exquisite table-talk goes on.

The nifty porcelain tea-cups (not those over-sized mugs you usually get as safety awards from your company), which have been around for centuries, often have this wide-top profile too. The less-advanced cylindrical glasses came into general use as tea-serving ware only in the nineteenth century simply because of their lower cost and were traditionally used by men, whereas the artful china tea-cups were politely left for the better half of the human race. Over time though, the inferiority of the tea-glasses was improved a bit with the invention of glass-holders (back then often decorated with their owner's monograms).

Fig. 24.4: Azeri people prefer drinking tea from Armudi glasses (in English ''Armudi'' means ''Pear'').

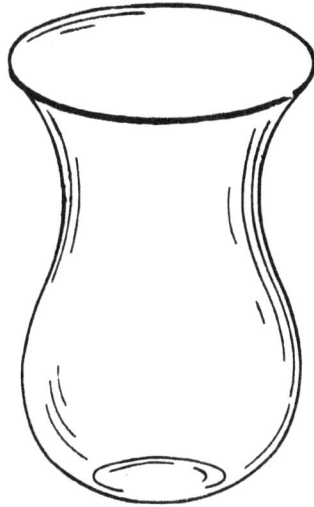

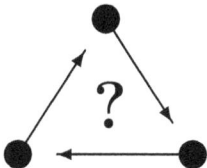

Can you think of what physical criteria material for glass-holders should satisfy? Would, say, aluminum and silver be good candidates?

Chapter 25

The physics of good coffee*

See how easy it can be to make a man happy:
a nice cup of coffee does it all.

Edoardo De Filippo, *"These Ghosts."*

The traveler who is used to going from country to country has probably observed that, in this century of globalization and multinational monopolies,
the very same drinks are being offered in New York or Kathmandu while
coffee, on the other hand, is still very much a regional drink. To drink a
coffee is a very different experience depending on whether you are in Turkey
or Egypt, Italy or France, Finland or America. If you ask for a coffee in a
bar in Naples, it will be served in a small elegant cup not much larger than
a thimble – at the bottom of which a viscous, nearly black liquid covered
with an inviting foam is slowly moving around. On the other hand, if you
order a coffee in Chicago, you will get a styrofoam container filled with a
half-liter of hot brown water. Here we do not wish to judge which of the
two coffees tastes better or makes you feel better; instead, we will simply
discuss the various methods employed in preparing coffee, and the relevant
physical processes involved in those methods.

*This chapter was originally published in English (translated from Italian by E. Fortin)
under the same title as an article in *Physics in Canada*, Volume 58, No. 1, pp. 13–17
(2002).

25.1 Boiled coffee

This method of preparation is very ancient and has extended nowadays up to Finland and the northern regions of the Scandinavian countries. Roasted coffee is coarsely ground, poured into water (about 10 grams of coffee per $150-190$ ml of water) and boiled for about 10 minutes. Without filtration, the drink is served directly into cups where it is left to settle for a few minutes. There is no interesting physical process involved here, and the authors will refrain from discussing the taste of such coffee.

25.2 Filter coffee pot (dripper)

This type of coffee-maker is quite common in America, in northern Europe, in Germany, and in France. Its principle of operation is very simple and the process lasts $6-8$ minutes. Coarsely ground coffee is poured into a conical paper filter. Boiling water poured from the top"washes" the coffee grains, penetrates across the filter and is collected in a glass container. As a result, one gets a light coffee since only a few of the oils make it through the dense filter. In addition, the coarsely ground coffee granules and the lack of pressure do not favor the complete extraction of all the flavors and aromas present in the coffee. In America, the norm is $5-6$ grams of coffee per $150-190$ ml of water, while in Europe, it is 10 grams of coffee per cup of water.

25.3 Turkish coffee

We continue our discussion with a process of preparation known as "coffee a la Turk". In this case, the coffee beans are ground to a fine powder and this powder mixed with sugar is poured into a conical shaped metal container (usually copper or brass) which is called an "ibrick", Fig 25.1. Cold water is then poured into the ibrick and the container is immersed in red hot sand.

In another recipe, the powdered coffee is put into the container which has already been filled with boiling water (if you do not have easy access to hot sand, you can use a gas burner or electric stove). The heat transported from the sand across the bottom and the sides of the ibrick generates convective currents while heating the liquid: as it moves, the hot liquid brings

Fig. 25.1: ''Ibrick''
for making Turkish
coffee.

up some of the coffee powder on the container's inner surface. Thanks to surface tension, a "crust of coffee" is formed on the surface. Little by little, the contents of the ibrick reach the boiling point, and bubbles start to break the crust forming a kind of foam. At this point, the process is interrupted by removing the ibrick from the sand so as not to damage the taste of the coffee. This process must be repeated twice more until a thick layer of foam has formed. The contents are then poured into small cups and one waits for the coffee powder to settle on the bottom. At the end, one obtains a dense and tasty beverage (especially if a small quantity of water is used).

25.4 Instant coffee

The pressures of present-day living have created this type of coffee. It is prepared from a real coffee which is evaporated at high temperature and low pressure. The resulting powder is preserved in a vacuum container where it can keep for a long time. To prepare the drink, the powder is simply thrown into hot water.

25.5 The Italian moka

The "moka" is the most commonly used coffee maker in Italy. It is made of three parts: the base where the water is to be heated, a cylindrical metallic filter where the finely ground coffee powder is deposited, and the top section which has the approximate shape of a truncated cone, where the finished coffee arrives, Figure 25.2. The filter is the heart of the moka: below the filter is a fixed metallic funnel which draws very near the bottom of the base container. The moka does not leave much room for inventiveness in preparation. Instead, in contrast to other methods of preparation, the recipe determined by the pot's design must be followed.

Fig. 25.2: The Italian moka coffee maker

The process of making coffee with the moka is very interesting. The ground coffee is gently pressed into the filter and cold water is poured into the base container. The moka is closed by tightening the base and the top part around the filter which covers the water contained in the base. A rubber ring seal insures the tightness between the two parts. The water is heated on a weak flame and, consequently, the vapor pressure over the surface of the water increases rapidly, forcing the water to rise within the stem of the funnel leading through the coffee powder contained in the filter. The coffee then rises through a thin tube to the upper part of the pot. At this point, the coffee may be served in small cups.

This method appears simple and easy to understand. But what is the moving factor behind the process? Certainly the heat from the flame. Initially, the water is heated in a closed space – where the water itself occupies most of the available volume. The water quickly reaches $100°\,C$ (boiling temperature at sea level, that is $212°$ Fahrenheit) and the saturated vapor pressure above the water reaches 1 atmosphere. Providing further heat results in an increase in both water temperature and saturated vapor pressure: water and vapor remain in equilibrium while temperature and pressure keep on increasing. (The saturated vapor pressure as a function of the temperature may be found in Table 24.1, page 214). On the other hand, the external pressure (above the filter) remains equal to the atmospheric pressure. The saturated vapor which is at a temperature slightly above $100°\,C$ acts as a spring, pushing the boiling water across the coffee powder. In this way, the extraction of all the flavors and aromas – the essence of the coffee – and other components, transforms the water into a tasty beverage.

Obviously, the taste depends on the quality of the coffee powder, the water temperature, and on the time required for the water to cross the filter. The secrets for the preparation of the coffee mixture are those of each producer, based on talent, work, and long experience. What determines the transit time of the water across the filter can be understood without resorting to industrial espionage – relying solely on the basis of physical laws.

In the middle of the nineteenth century, two French engineers, A. Darcy and G. Dupuis, performed the first experiments on the motion of water through sand-filled tubes. This research represents the start of the empirical theoretical development of filtration which is successfully applied today to the motion of liquids through solids having interconnected pores and cracks. It was this Darcy who formulated the so-called law of linear filtration which today bears his name. That law relates the mass of fluid/sec (Q) going through a filter of thickness L and cross section S with the pressure difference ΔP across the filter:

$$Q = \kappa \frac{\rho S}{\eta L} \Delta P, \tag{25.1}$$

where ρ and η respectively represent the density and the viscosity of the fluid, κ, (the constant of filtration) does not depend on the characteristics of the liquid, but only on the porous medium.

In the original work of Darcy, the pressure difference ΔP was assumed to originate exclusively from the force of the weight. In that case, $\Delta P = \rho g \Delta H$, where g is the acceleration of gravity and ΔH represents the height difference between the two extremities of the filter which are supposedly vertically mounted. From Eq. (25.1), κ has the dimensionality of the area (meter^2). The values of κ are in general very small. For example, in a large grain sandy terrain, the value of κ is of the order of $10^{-12}-10^{-13}\,m^2$, and in a terrain of pressed sand $\kappa \sim 10^{-14}\,m^2$.

Let us try now to apply Darcy's law to the study of our moka coffee maker. For example, it would be interesting to know the temperature reached by the boiling water within the base of the moka. This temperature can be obtained from the dependence of the boiling temperature on the pressure and Darcy's law:

$$\Delta P = \frac{m}{\rho S t} \frac{\eta L}{\kappa}. \tag{25.2}$$

The usual thickness of the filter in a moka for three persons is $L = 1\ cm$, $S = 30\ cm^2$ and $m = 100\ g$ of coffee is prepared in $t = 1$ minute. For the coefficient of filtration, we can use that of a terrain of coarse sand $\kappa \approx 10^{-13}\ m^2$, $\rho = 10^3\ Kg/m^3$. As for the viscosity, one must be careful because it is temperature dependent. In the tables of physical quantities, $\eta(100^\circ C) = 10^{-3}\ Pa \cdot s$ from which $\Delta P \approx 5 \cdot 10^4\ Pa$ follows. From Table 24.1 one sees that the corresponding water boiling temperature is $T^* \sim 110^\circ C\ (230^\circ F)$.

We have thus understood the physical basis of the coffee preparation process in an Italian moka. However, there are obscure aspects in that machine which can transform it into a real bomb, a menace for the kitchen walls and ceiling, and of course for the safety of whoever may be in the vicinity. Why and how can it happen? Clearly, such a disaster can only occur if the safety valve located in the base of the coffee maker is obstructed (old coffee makers are dangerous!), and at the same time, if the normal flow across the coffee filter is blocked off. For example, too fine a coffee powder compressed too hard could become impenetrable for the water. If both of these conditions occur, the pressure in the base can increase to a point where the thread upon which base and top are screwed breaks down.

Let us now try to explain what happens when the coffee powder is packed too tightly. We must remember that Darcy's law is empirical in nature. At the microscopic level, however, the filter can be imagined as a system of interconnected capillaries of various sections and lengths. The validity of Darcy's law requires a laminar flow of the liquid across the filter. In reality, various irregularities in the capillary structure result in the formation of vortices and consequently a transition to the turbulent mode. These dissipative phenomena require an increase in pressure in order to keep a constant flux across the filter. Another factor that determines the lower limit for the transparency of the filter is related to the surface tension. In fact, it is known that in an ideal capillary of section r, one must apply a pressure difference $\Delta P > 2\sigma/r$. Taking the capillaries' radius as $\sim 10^{-4}\ m$ and remembering that σ is about $0.07\ N/m$, one obtains a threshold of $10^3\ Pa$ for the filter's transparency. This value is about one

order of magnitude lower than the pressure difference across the filter at
$110°\,C$. It is clear that, if too fine a coffee powder is packed too hard within
the filter, the effective radius of the capillaries could become much smaller
than the $10^{-4}\,m$ previously estimated. Under these conditions the filter
could, in practice, turn "opaque" or impenetrable.

Coffee pots transformed into bombs can indeed be quite dangerous. Let
us start from the worst situation: the filter and safety valve are blocked
and 100 grams of water are heated in a volume not much greater than
that occupied by the water. For temperatures near the water critical phase
(where the density of the vapor and of the water are comparable), at a
temperature of $T = 374°\,C = 647\,K$ $(T = 705°\,F)$, all of the water becomes
vapor. In principle it is possible to heat the moka further, but at higher
temperatures, the pot will start to glow (this has not been observed yet)!
Thus, for a reasonable calculation one can assume a final temperature of
$T = 600\,K$ $(T = 620°\,F)$. The vapor pressure in the base of the moka can
be easily estimated using the ideal gas equation:

$$PV = \frac{m}{\mu}RT. \tag{25.3}$$

Since $m = 100\,g$, $V = 120\,cm^3$, $\mu = 18\,g/mole$, $R = 8.31\,J/(mole \cdot K)$
we can find $P \approx 10^8\,Pa = 10^3$ atmospheres. The associated energy is
impressive, reaching

$$E = \frac{5}{2}PV \sim 70\,KJ \tag{25.4}$$

and the explosion could thus throw moka parts at speeds of up to hundreds
of meters per second. From this calculation, it is obvious that the explosion must occur much before the temperature of $600\,K$ is reached. But it
demonstrates the strong forces developed within the moka with excessive
heating: forces sufficient not only to dirty up the kitchen with coffee but
also to be the source of other problems. A curious effect related to the explosion is the following: if you are near the moka when it explodes, and you
are lucky enough not to be hit by pieces of metal traveling at high speeds,
you will, in all probability, be hit with overheated vapor jets and coffee
powder. Your first reaction will be to quickly remove your wet clothes to
avoid being scalded. A moment later, you may realize, with great surprise,
that you have a feeling of coldness rather than heat.

The explanation for this surprising effect is rather simple: the vapor
expansion following the explosion is very rapid, not allowing time for a

heat exchange with the ambient. In other words, the vapor expansion is adiabatic and with good approximation obeys the related ideal gas laws:

$$TP^{\frac{1-\gamma}{\gamma}} = const, \qquad (25.5)$$

where $\gamma = \frac{C_P}{C_V}$. Here C_P and C_V are the molar heat capacities of water at fixed pressure and fixed volume respectively.

As is well known from molecular physics, C_P and C_V can be expressed by means of the number of the degrees of freedom i. Indeed, as we know from the law of the equal distribution of energy, for each degree of freedom at fixed volume there is a corresponding contribution to heat capacity equal $R/2$, while $C_P = C_V + R$. For a molecule of water $i = 6$ and $\gamma = \frac{i+2}{i} = 4/3$, hence

$$T \sim \sqrt[4]{P}. \qquad (25.6)$$

Assuming the initial temperature and pressure of the vapor to be $500\,K$ $(440°\,F)$ and $10\,atm$ respectively, one sees that, upon reaching a pressure of one atmosphere, the gas may attain a temperature as low as $100\,K$ $(-280°\,F)$.

In conclusion, coffee prepared in a moka is strong and aromatic but does not reach the quality of an espresso served in a good bar. The main reason is the high temperature of the water forced through the filter by the vapor. Here, then, is some advice to improve the quality of coffee prepared with the moka: warm the coffee pot slowly; in this way the water will traverse the filter slowly and, at the same time, excessive heating of the vapor in the base container will be prevented. Finally, a very specific coffee can be prepared with a moka in an alpine refuge where the external pressure is less than one atmosphere (for example, at an altitude comparable to that of Mount Everest, the water boils at about $70°\,C$ ($\approx 160°\,F$)) and the superheated water reaches only a temperature of $85-90°\,C$ (the best for coffee preparation).

25.6   The ancient coffee maker – "la Napoletana"

This coffee maker reminds us somewhat of the moka with the main difference being that the "motor" forcing the liquid through the filter is gravity rather than vapor pressure. The coffee pot consists of two containers, one above the other and separated by a filter filled with coffee, Figure 25.3.

When the water from the lower container reaches the boiling point, the coffee pot is inverted. The filtration then occurs under the action of the pressure of a water column of a few centimeters (ΔP on the filter does not reach $10^3\,Pa$). The coffee preparation process is slower here than for the moka. It is interesting to make an experiment and prepare the coffee in the two different pots and verify Darcy's law where the preparation time is inversely proportional to the pressure, the validity of which was stated above. According to the connoisseurs, a coffee prepared "alla Napoletana" is better than that made in a moka: here the filtration process occurs more slowly and the flavor of the coffee is not damaged by contact with super-heated water. Unfortunately, the pace of modern life does not leave us time for calm discussions while waiting for the coffee "alla Napoletana". This luxury remains only in the paintings showing the Neapolitan life of the past and in the works of Edoardo di Filippo.

Fig. 25.3: ''La Napoletana'' against the background of Vesuvius.

25.7 The "Espresso"

Not all Neapolitans were so patient however. It seems that in the XIXth century, a resident of the Kingdom of the Two Sicilies, who could not wait patiently for the coffee "alla Napoletana", managed to convince an engineer friend from Milan to devise a new type of coffee maker that could prepare a good, dense, aromatic coffee in less than half a minute. Nowadays, a cup of good coffee involves a treasure box of secrets about the growth and harvesting of coffee beans, the preparation of the mixture, and, obviously, the roasting and grinding of the beans. Behind the art of coffee making stands a sophisticated technology: the espresso machines are much more complicated than the simple coffee makers described earlier.

Normally, espresso machines are found only in bars and restaurants but, for enthusiastic coffee lovers, there also exists a home version of the machine, Figure 25.4. In the professional type of machine the water, at a temperature of 90–94° C, is pushed at a pressure of $9-16$ atmospheres across the filter containing finely ground coffee. The whole process lasts $15-30$ seconds and provides one or two small cups ($20-30$ ml) of espresso. The mechanism responsible for the passage of the liquid across the filter can also be explained by Darcy's law as for the moka. However, the pressure here is 10 times greater while the temperature is around 90° C. A higher pressure increases the speed of travel across the coffee powder while a lower temperature ensures that the unstable components contributing to the coffee flavor are not decomposed. Although this must appear strange, the espresso contains less caffeine than the coffee prepared with the moka because the contact between water and coffee powder is shorter ($20-30$ seconds as compared to several minutes for the dripper). The first espresso machine was presented in Paris in 1855. In the modern machines used in bars and restaurants, the water reaches the required pressure with the help of a pump. In the classical machines, one first raises the lever to introduce the water contained in the heating cylinder situated within the vessel under the filter. Successively, the same lever is lowered to push the water across the filter. The pressure applied to the liquid is thus provided by the force of the arm multiplied by the lever action.

25.8   Variations on the Espresso theme

With an espresso maker and a good coffee mix, it is possible to be inventive. For example, in Italian bars, you can drink: a concentrated *"caffè ristretto"* ("restricted" coffee) which is prepared with a standard quantity of coffee but with less water; a *"caffè lungo"* ("long" coffee) prepared with a normal quantity of coffee but with more water; a *"caffè macchiato"* ("dappled" coffee) which is espresso with a dash of milk; a *"caffè corretto"* ("corrected" coffee) which is espresso with a liquor, and other 50-60 variations on the espresso theme.

Fig. 25.4: Even the
house version of the
espresso machine looks
impressive.

Especial discussion requires a *"cappuccino"* which is espresso in a medium sized cup to which is added milk "beaten" with the vapor in order to obtain a light, frothy foam. A good barista can pour the milk on the coffee

Fig. 25.5: The scene
from''Swan lake''
painted at the surface
of cappuccino by the
winner of the Italian
Barista Championship
Pietro Vanelli.

so as to write the first letter of your name on the surface, while the Barista Championship winner can paint the scene from "Swan Lake" (see Figure 25.5). Finally, one can simply add a bit of cacao to the foam.

They say that nowadays in Naples, a few bars still serve the *"caffè prepagato"* ("prepaid coffee"). A well-dressed gentleman accompanied by a lady enters a bar and orders three coffees, two for themselves, and one "prepaid". After a short time, a poor man enters the same bar and asks,

"Is there a *caffè prepagato?*" And then the barman pours him a cup of free coffee.

Naples always remains Naples... and not only in the Toto movies.

25.9 Air humidity and degree of coffee beans grinding for espresso

Today, in the age of globalization, the borders between countries are in many respects erased and the characteristic national products are gaining world status. One of them is the Italian espresso. Thanks to the development of technology, a relatively inexpensive coffee machine and corresponding capsules with various types of coffee today can be bought in almost any country of the world. This allows you to drink a cup of quite good *"caffè ristretto"* or *"caffè lungo"* as in the English province, traditionally preferring tea, and in Turkey, which has its own original coffee culture. Of course, the standardized taste of coffee from the capsule cannot be compared with what you can enjoy in coffee shops in Naples or Trieste. Nevertheless, the capsule technology definitely improved the standard requirements for this drink worldwide.

To achieve the top mastery in coffee brewing, one should tirelessly work on the selection of varieties of its grains during the creation of the mixture, their proper storage in a specially selected atmosphere, unique technologies for their toasting, and strict observance of the shelf life. Perfection of coffee makers has led to the invention of devices that satisfy the human needs in this drink from preparing the morning portion of coffee in a bachelor's house to serving a thousand people a day in the large institutions. Today hundreds of thousands of people are engaged in this huge coffee industry. However, with all this, the presence of a luxurious chrome Italian machine and a mixture of coffee beans of the best brands in expensive bars of Moscow, Paris or New York guarantees you a high price, but not the quality of this drink.

Let us talk now about the "espresso" brewing not as about technology, but as an art, which is ultimately based on the laws of physics. The final intermediary between the coffee machine and the consumer is the *"barista"*. A lot depends on his art and devotion to the profession. And it is not just that he regularly cleans the coffee machine of sediment, warms it up early in the morning and pours out the top ten cups without offering them to the first customers. It is also necessary (but not enough) to take care of the cleanliness of the grinder, grind 7-9 grams of a good mixture of coffee beans per serving immediately before brewing, and much more. Among the secrets of the mastery of Italian *"barista"*, the important element is attentive observation of changes in air temperature and humidity, with appropriate adjustment of the degree of grinding of grains. At first glance, their recipe for quality coffee making seems paradoxical: with increasing humidity or air temperature, coffee beans should be grounded more roughly. Let us try to understand the physical reasons for this advice.

To do this, we rewrite Darcy's law (25.2) in explicit form for the filtration time:

$$\tau = \frac{1}{\kappa} \frac{m\eta L}{\rho S \Delta P},$$
(25.7)

i.e., relating the time of preparation of a cup of coffee τ to the mass of the obtained beverage (m), the viscosity of water (η), the sizes of the filter (S, L), and the difference of pressure on it (ΔP). The optimum pressure and temperature for the espresso brewing are set in the coffee machine in advance and the *"barista"* does not change them in the process of work. The viscosity of water also remains unchanged in Eq. (25.7): it depends only on the temperature, which we assume to be constant. The mass of the drink in the cup can vary depending on the customer's request - from 25 grams in the case of a *"caffè ristretto"* order, up to twice as much in the case of a request for a *"caffè lungo"*.

Namely by the filtration time the barista judges the quality of the drink: for the *"caffè ristretto"* this time should be 18-25 seconds, for *"caffè lungo"* it is twice as much. If coffee "falls" into the cup in a shorter time, then such an underexposed drink turns sour, with a light loose *"crema di caffè"*(foam). Overexposed coffee, on the contrary, has a dark-colored *"crema"* and turns bitter.

As one can see from Eq. (25.7), the only parameter by which the *"barista"* can affect the brewing time τ is the porosity coefficient κ of the coffee powder in the filter. Varying it, the *"barista"* achieves the necessary filtration time, despite the change in temperature and humidity of the ambient air.

Imagine that the coffee shop is located in the open air at the "South

Pole Scientific Station" in Antarctica.[a]   At a temperature of $-40°\,C$ to $-50°\,C$ and polar conditions, the vapor density in the ambient air (absolute humidity) is negligible. The smallest particles of coffee (of the size r_l) formed during the grinding process in the coffee grinder are electrified (free electrons from the grinder's knives transfer to them).[b] The fact that the grounded coffee is really electrified is evident since this powder sticks, say, to a coffee grinder.

The electrostatic forces that arise between them have the character of attraction and lead to formation of effective agglomerates of these particles, whose sizes are noticeably larger $R_a \gg r_l$. In the subsequent filtration, it is these composite agglomerates that represent obstacles to water leaking through the filter: water becomes a coffee drink when washing them. Thus, it is the size of the agglomerate particle R_a that determines the value of the porosity coefficient κ, and hence the brewing time of the coffee portion.[c] Thus the "barista selects the size of the grind so that the time τ would belong to a range of 18-25 seconds.

We will now transfer our coffee shop to the equator, for example, to Singapore. Here the air is saturated with water vapor, the relative humidity reaches 90%, the temperature is the same $40°\,C$ but with the sign "+". As a result, absolute humidity increases hundreds of times in comparison with the icy desert of Antarctica. Therefore, if we choose as the characteristic grinding size r_l, also here, the agglomerates that make up the coffee powder in the filter will turn out to be much smaller than R_a. Indeed, the tropical air is saturated with moisture and the electrostatic forces are

---

[a]The Southernmost habitation on Earth.

[b]The reason for this process is simple, it is exactly the same as the electrization of hair when combing a plastic comb or synthetic clothing when it is worn. This is friction, in which a certain amount of charge is transferred from one body to another. As early as 1733, the French scientist Charles François de Cisternay du Fay (14 September 1698 16 July 1739) after carrying out numerous experiments, proved that all the types of electricity known at that time, that is, of electricity of different origins - the celestial (lightning), the animal (obtained from creatures, for example, from electric acne) are reduced to two types. These are "the glass one", obtained by rubbing glass on silk and "the resin one", formed when rubbing the resin on the wool. After the experiments of Benjamin Franklin, 15 years later, it became clear that this division corresponds to our today's positive ("glass") and negative ("resin") electric charges.

[c]Provided that the *"barista"* compacts the powder in the filter always with the same mechanical force.

substantially weakened.[d] Therefore, the coefficient of porosity of the coffee powder ground in a warm and humid environment will be substantially less than in cold and dry air. The filtration time (25.7) will increase, the coffee will be overexposed. To avoid this, a good *"barista"* changes the grind size of coffee beans almost every half hour, depending on changes in humidity and air temperature.

Another important factor is that water is not a bad electrical conductor and so the charges are redistributed much faster in slightly damp coffee than in the dry one. Hence in humid air, the coffee beans should be grounded more roughly than in dry air.

---

[d]Recall the dielectric permittivity ϵ of the medium in the denominator of the Coulomb law. For water $\epsilon = 81$.

Chapter 26

"Nunc est bibendum": physicists talk around glasses of wine

Oh, beloved child, bring the precious chalices,
for the son of Zeus and Semele
gave to human being the wine
to let them to forget their sorrows

Alcaeus[a] *""*

Wine is one of the most civilized things in the world
and one of the natural things of the world that has been
brought to the greatest perfection, and it offers a greater
range for enjoyment and appreciation than, possibly, any
other purely sensory thing which may be purchased.

Ernest Hemingway,[b] *"Death in the Afternoon."*

26.1   On the origin of wine and the methods of wine making

What can we add to the idea of wine created over the centuries by poets,
writers, journalists and oenologist? The epigraphs to this chapter were
written over a 2,500 year period: the first one comes from one of the earliest
poets of our civilization, the second from one of the most popular modern
writers. These remarks praise wine and are the manifestation of the love of
land and for one of its most valuable fruits, transformed by labor and skill

---

[a] Alcaeus (620–580 B.C.), Greek poet.
[b] Ernest Hemingway (1899–1961), American writer.

into a divine drink, playing a special role in the life of human beings. These remarks are also somewhat related to what we are going to talk about later on.

The origin of wine can be traced back to prehistoric caves, where archeologists have found records of grapes. Thousands of years B.C., the fruits of the grapevine, which presumably originated in India, were known. According to a legend which would be hard to verify, wine was discovered at the court of a Persian king. A distressed lady living in the royal palace decided to kill herself. Following the suggestion of a priest-like man appointed to perform religious rites she drank the strange-looking liquid being formed at the bottom of large bowls filled by grapes. The result was unexpected: her depression was gone, gloomy thoughts gave way to happiness. This experiment highlighted the felicitous qualities of the drink and, after testing it personally, the king introduced the wine for regular consumption. Since that moment, the status of wine had been on the ascent and it also became a sacramental drink. It is well known that in the Christian tradition, wine is part of the ritual for the Holy Communion. In the Jewish ritual of Kiddush, the ceremony of blessing is performed with wine and bread. Arabic legends are claimed to state that Adam was familiar with the vines, and it was the grape rather than the apple that was the "forbidden fruit". The migration of Aryans from India to the west around 2500 B.C. led to a spread of these methods throughout the ancient world. It is certain that wine was being produced in Sicily about 2000 B.C., probably due to colonization by Greeks and Egyptians. Some time later, Sabinis, who lived on the territory of modern Italy (recall the painting "The Rape of the Sabines" by N. Poussin) and Etruscans, who populated modern Tuscany 3000 years ago, started making wine. It is known that in ancient times, people acknowledged the healing properties of wine, and also used it as an antiseptic remedy for treating wounds. Wine-based medical prescriptions were found in Egyptian papyruses. Hippocrates prescribed wine as an anti-inflammatory remedy. Speaking of Romans, we may remember how Horatio and Virgil eulogized wine in many of their compositions. In 42 B.C., Columella wrote an excellent instruction book on viticulture. In the first century A.D., winemaking was widely spread throughout the Roman Empire. In his struggle against overproduction, Emperor Domitian prohibited an increase of the area of grape cultivation in the Roman Empire and ordered that the vineyards in the provinces be cut by half. By 200 A.D., the crisis in agriculture related to the Barbarian invasions led to the

decline of viticulture, and soon it completely disappeared due to the advent
of Muslims in the historic arena, for the use of alcohol is prohibited by the
Koran.

In the Middle Ages, wine was sporadically produced in castles or in
monastic minsters. Only the period of the Renaissance, i.e., beginning in
the sixteenth century, in Europe restored the mass production of wine. As
Hemingway said, the progress in wine making, along with the revival of arts
and culture, highlights the development of civilization and liberation from
barbarism. Modern oenology – the science of wine making – originated at
the beginning of the twentieth century, due to dramatic events in viticulture: the mass migration of people caused the transfer of unusual parasites
into traditional wine-making areas, which caused the destruction of vines.
Some of these parasites such as powdery mildew (*Uncinula necator*) and
downy mildew (*Plasmopora viticola*) were repelled by chemicals based on
sulphur and copper. A notorious phylloxera (*Phillossera vastatrix*), however, proved to be resistant to pesticides. To get rid of it, oenologists
turned to genetic methods. The grape vine brought to America by European settlers long ago was gradually evolving in the new environment. Its
roots had developed an external protective tissue, which made it immune to
phylloxera. Therefore, to save the best European plants, the *Vitis vinifera*,
oenologists began grafting onto them well-weathered species native to the
United States (*Vitis riparia, Vitis rupestris*, and *Vitis berlandieri*). Their
offsprings are literally at the root of modern European viticulture. Old
European vines based on *Vitis vinifera* survive only in several small areas
in Europe: in Jerez (Spain) and in Colares (near Lisbon), where sandy soil
prevents the growth of phylloxera, and also in Moselle or in Douro areas,
where shale soils play an analogous role. Today the battle against diseases
of the vines is a primary concern for the oenologists, though sometimes
parasites can be useful. For example, the gray mold *Botrytis* is regarded as
a beneficial kind of mold: in Sauternes frequent fogs in the fall encourage
the proliferation of this infection. In the most infected berries, the mold
absorbs up to 20 percent of the moisture, thus increasing sugar content in
the grape juice and creating a unique taste in Sauternes wines,[c] delicious
in combination with Roquefort[d] and pâté de foie gras.[e]

---

[c]Sauternes (France).
[d]Roquefort – famous French cheese with lots of green and blue mold.
[e]Pâté de foie gras – paté made of the enlarged liver of a specially fed goose.

After the grapes are harvested and pressed the process of fermentation begins. Wine berries have a special property: it is enough to scratch their skin and leave them for a short time in grape juice and a natural process of fermentation – conversion of sugar into alcohol – will begin. This process takes place due to the presence of *Saccharomyces ellipsoideus*, the yeast derived from the grape skin. Sometimes yeast cultures are added to the grape juice; this method is called "induced fermentation". During the fermentation process, sugar is converted into ethyl alcohol, carbon dioxide, glycerin, acetic and lactic acids, and many other substances. We mention here only the substances essential to our pending story. The attempt to describe in detail how to create good wine would be as hard as the attempt to describe in detail how to paint a famous masterpiece of art or to provide instructions on carving a masterpiece from a marble slab. Here we will confine ourselves to the description of certain physical phenomena occurring in wine making. Grape skin contains yeast cultures that convert juice into both wine and vinegar; when left unguided the yeast cultures can spoil the beverage. Thus an oenologist once said that the Lord had not been planning to transform grape juice into wine: vinegar should have come out in the end. To avoid such an undesirable outcome and to streamline and harness the fermentation process is the responsibility of man. Modern methods of fermentation based on the knowledge of chemical processes and high technology offer better results than home wine-making. Today, for example, to produce high-quality wines oenologists turn to limited areal density of plants yielding better and more precious grapes. Earlier, wine makers preferred planting tall plants, as they were more convenient for harvesting. Now, to improve the quality of the wine, short plants are grown, with berries getting extra heat from the sun-warmed soil. In the past, the adverse effect of acetic yeast was controlled by certain chemical components. Today, alternative methods based on fine filtration and deep refrigeration are frequently used.

Fermentation is an exothermal process that is accompanied by a marked generation of heat. As a result, uncontrolled production of wine may lead to a temperature rise in must to temperatures as high as $40-42°\,C$. In this case, the exquisite and volatile flavors of fruits and flowers that could create a unique bouquet are lost. In modern production, the fermentation is performed in double-walled stainless steel tanks reminiscent of Thermos flasks. Coolant circulating between the walls keeps the temperature of must low – no more than $18°\,C$, thus helping to preserve the wine flavors. However, this method requires a longer time for fermentation; it takes up to three weeks instead of $7-8$ days to complete fermentation at low temperature under natural conditions.

Fine filtration and cooling also help to clear wine of certain sulphites which had been added previously to slow down the acidification. Two factors play a role in this process. Each additional percent of alcohol lowers freezing temperature for wine by $0.5°\,C$ relative to $0°\,C$. Thus, wine which contains 12% of alcohol will freeze at about $-6°\,C$. At the same time, sulphites form organic complexes with the peptides found in wine. These complexes form small crystals that can be filtered off even before wine is frozen. A quick view of the previous methods explains why a contemporary winery more and more resembles a research laboratory. Low temperatures also help wine makers increase relative sugar content, so grapes are slightly frozen before pressing. Some water is condensed into ice and the sugar content is higher in the residual juice (the so-called *ice-wine* in Germany, Austria, and Canada). Another way to increase sugar concentration is to dry grapes (Erbaluce, Caluso) for sweet wine manufacturing: the so-called passito (Sicily and other regions of Italy), muscat (Crimea) and other wines. Let us now explore what we regard as several interesting physical phenomena related to alcoholic beverages and their consumption.

26.2   Tears of wine

If we swirl wine in a glass, we can observe an interesting phenomenon: it often leaves a kind of coating on the inside of the glass, that separates into viscous-looking rivulets, called legs or tears (see Figure 26.1). These legs slowly slide down the glass, returning to the wine's surface. Sometimes it is believed that the presence of the tears indicates the high quality of the wine. Connoisseurs of wine like to discuss the tears at the table using terms like glycerin and corpulence (full-body).

Let us analyze the physical origin of this phenomenon. First of all, let us note that in order to inspect "wine tears" in a strong enough alcohol-water solution (more than 20% of alcohol) there is no need to swirl or shake the glass. The unusual effect of mass convection up the wall in the form of a thin film and its counter flow down as "wine tears" is observed, in this case, even in a standing glass. In hydrodynamics the first part of this phenomenon is known as the Marangoni effect. It consists of the motion of

Fig. 26.1: Wine tears
on the walls of a
glass.

a film boundary in the direction opposite to the gravitational force due to the appearance of a surface tension gradient over the height of such a film. The point is that alcohol vaporizes from the film's surface faster than water. The surface tension of water is larger than that of alcohol. Consequently, nonhomogeneous vaporization of the alcohol from the surface of films of varying thickness (along the wine surface in glass) results in the appearance of an alcohol concentration gradient and hence to the gradient of the surface tension. This inhomogeneity leads to the generation of a force pulling the film up the glass wall. The details of such boundary motion were studied in 1992 in the experiments conducted by French physicists Fournier and Cazabat.[f] They found a surprising relation between the film boundary displacement (climb) L and time t:

$$L(t) = \sqrt{D(\varphi)t}.$$

Corresponding dependence in the coordinates L versus $t^{1/2}$ is shown in Figure 26.2.

In fact, such functional dependence was already well known in physics. It was first derived theoretically by Einstein and Smoluchovski, and it established the relation between the effective particle displacement in its diffusive motion and the diffusion time. The coefficient D in the Einstein-Smoluchovski formula is a so-called diffusion coefficient and it is defined by means of the particle velocity and its mean free path. In their experiments, the French scientists also discovered the nontrivial dependence of their diffusion coefficient $D(\varphi)$ on the solution's alcohol content (proof) φ (see Figure 26.3).

As was already mentioned, when the alcohol concentration in solution was less than 20% ($\varphi < 0.2$), very little spontaneous film climb up the

---

[f] J. B. Fournier, A. M. Cazabat, "Tears of Wine", *Europhysics Letters*, **20**, 517 (1992).

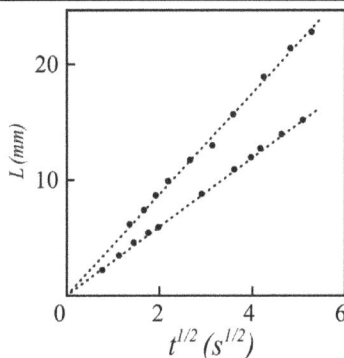

Fig. 26.2: Dependence
of the wine film
boundary displacement
on time.

glass wall was observed. Strikingly, in contrast to intuitive expectations, the largest values of the coefficient D were obtained for the relatively dilute solutions, even though the alcohol evaporation is going slower. Looking at the experimental data (Figure 26.3, upper curve), one may notice that the diffusion coefficient D reaches the minimum at $\varphi = 0.9$ and surprisingly does not disappear even when $\varphi = 1.0$, i.e., even pure alcohol climbs up the glass wall. The authors attributed this phenomenon to the absorption of water vapor from the atmosphere by the pure alcohol, which results in the appearance of a small surface tension coefficient gradient. Their assumption was verified by repeating the same measurements for pure alcohol in a dry atmosphere. The corresponding coefficient D was found to be smaller and monotonically decreasing with the growth of φ, and is reduced to zero at $\varphi = 1$ (see Figure 26.3, lower curve).

Now let us move to a discussion of the wine tears formation. The first and most evident reason for the formation is the force of gravity, which counteracts the surface tension forces and tends to bring the runaway liquid back. Nevertheless, the so-called Raleigh instability is responsible for the nonhomogeneous character of this return (in the form of separate rivulets). The insight into this complex phenomenon can be gained through examples. When a boat floats onto a lake it generates a wave perturbation of the water surface. As time goes by, these waves decay and the water surface returns to its initial state. One can demonstrate that if the lake had been turned upside-down i.e. the gravitational force had been directed bottom-to-surface rather than surface-to-bottom, the smallest deviation of the surface from the horizontal line would grow up to a squall and destroy

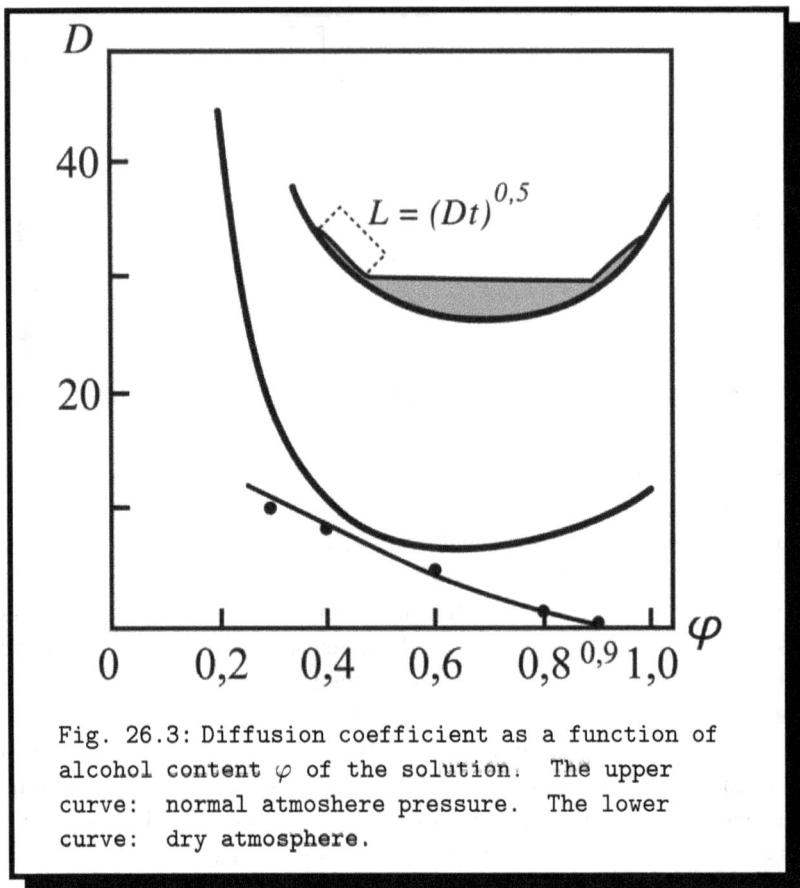

Fig. 26.3: Diffusion coefficient as a function of
alcohol content φ of the solution. The upper
curve: normal atmoshere pressure. The lower
curve: dry atmosphere.

the boat instead of decaying peacefully. This instability is responsible for the wine tears' formation: the film border, which scrambles up due to the Marangoni effect, turns out to be unstable with respect to small perturbations. Due to random inhomogeneities, small droplets appear on the glass surface. They go down engulfing liquid from border domains and leave beautiful arches behind (see Figure 26.1).

Hence, the more alcohol the solution contains, the less pronounced the Marangoni effect turns out to be, the weaker the binding effect of the surface tension on the wine tears' formation, the more wine weeps appear and less the distance between the streamlets is also to be observed. Nevertheless, this dependence turns out to be so weak that it does not allow the use of wine tears as a reliable indicator of a wine's proof without looking at the bottle label. With regard to glycerol ($C_3 H_8 OH$), the content of this sweet alcohol in wine is normally very low (it is of the order of 1–2%) and despite the fact that it certainly affects the taste of wine, it has almost no impact on its proof. Regular ethyl alcohol determines wine's characteristics and contents. It should be mentioned that low ethyl alcohol wines are less viscous, hence it seems that the glycerol content does not even determine the wine viscosity. Thus, the observation of the wine tears or even their flow does not provide grounds for serious judgement on the glycerol content in wine.

26.3 Champagne and its bubbles

Strictly speaking, only a sparkling wine produced by a special process in the French province of Champagne has the legal right to carry the name Champagne. The rest are just different sparkling wines, and they are many. Sparkling wine is produced by a special type of fermentation; it can be made from any grape variety, although, as a rule, only three varieties are used: Pinot Noir, Pinot Menier, and Chardonnay. Of these grapes, only Chardonnay is actually white. If the label indicates Blanc de Blanc (white from the white), it means that the sparkling wine was manufactured from pure Chardonnay. Blanc de Noir (white from the black) means that Chardonnay was not used for this wine. There is also Champagne rosè. There are two ways of creating Champagne rosè: either to "correct" white wine with dry red wine, or to keep the unfiltered juice together with skin for a certain period of time, then to filter it off and proceed with making Champagne.

26.3.1 *Méthode Champenoise (Champagne method)*

If you see these words on the label, it means that the wine was produced by the following method: in Champagne, grapes are harvested manually to avoid damaging the berries, thus the yeast from the grape skin will not get into the juice. Grapes are pressed twice. After the first pressing, 80 percent of the juice is extracted, and Champagne produced from it will be called Cuvee. Champagne produced from the remaining 20 percent will be named Taille; the best houses of Champagne do not make it, they only produce Cuvee. There are several ways of saturating wine with carbon dioxide.

The method we are dealing with begins with white wine fermentation as above. Then the wine is combined with reserved wines from prior years and bottled.[g] Later some grape sugar and yeast are added into every bottle and the bottle is closed with a temporary cork resembling a beer cap. The process of secondary fermentation begins. Champagne ages in the bottle for several years: usually up to three, although for special occasions it may stay there for up to six years. The law says that Champagne has to age in a bottle for no less than a year. The next step is to remove the yeast from Champagne. In this process, the bottles are placed into special racks where they are turned and inclined at $15°$ per day so that the dead yeast settles on the cork. The last operation is the evacuation of the yeast deposits and the replacement of the temporary cork with a permanent one. This process is not easy, because by that time Champagne is supersaturated with carbon dioxide. It takes a real virtuoso of *Méthode Champenoise* to open the bottle without shaking it and to replace the temporary cork with the yeast deposits on it with a permanent cork. Modern cryogenic technology has streamlined production: the neck of the bottle is frozen, the deposit is removed, and the bottle is corked again.

26.3.2   *Bubbles and foam*

We all know that sparkling wine is sold in special bottles that can resist high pressure. The reason is that a large amount of carbon dioxide in a metastable state can be released if the external conditions change. Thus, for instance, there are different ways to open a bottle of Champagne, depending on the purpose, skill, and temper of the person who is trying to open it. First, you can slightly let the gas out of the bottle, slowly pull the cork out, and pour the drink into the flutes without spilling a drop; or you can let the cork hit the ceiling to mark a New Year and spill half of the bottle as foam. That is how Michael Schumacher, Formula-1 winner, celebrated his victory. The secret is simple – to get a lot of foam you should vigorously shake the bottle before opening it. As a result, the amount of gas that is present in the neck of the bottle will mix with Champagne and form numerous bubbles. When the cork shoots up, the pressure inside the bottle sharply falls down and the bubbles formed earlier serve as the centers of emission

---

[g]A mixture of new and old wines is called dosage. Depending on the amount of added sugar, Champagne will vary in dryness or sweetness and will be labeled Brut (dry), Extra Sec (extra dry), Sec (medium sweet), Demi-Sec (sweet), Doux (very sweet).

of the diluted carbon dioxide in all the volume of the liquid. Thus, a bottle of Champagne serves as a fire extinguisher and shows the emotions of the Formula-1 winner. Champagne bubbles may entertain not only the viewers of Formula-1. Undergraduate physics majors may ignite the curiosity of language majors of the opposite sex by casting a chunk of chocolate into a glass of Champagne or mineral water. The reader himself can try it out and see that the chocolate chunk will go down, then resurface some time later, and then submerge again. These oscillations will take place several times. We are leaving this little riddle for the reader to explain and move forward to discuss the other less obvious properties of Champagne bubbles.

Speaking about Champagne, physicists cannot overlook its unusual acoustic properties. In Chapter 9, we already touched upon this topic, explaining how Champagne bubbles reduce the crystal tingle of glasses. Here we will concentrate on the other phenomenon: particularly "perlage" *i.e.* hiss of the bubbles bursting at the Champagne surface. The fizzing sound of the fresh, just-poured-in-glass, Champagne is caused by the emergence of miniature avalanches in its foam. The hissing sound is the sum of multiple separate spontaneous bubble shots. The specifics of such micro-explosions were studied by Belgian physicists.[h] If the number of such shots per second had been constant we would have heard "white noise", like the one produced by an untuned radio. This kind of a hiss occurs in acoustic processes when the uniform spectrum contains a broad range of frequencies. Careful examination showed that the acoustic spectrum of the bursting Champagne bubbles has nothing to do with white noise. The measurements performed by a sensitive microphone showed that the signal intensity depends considerably on the frequency. This dependence developed due to the fact that the bubbles burst, not independently, but in a collective way, affecting each other. Every shot lasts of the order of 10^{-3} second. Some of them occur in fast consecution one by one and their noise runs into a unique, audible, sound signal. Some bubbles burst independently and they can hardly be heard. The foam blows off as the liquid in the bubble walls falls down under the action of the gravitational force, making the walls too thin for the bubble to survive. Instead of Champagne (whose foam blows off too quickly) scientists studied this process in soapy water, where the foam dissolves very slowly. Such a convenient object allowed them to study

---

[h]N. Vandewalle, J. F. Lentz, S. Dorbolo and F. Brisbois, "Avalanches of Popping Bubbles in Collapsing Foam", *Phys. Rev. Lett.* 86, 179 (2001).

the time dependence of the length of intervals between two subsequent bubble shots. They discovered this dependence to be of the power character.
This means that the characteristic scale of such an interval just does not
exist: the time between two such subsequent events can be an arbitrary –
from milliseconds to several seconds – and there is no way to predict the
duration of the interval obtained as a result of this measurement. Similar
power laws turn out to be characteristic for many natural phenomena like
earthquakes (which have no characteristic amplitude), mud flows, and solar
flares. Usually they take place in systems where the interaction between
the elements plays an important role in the behavior of the system as a
whole. Thus, a small pebble falling down a slope that is prone to avalanche
can either provoke a terrible disaster or just simply roll down.

26.4   "Bread" wine – vodka

Grape vines do not grow well in the North. There, wine is replaced by other
beverages produced by distillation of the must of apples (e.g., Calvados in
Northern France), plums and apricots (e.g., Slivovica brandy in Bulgaria
and the Czech Republic), and, finally, of different sorts of grain (e.g., whisky
in the United Kingdom and vodka in Russia). In old times in Russia, vodka
was called Bread Wine $N21$ (Smirnoff). Now it is consumed at dinner in the
same way as grape wine in France and Italy. Have you ever considered why
all the previous beverages have about 40% alcohol content (ethyl alcohol,
$CH_3 CH_2 OH$)?

The first reason can be found in a legend based on physical evidence. At
the end of the XVIth century in Russia there was introduced a government
monopoly on vodka production, which made the government responsible for
its quality. To increase their profits saloon keepers began diluting vodka,
which got them into hot water with their customers who were unhappy with
the low quality of their favorite beverage. To put an end to these practices,
Ivan the Terrible issued a decree which allowed the patrons "to beat the
saloon keeper to death" if the surface of the vodka he was serving could not
be set alight. It was found that 40% alcohol content is the threshold when
alcohol vapor can still burn near the surface at room temperature. Naturally, saloon keepers used this lowest level of alcohol content thus reaching
a compromise between the lure of profit and concern for their health.

There is one more reason for this "magic number". Volume expansion is a well-known phenomenon: when temperature rises, the volume of the objects increases, and when temperature goes down, it decreases, respectively. Most substances (including alcohol) obey this law. Water is different: below $4°\,C$ its volume starts growing as temperature declines; when freezing, its specific volume jumps up 10 percent! That is why bottled water should not be left outside at temperatures below freezing: frozen water will break the bottles but vodka can be left outside. In Siberia, boxes of vodka bottles are left outside and the bottles stay intact. The explanation consists of two points. First, the presence of such a noticeable alcohol content in the water hinders its crystallization, and thus its specific volume does not jump as the temperature decreases. Second, at volume ratio of 4:6, the total coefficient of volume expansion is close to zero: water "deviation" is compensated by alcohol's "normal" behavior. It is enough to take a look at the tables in the handbook and to compare the coefficients of thermal expansions of water $(\alpha_{H_2O} = -0.7 \cdot 10^{-3}\,\text{grad}^{-1})$ and alcohol $(\alpha_{C_2H_5OH} = 0.4 \cdot 10^{-3}\,\text{grad}^{-1})$.

The third reason for using 40% was a discovery attributed to the well-known Russian chemist Dmitrii Mendeleev. He is believed to have proved that 40% vodka does not change its alcohol content when left in the open. Thus, a shot of vodka left overnight on the table will be found in the morning filled (unless having evaporated) with the same beverage (which will not happen in the case of wine or champagne) and may help to ease a hangover headache. A curious reader could try to use the results shown in Chapter 20, where the mass loss rate from the liquid surface was estimated for alcohol and water molecules. The result of this calculation gives 55% as the concentration of alcohol for a stable solution, which considerably exceeds the value proposed by Mendeleev and successfully used in production. The point is that the molecules of alcohol and water interact in solution and therefore they cannot be considered in the "gas model" as two non-interacting and interpenetrating liquids.

The fourth reason for vodka being 40% proof is related to the viscosity jump which takes place in the vicinity of this concentration in alcohol-water solution. In the dilute solution, alcohol molecules share the environment with the water molecules but practically do not interact with them. Around a concentration of 40% the solution is so dense that they begin to form long one-dimensional chains, i.e., the process of polymerization begins. As a consequence, the viscosity changes drastically and the solution improves its organoleptic properties. The last reason known to the authors is related to the calibration procedure, i.e., to the beverage quality control in the process of home wine-making. When a piece of lard is placed into alcohol-water solution, it will be in the state of indifferent equilibrium only at 40% content of alcohol. In a stronger beverage it will sink, in a weaker solution, it will float on the surface.

26.5   The role of wine in the prevention of cardio-vascular diseases: the French paradox (or Bordeaux effect)

We would like to precede this section by reminding the reader that:

⚠   **According to US legislation the use of alcoholic beverages is prohibited for persons under 21 years of age.**

The necessity of such a rule is due to the negative effects of alcohol that are not mentioned here.

Figure 26.4 presents the mortality rate caused by cardio-vascular diseases (per 100,000 of the population) in different countries of the world versus the average animal fat consumption (in kilocalories per day). The correlation is obvious: the more animal fats consumed (cholesterol), the higher the mortality rate from cardio-vascular diseases (an almost linear increase). Nevertheless, there is a point that does not correlate. This point indicates the position of France: the French consume a noticeable amount of fats, but their mortality rate from cardio-vascular diseases is relatively low. Indeed, as we can observe, the French eat more fat than the British, although their mortality rate from heart attacks is almost four times lower.

This statistic was obtained from the MONICA project in 1992 and published in *The Lancet*. It was the CBS (Columbia Broadcasting System) anchor, however, who was the first to get behind these data. In 1991, he made them public under the catchy title of the "French paradox", which later also became known as the "Bordeaux effect". From the moment it was discovered, this anomaly was accredited to the regular consumption of noticeable amounts of red wine by the French, specifically in the Bordeaux province. Further scientific research in the other red wine producing areas led to the unambiguous conclusion: "red wine consumption results in the noticeable lowering of the risk from cardio-vascular diseases". What is the reason behind this? This question was not easy to answer. The point is that red wine contains about 2,000 different substances: various

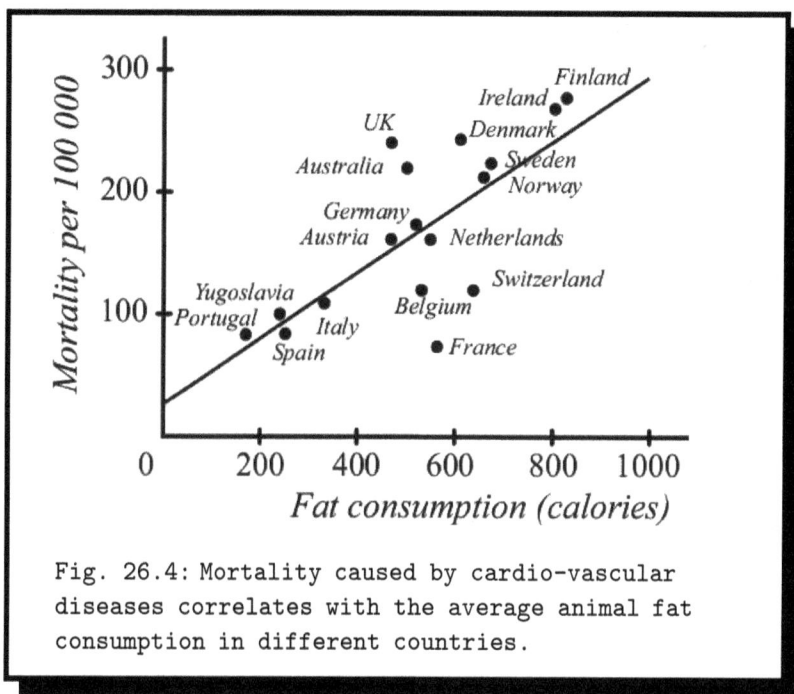

Fig. 26.4: Mortality caused by cardio-vascular diseases correlates with the average animal fat consumption in different countries.

acids, phenols, vanilla, and traces of almost all known minerals. Polyphenols, whose content in red wine is about 1 gram/liter, and Phytoalexin, which is found in grape skin, generated a special interest of the researchers. It was discovered that Phytoalexin contains trans-resveratrol which, being a strong anti-oxidant, prevents brain cells from aging. Further research showed that Polyphenols have an anti-lipoprotein impact, reduce the influence of Endothelin-1 (the major culprit of cardio-vascular diseases), and prevent the formation of platelets in the arteries. In this study we are not going any further into the jungle of medical science. Here we will try to estimate a healthy daily dose of red wine based on the physical methods of research. On the basis of the data from the MONICA report, we can assume that the probability of cardio-vascular diseases decreases with red wine consumption in accordance with an exponential law (as often occurs in nature):

$$I = I_0 \cdot e^{-b/b_i},$$

where I_0 is the probability for a non-drinker to contract the disease (e.g. for an Englishman, Figure 26.4) and b_i is the characteristic constant that can be estimated around a bottle a day.

It is obvious however, that one should not be overly enthusiastic about this method of heart attack prevention: too much alcohol can lead to serious diseases like cirrhosis of the liver. The danger is easier to estimate through vodka consumption: regular consumption of several glasses of vodka per day will have adverse health effects and will over time increase the risk of cirrhosis. Below we will again create a model based on an exponential function:

$$C = C_0 e^{b/b_c},$$

where C_0 is the probability of cirrhosis in a non-drinker, and the constant b_c can approximately be assumed around three bottles of wine a day. Adding the probabilities of both diseases one has

$$W = I_0 \cdot e^{-b/b_i} + C_0 e^{b/b_c}.$$

"Optimum" wine consumption can be estimated to occur at the minimum in W. Differentiating and equating the derivative to zero

$$\frac{dW}{db} = -\frac{b}{b_i} I_0 \cdot e^{-b/b_i} + \frac{b}{b_c} C_0 e^{b/b_c} = 0,$$

we discover that the corresponding amount of wine is determined by the expression

$$b^* = \frac{b_i b_c}{b_i + b_c} \left(\ln \frac{b_c}{b_i} + \ln \frac{I_0}{C_0} \right). \tag{26.1}$$

The probability of getting a heart attack is 5-6 times greater than the probability of suffering from cirrhosis. Substituting these data in the Eq. (26.1), we find that half a bottle, i.e., two 2 glasses of red wine is the "optimum" amount of daily consumption. This is the amount of wine that Tuscany farmers drink regularly, for instance.[i]

---

[i]Note that such generous daily consumption of wine is not a flight of imagination. In Heidelberg castle e.g., visitors can still find the world's largest barrel from which wine was pumped to the big dining hall by a special system of manually operated pumps. Average per person wine consumption in this castle, including children and old people amounted to 2 liters per day. The court jester, midget Perkey, would regularly drink

26.6   Quality estimation and attribution of wine: SNIF-NMR method

When discussing the merits of responsible consumption of red wine endowed with certain properties, we must be sure that its origin matches what is written on the label. Contemporary physics offers the most accurate, NMR based methods to determine the origin of wine: SNIF (Specific Natural Isotope Fraction). Nuclear magnetic resonance (NMR) basically means the following: nuclei (for example, the protons in a hydrogen atom) have a small magnetic moment, a kind of microscopic compass. When placed in a homogeneous constant magnetic field **H** these magnetic moments precess around the direction of the field with a frequency ω_L proportional to the strength of H. Now let us consider what happens if one applies also a small radiofrequency (RF) field by means of a coil connected to a current generator and with an axis perpendicular to the field **H**. When the frequency of the RF field is just ω_L (now you understand why one speaks of "resonance"), electromagnetic energy is absorbed by the nuclei and consequently they reverse the direction of the magnetic moments with respect to the field **H**. This simple vectorial description is not far from the real phenomenon, although it is actually based on a complex quantum behavior of the elementary magnetic and angular moments of the nuclei (for more details see below in Chapter 28). The point we must emphasize here is that electronic equipment allows us to measure with very high precision the value of the magnetic field **H** at which the resonance occurs from the so-called "NMR resonance lines" of the absorption spectrum. The value of the resonance field is actually the local one at the nuclear site, namely the one resulting from the external field **H** and from small correction terms (very small indeed but still detectable) due to the electronic currents. Thus, it is clear that the resonance lines occur at different values of the electromagnetic irradiation frequency, depending on the environment of electrons around the nucleus. For instance, in ethyl alcohol, $CH_3\,CH_2\,OH$, one may expect resonance lines of intensities in the ratios $3 : 2 : 1$ that occur at different frequencies in correspondence to the protons belonging to the molecular groups CH_3, or to CH_2 or to OH. Thus the resonance spectrum becomes a kind of snapshot of the molecular configuration. The SNIF method was devised by Gerard and Maryvonne Martin in Nantes in the 1980s, initially with the purpose of con-

---

12 bottles a day. Cirrhosis of the liver was not the cause of his death. He died from dysentery after he drank a glass of untreated water, because he lost a bet.

trolling the eventual enrichment of the wines by addition of sugar. In 1987, a company called Eurofins Scientific was founded and an *ad hoc* database was started. Today this database contains the NMR spectra of a large number of wines from France, Germany, Italy, and Spain. The SNIF method has been accepted by the European Community (1989) and by the Organization International de la Vigne e du Vin (OIV). It has also been acknowledged as the official method by the Association of Official Analytical Chemistry (AOAC) in the USA and Canada and awarded "The Method of the Year" by the AOAC in 1996. Nowadays, the SNIF method makes it possible to single out ethyl alcohol with the same chemical structure but different botanic origin. It is possible to find out whether or not a wine originated from the must of a particular vine of a particular region. The method is based on the fact that, as a consequence of the various processes of photosynthesis, of the plant metabolism, and of the different geographical and climatic conditions, the fraction of deuterium with respect to hydrogen is different from area to area and from vine to vine. The fraction of deuterium (D) with respect to hydrogen (H) is usually measured in part per million (ppm). While it is around 16,000 on Venus and 0.01 in the troposphere, on the earth the fraction D/H is around 90 at the South pole and 160 at the equatorial belt, and displays a relatively large variability. A further possibility for the characterization is related to the distribution of deuterium among different molecular groups. Again, for ethyl alcohol one can find $CH_2D - CH_2 - OH$ or $CH_3 - CDH - OH$ or $CH_3 - CH_2 - OD$. The percentage (D/H) for each molecular group can be estimated from the NMR spectra of deuterium, since the resonance lines occur at different frequencies because of the corrections to the external magnetic field due to the different electronic currents induced in each of the groups. From the NMR signals, compared by means of a suitable computational program with the spectra collected in the database, one can find out if sugar has been added to wine, determine the geographical origin of mixed wines versus the one claimed on the bottle label, etc.

PART 4

Windows to the quantum world

Finally the time has come to tell you about strange laws ruling in the world of microparticles. You will learn how these laws reveal themselves through "superphenomena" which take place at very low and not so low temperatures. We shall try to make things as easy as possible. Still this will not be small talk, for any real comprehension appeals to the mathematical language of modern physics. But we hope that you will appreciate the marvels of this unbelievable, surprising, and promising world.

Chapter 27

The birth of new physics at the turn of the century

Einstein said that if quantum mechanics were correct then the world would be crazy. Einstein was right - the world is crazy.

Daniel M. Greenberger.

In the second half of the 19th century production and manufacturing industries were rapidly developing in Europe, whereas in fundamental science, as well as in arts, the feeling of closure and exhaustion of existing paradigms[a] was under way. In physics, the flawless classical mechanics and thermodynamics, optics and electromagnetic theory have already taken shape. "The most important fundamental laws and phenomena of physical science are known already. These laws have been proven with such authenticity, that the possibility of their changes due to the newly discovered knowledge is bordering on improbable" Albert Michelson, the American physicist, said in 1899. "The future role of researchers will be reduced, — he would say, — to verification of the obtained results to the sixth digit after a period". Soon, evidence proved that he was wrong.

In art, sculpture, architecture and music, classicism has achieved the undisputed perfection. To further explore the properties of nature beyond

---

[a]Paradigm (from ancient Greek $\pi\alpha\rho\alpha\delta\epsilon\iota\gamma\alpha$ "example, model, sample") - is a set of values, methods, approaches, technical skills and means, recognized by the scientific community within the frames of established scientific tradition at a certain period of time. Classical art adopted the term "paradigm" as a model to emulate, an ideal form, a universal prototype for all art forms. In the broader meaning: the amalgamation of factors, defining the idiosyncrasy of a certain historic art type. Since the 1920s the term "paradigm" was gradually replaced by its modernistic equivalent: "world image".

direct human perception, and to express through art the subtle variations of human feelings it was necessary to broaden these paradigms.

Thus, at the end of 1860s - beginning of 1880s the Impressionism,[b] a new art movement, originated. Its main goal was to convey the subtle, easily changeable impressions. Impressionism was based on the latest discoveries in optics and color theory. Impressionism produced the most striking effect in painting, where special accent was made on conveying color and light. Another new art trend - Art Nouveau - determined the withdrawal from the classicism in art. This movement featured the abandonment of straight lines and angles, and the transition to softer natural lines as well as the interest in new technologies.

Fig. 27.1: Ludwig Boltzmann (1844-1906), Austrian physicist, one of the founders of statistical physics and physical kinetics. Boltzmann's students remember him as a brilliant lecturer. When Paul Ehrenfest, a future prominent theorist and at that time a student, presented a report on one of Boltzmann's publications, he exclaimed: ''I wish I knew my publications so well!''.

New ideas and methodology keep coming to physics. Thus, since the 1870s, Austrian physicist Ludwig Boltzmann, American mathematician and

---

[b]fr. *impressionnisme*, from impression.

physicist Josiah Willard Gibbs and British scientist James Clerk Maxwell formulated the so-called statistical method in order to describe the properties of physical systems, containing a great number of particles moving according to the law of classical mechanics. This method allows one to obtain necessary information about the properties of such systems without solving a great number of differential equations of classical mechanics, created for every particle.[c] Voluntary rejection of a detailed deterministic

Fig. 27.2: Josiah Willard Gibbs (1839-1903). American mathematician and physicist, one of the founders of chemical thermodynamics and statistical physics. His name is associated with such terms as ''Gibbs' paradox'', ''Gibbs' canonical, micro-canonical and macro-canonical distributions''.

description for the sake of a much less informative probabilistic description allows one to create a new science - statistical physics. Within its framework, it is possible to define the fundamental laws emerging during the particle growth in the system and determining the behavior of gas, liquids, hard matter, etc., without getting into extra details. To that end, qualitatively new physical entities are being introduced, e.g. the entropy, which is characterizing a degree of systemic disorder. Also, earlier introduced phenomenological concepts are being confirmed (such as temperature, power). It makes possible not only to understand the processes going on in such sys-

---

[c]It should be noted, that the XXIst century physics due to the advent of supercomputers, is getting back to this earlier rejected idea in the form of the so-called "molecular dynamics" method.

tems from the same position, but it also makes predictions important for practical implementation. It should be pointed out, that the indeterminism of statistical physics, its probabilistic nature of the description of multiple-particle system properties, are still a matter of choice for a researcher as well as the matter of choice for the method of interpretation of reality by impressionists in the art of painting.

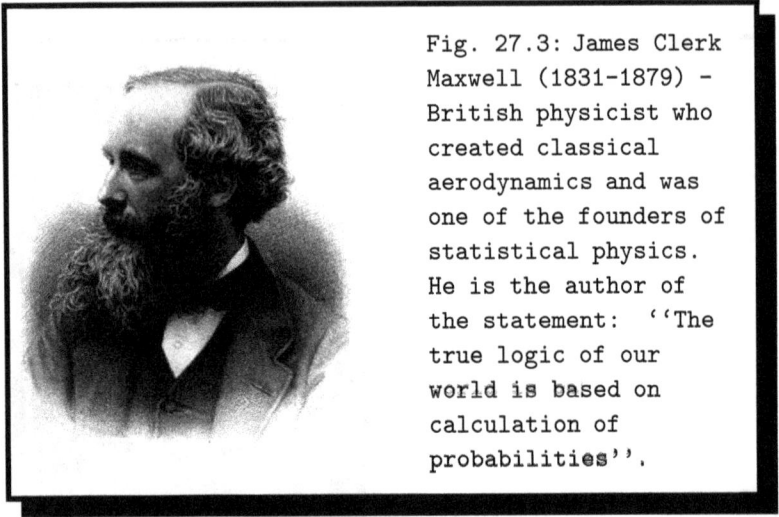

Fig. 27.3: James Clerk Maxwell (1831-1879) – British physicist who created classical aerodynamics and was one of the founders of statistical physics. He is the author of the statement: ``The true logic of our world is based on calculation of probabilities''.

But when we attempt to explore the physics of the microworld the picture may be very different. Leucippus, the ancient Greek philosopher (5th century BC), was the first to describe the atom — the elementary, indivisible, tiny particle of matter. According to Leucippus, atoms are absolutely solid, infinitely versatile in shape and size, and exist in vacuum in the form of an uninterrupted chaotic motion, colliding as a part of this process and forming unique vortices. "Everything occurs for a reason and by necessity, because vortex is the reason for incipient formation of all things" — these vortices serve as building blocks for all things found in the Universe.[d]

Democritus, the pupil of Leucippus, was so actively promoting and further developing his teacher's ideas, that with time it became unclear where

---

[d]Let us remember these words were said by an ancient Greek philosopher. Later we will get back to vortices and vortex matter in this book.

Fig. 27.4: Leucippus of Abdera or Miletus (500-440 BC) - Greek materialist philosopher, one of the founders of the theory of atomism. Of all his works only one line from his book ''On Reason'' survived: ''Nothing occurs at random, but everything for a reason and by necessity''.

the borderline between their views lied. This is the reason why today when we speak of atoms the name of Democritus is the first to be mentioned.

Fig. 27.5: Democritus (460-370 BC) - ancient Greek philosopher, one of the founders of the atomic theory of the Universe. In ancient times Democritus was famous not only for his prominent studies, but also for perfection of his literary style.

In 1897, Joseph John Thomson discovered the elementary particle with negative charge — the electron. In several years this discovery allowed him,

Fig. 27.6: Joseph John Thomson (1856-1940). British physicist known in the history of science for his discovery of the electron. He created a model of the atom shaped as a ''plum pudding''.He delivered a popular lecture in Cambridge University about the physics of golf.

following in the footsteps of Lord Kelvin, to create his model of the atom, in which he suggested that "...atoms of elements consist of several negatively charged corpuscles contained in a sphere with a uniformly distributed sea of positive electric charge...". This model is known in the history of science under the nickname of "plum pudding", but it was a short-lived model. In 1911 Lord Rutherford discovered a very small compact nucleus containing the whole positive charge by deflecting alpha particles passing through a thin gold foil. Rutherford's discovery made Thomson's "plum pudding" a part of the history of physics. In his turn, Rutherford pioneered a "planetary" model of the atom with negatively charged electrons rotating in the vacuum around a positively charged nucleus. The problem is that, while rotating along circular orbits, charged particles, in accordance with the laws of electromagnetism, should emit energy and finally fall into the nucleus.

To eliminate this paradox, in 1913, Danish physicist Niels Bohr suggested mysterious postulates, which "exempted" the electron from its duty of emitting energy during its rotation along certain electron selected orbits, whereas the light of certain wavelengths can be radiated only during electron transfer between them. Circumferential motion with transient radii, specified by Bohr, were prohibited. These postulates allowed Bohr to explain single-curved spectra observed in different gas absorption, and even

Fig. 27.7: Ernest
Rutherford (1871-1937)
- New Zealand
physicist, one of the
founders of nuclear
physics, Nobel Prize
winner in Chemistry
(1908). ''If a
scientist cannot
explain a cleaning
lady, who is taking
care of his lab, the
rational of his study,
it means that he
himself does not
understand what he is
doing''.

predict new series of such spectra; however their meaning remained a mystery.

Around the same time, when J. J. Thomson was studying in Cavendish Laboratory of Cambridge University the deviation of cathode rays in electric and magnetic fields, which would soon lead him to the creation of a hypothesis of the existence of matter at the subatomic level, German physicists were trying to find response to the question: what is the connection between temperature, color palette and intensity of the light emitted by a hot iron rod (later it was named the problem of absolute blackness of the body)? This problem seems trivial, but its solution was of primary importance for a country newly created in 1871 (Germany), which was struggling to build a strong industry to win the competition with England and USA. However, every effort within the frames of classical electromagnetism and thermodynamics, would lead to non-valid results and contradict available experimental data. This problem was solved by Max Planck — professor of Berlin University, classical, pedantic German scientist. The price he had to pay for it was the idea of energy quanta, which he introduced into physics

Fig. 27.8: Niels
Henrik David Bohr
(1885-1962), the
Danish physicist and
philosopher, Nobel
Prize winner in
Physics (1922) for the
foundational
contributions to
understanding atomic
structure and quantum
theory. In response
to well known Einstein
saying ''God does not
play dice'', Bohr
said: ''Please, stop
telling God what he
should be doing''.

to save it from the so-called "ultraviolet catastrophe".[e] Max Planck later put it as "In short, what I have done can be described as an act of despair ..."

---

[e]Later this law received a bleak name of "ultraviolet catastrophe" or "Rayleigh-Jeans catastrophe". It was discovered within the framework of classical electromagnetic and thermodynamic theory, and provided a response to the question about dependence of spectral density of electromagnetic radiation of a heated body on frequency of radiation and its temperature. The truth of this relationship would make thermodynamic equilibrium between the matter and radiation impossible, because, according to this relationship, all thermal energy should have been transferred into the energy of the short wavelength part of the spectrum. Among other things, it would mean the cooling of the Universe down to absolute zero due to short wave electromagnetic radiation. Max Planck pointed out the inapplicability of concepts of classical physics to this domain of energies, and while introducing a categorical concept of energy quantization, found a correct, internally consistent expression for spectral density of electromagnetic radiation. Nevertheless, for long wavelength part of the spectrum, Planck's formula is transformed to the Rayleigh-Jeans radiation law developed within the framework of classical physics.

Fig. 27.9: Albert Einstein (1879-1955) one of the most prominent physicists of the XXth century, the author of special and general theory of relativity, Nobel Prize winner in Physics in 1921. Many statements by Einstein became catch phrases. Here is one of them: ''You see, telegraph cable resembles a very long cat. You pull its tail in New York, and his head is meowing in Los Angeles. See? Radio works the same way: you are sending signals from here, and they are received there. The only difference - there is no cat''.

The next step on the way to understanding discreteness of microscopic physics was made by Albert Einstein. In 1905 he published the article explaining the then mysterious phenomenon: electronic emission under the electromagnetic radiation (light, in particular). Alexandre Becquerel first observed it in 1839 in electrolytes, and later, in 1873, Willoughby Smith —

in his experiments with selenium, and in 1887, Heinrich Hertz — with zinc. Einstein explained the enigmatic properties of photo-current[f] by reverting to the old Newton's theory of corpuscles rejected by Huygens. Only the hypothesis that the light behaves like a particle possessing impulse and energy will allow us to explain the necessary energy concentration required to extract an electron from a volume of metal. Later on, in 1909, Einstein announced to the public: future studies of radiation will allow to discover the synthesis of particles and waves.

Fig. 27.10: Louis -Victor -Pierre - Raymond, 7th duc de Broglie better known as Louis de Broglie (1892-1987) - French theoretical physicist, one of the founders of quantum mechanics, Nobel Prize winner in Physics in 1929.

Another "brick" was laid into the construction of quantum physics by a 31-year old French duc and German prince Louis de Broglie. Later he wrote: "In 1923, after long and lonely speculations it suddenly occurred to me that the discovery made by Einstein in 1905 should be summed up and expanded to include all particles, in the first place - electrons". De Broglie challenged himself with a simple question: if the light waves can behave like particles, why then such particles as electrons cannot behave like waves? Einstein characterized his work as "the first timid step towards solving one of the most intricate puzzles of modern physics".

---

[f]Proportion between its density and light stream; linear growth of maximum electron kinetic energy with frequency and its independence from light intensity; for every substance the presence of its own red borderline of photo-effect.

Several years later the American physicists Clinton Davisson and Lester Germer confirmed the brilliant de Broglie hypothesis observing the diffraction of electrons scattered from a metal surface, i.e., demonstrating their wave nature. Putting wave-particle duality on a firm experimental basis, this experimental finding represented a major step forward in the development of quantum mechanics.

Finally, in the end of 1926, Erwin Schrödinger summarized De Broglie's wave-corpuscle dualism from free particles to the particles moving in an external field. After coming back from a ski trip, he presented for consideration of his colleagues the equation which later became the main instrument of microworld mathematical description. It determined a certain function describing the wave process with variable wavelength. Its meaning was still unclear, but the equation had already contained the long expected discreteness, which was predicted separately by Planck, Einstein, and Bohr! Thus, discrete levels of electron energy in the atom appear here naturally, and they are related to its finite size. In a certain sense, the atom can be compared to a violin: a violin creates sounds of certain wavelengths, the same way as the atom in Bohr's orbits should contain a whole number of wavelengths, which could be assigned to an electron moving in the nucleus field.

Fig. 27.11: Erwin
Rudolf Josef Alexander
Schrödinger
(1887-1961) Austrian
theoretical physicist,
one of the founders of
quantum mechanics,
Nobel Prize winner in
Physics in 1933.

Gradually, as a result of long and complicated discussions between the best minds of that time, the philosophical kernel of newly created quantum physics has been developed. Microworld cannot be described in customary, every day observed and easily measurable terms of coordinates, trajectory, impulse. It can be described only in terms of probability and the square of the complex wave function modulus, which is a part of the Schrödinger equation, and determines the probability distribution of quantum particle's presence in space. There are pairs of physical values: coordinate and momentum, energy and time, angular momentum and phase, which cannot be determined simultaneously with precision. Indeterminism was not introduced into physics of the XXth century as a result of voluntary rejection of accurate system description for the sake of convenience, but rather as an inevitable property of the microworld. But, nevertheless, when the problem was formulated correctly quantum mechanics allows us to receive absolutely accurate answers: how to produce a laser with a certain wavelength, to what level of energy should the particles be scattered to detect a Higgs boson, what should be the mass of hemispheres of radioactive element for a nuclear bomb to explode.

In the following chapters we will demonstrate how quantum mechanics "works" in modern physics and everyday life.

Chapter 28

The uncertainty principle

Coordinate and momentum resemble the male and fe-
male silhouettes in an antique barometer. For either of
them to show up, the second must disappear.

Werner Heisenberg.

In 1927 the German physicist Werner Heisenberg[a] discovered one of the
most fundamental and astonishing concepts of quantum mechanics: the
"uncertainty principle". It is so different from our ordinary experience and
at the same time it is so essential for the description of the properties of
the microworld that we start our journey namely from this point.

Suppose that watching some body we managed to determine the projection of the momentum onto the x-axis with the precision Δp_x. Then
we shall not be able to measure the corresponding x-coordinate with pre-
cision greater than $\Delta x \approx \hbar / \Delta p_x$, where $\hbar = 1.054 \cdot 10^{-34} \, J \cdot s$ is Planck's
constant.

At first, this relation looks perplexing. Remember that according to
school textbooks Newton's laws allow one write down equations of motion
of a body and to calculate time-dependencies of the coordinates. Knowing
those one can compute the velocity $\vec{v}$ (that is the time derivative of the
coordinate $\vec{x}$), the momentum $\vec{p}$, and its projections. It looks as if we
have established both coordinates and momenta and there is no uncertainty
principle. Indeed, this is a fact in classical physics,[b] but the situation
changes in the microworld, Figure 28.1.

---

[a]W. K. Heisenberg, (1901–1976), one of the founders of quantum mechanics; 1932 Nobel
Prize in physics.
[b]Of course the uncertainty relation holds in the macroworld too. But there it is not

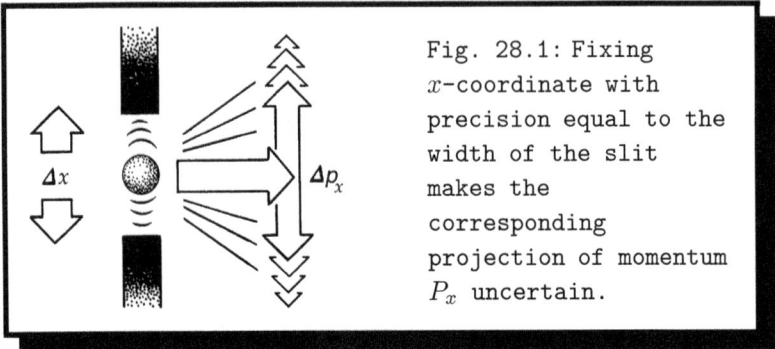

Fig. 28.1: Fixing
x-coordinate with
precision equal to the
width of the slit
makes the
corresponding
projection of momentum
P_x uncertain.

28.1   Momentum and coordinate

Imagine that we want to trace the motion of an electron. What should we do? The human eye can hardly do this. Its resolving ability is too weak to discern an electron. Then, let us try a microscope. The resolution of a microscope is limited by the wavelength of light used for the observation. Wavelengths of the usual visible light are of the order of $100\,nm$ ($10^{-7}\,m$) and one cannot see particles smaller than that in a microscope. Sizes of atoms are of the order of $10^{-10}\,m$ and there is no hope to discern them, not to mention a single electron.

But let us fantasize. Suppose that we managed to construct a microscope which exploits electromagnetic waves of smaller length, say, X-rays or even γ-rays instead of visible light. The harder the γ-radiation we use the shorter the corresponding waves and the smaller the objects which can be detected. This imaginary γ-microscope seems to be an ideal instrument capable of measuring electron positions with a desired accuracy. And what about the uncertainty principle?

Think over this hypothetical experiment (some physicists like the German word *Gedankenexperiment*) once more. In order to get the information about the position of the electron, at least one γ-quantum must be reflected. One quantum carries the minimal amount of energy of the radiation that is equal to $E = h\nu = \hbar\omega$ where ω is the angular frequency of the field oscillations. Short waves have higher frequencies and their quanta carry bigger energies. But the momentum of a quantum is proportional to the energy. Colliding with an electron a quantum inevitably transfers to the particle

—————————————

restrictive and does not play a key role.

a fraction of its momentum. Because of that any coordinate measurement makes the electron momentum ambiguous. Rigorous analysis of the process proves that the product of the uncertainties cannot become smaller than Planck's constant.

You may think that the discussion referred only to this particular case. Or we have proposed the wrong device and there are more delicate ways to measure the coordinate without kicking the electron to a new state. Unfortunately this is not so.

The best scientists (including Einstein) tried to invent a thought experiment which could determine the position and momentum of a body with accuracy better than what the uncertainty principle prescribed. But all attempts failed. By the laws of nature this is impossible.

Our arguments may appear vague and there is no self-evident mental model at hand. Real understanding demands a serious study of quantum mechanics. But hopefully the preceding was enough to make a first acquaintance with the subject.

In order to mark the boundary between the micro- and macroworlds let us make an estimate. Tiny particles used in observations of Brownian motion[c] are about $1\,\mu m$ ($10^{-6}\,m$)in size and weigh less than $10^{-10}\,g$. Still these fragments of matter contain enormous numbers of atoms. The uncertainty principle tells us that for them $\Delta v_x\,\Delta x \sim \hbar\,/\,m \sim 10^{-21}\,m^2\,/\,s$. Suppose that we are going to fix the particle position with an accuracy equal to 1 percent of the size, $\Delta x \sim 10^{-8}\,m$. Then $\Delta v_x \sim 10^{-13}\,m\,/\,s$. This is a very small quantity and the reason for that is the small value of Planck's constant.

The Brownian velocity of such a particle is approximately $10^{-6}\,m\,/\,s$. Apparently the inaccuracy of the velocity coming from the uncertainty relation is negligible. It is less than one tenmillionth (0.0000001!) even for a body this small. And because of the $\hbar\,/\,m$ in the right side of the relation it is even less apparent for larger bodies. But if we reduce the mass (take for example an electron) improving at the same time the accuracy of measurements (let $\Delta x \sim 10^{-10}\,m$, be the atomic size) the uncertainty of the velocity becomes comparable to the velocity itself. For electrons in atoms the uncertainty principle carries full weight and may not be ignored. This leads to breath-taking consequences.

---

[c]The chaotic motion of small impurities in gas or liquids due to collisions with their molecules.

28.2   The probability waves

As discussed in the previous chapter, Niels Bohr allowed for electrons in his model of the atom to occupy only certain orbits with strictly defined energies. Electrons may change their energy only by jumping from one orbit to another. This "quantum" behavior explains many things and among those the atomic spectra and stability of atoms. Even now it is helpful in a simplified treatment of quantum effects. But it violates the uncertainty principle! Obviously, in contrast to the laws of the microworld both the coordinate and the momentum of an electron in orbit are definite, regardless of whether the orbit is quantum or classical.

A further development corrected this so-called semiclassical model. The actual behavior of electrons in atoms proved even more startling.

Suppose that we managed to find exactly where the electron is at a given moment. Is it possible to predict positively where it will be a bit later, to be definite, say, in a second? No, because as we know, the coordinate measurement has inevitably introduced an uncertainty into the momentum. Predicting where the electron goes is beyond the power of devices. What should we do?

Let us mark the spatial point where we have found the electron. Another mark will register the result of an analogous measurement performed on one more atom of the same type. The more measurements, the more marks. It turns out that although it is impossible to tell where the next one appears, the spatial distribution of the marks follows a pattern. The density of marks varies from point to point indicating the probability of detecting the electron.

We have to give up the idea of describing the motion of the electron in detail but we can judge the probability of locating it at different points in space. The behavior of an electron in the microworld is characterized by probability! The reader may dislike this strange suggestion. It is absolutely out of habit and contradicts our intuition and routine experience. But there is nothing we can do about the fundamentals of nature. The laws of the microworld are really different from those of every day. According to Einstein "God does not play dice" in order to predict the behavior of electrons. Alas, one has to face the facts.[d]

So, in the microworld the state of the electron is defined by the probability of finding it at various spatial points. In our pictorial model, the probability is proportional to the density of the marks. One may fancy that the marks form a sort of a cloud in which the electron lives.

---

[d]We must note that Einstein himself did not believe in the necessity of such gambling. He did not accept quantum theory until the end of his life.

But what regulates the structure of the probability clouds? You know that classical mechanics is ruled by Newton's laws. Quite similarly quantum mechanics has its own equation which determines the "spread" of an electron in space. This equation was found in 1925 by Erwin Schrödinger (see the previous chapter. Note that this was before the uncertainty principle clarified the reason for particle spreading. These things happen in physics.) The Schrödinger equation provides an exact and detailed quantitative description of atomic effects. But it is impossible to solve it without complicated mathematics. Here we shall cite some ready answers that illustrate electron spreading.

Figure 28.2 presents a scheme of an experiment on electron diffraction. The pattern of bands which appears on the screen is shown in the photograph. It is indubitably akin to a pattern of light diffraction. The result would be impossible to explain if electrons followed linear trajectories as prescribed by the laws of classical physics. But if they are smeared in space this is conceivable. Moreover, the experiment demonstrates that probability clouds exhibit wave properties. Probability waves such as the birth rate or crime waves are common in day-to-day life. The amplitude of the wave is maximal in the place of the greatest probability of an event. In our case it is most probable to find the electron there. In the photograph these zones are lighter in color.

The smearing of the electron in a hydrogen atom as obtained by exact mathematical analysis of several quantum states is portrayed in Figure 28.3. These are analogues of quantum electron orbits in the Bohr model of the atom. Again the probability to detect the electron is higher in the lighter regions. The pictures are reminiscent of the snapshots of standing waves in finite domains. Probability clouds are really magnificent! Besides, these abstract pictures do indeed define the behavior of electrons in an atom and explain, for example, energy levels and all that concerns chemical bonding.

The uncertainty principle makes possible an estimate of the dimensions of probability clouds without going into the particulars of their structure. If the size of a cloud is of the order of Δx then it makes no sense to speak of coordinate uncertainties greater than Δx. Consequently, the uncertainty of momentum of the particle, Δp_x, cannot be less than $\hbar / \Delta x$. The same expression determines within an order of magnitude the minimal momentum of the particle.

The smaller a cloud, the bigger the momentum and velocity of motion within the localization domain. It turns out that these general considerations suffice to make a correct estimate of the atomic size.

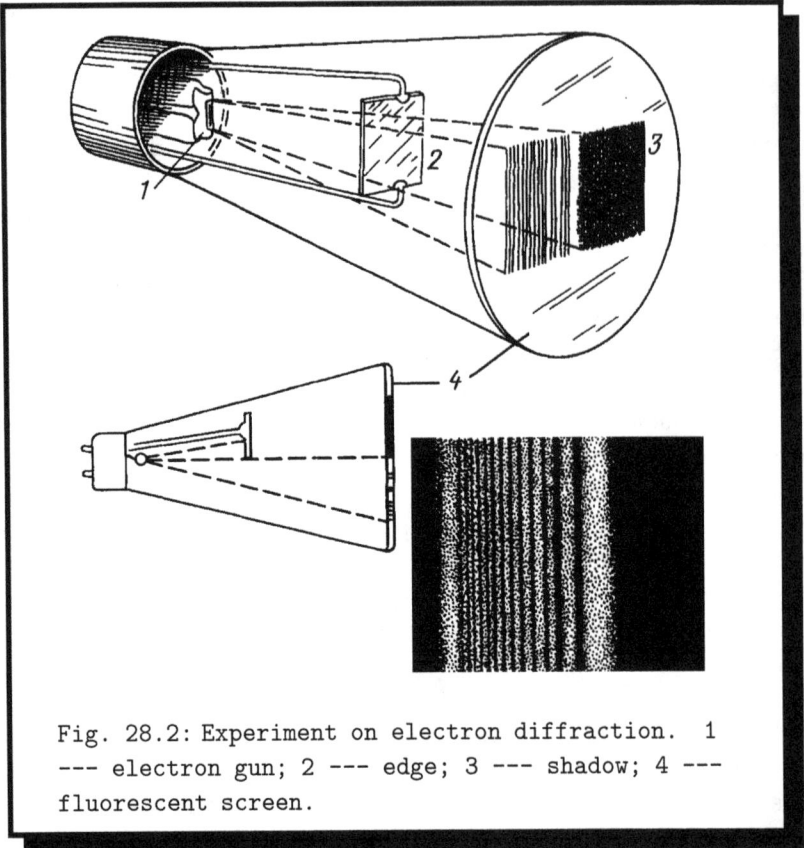

Fig. 28.2: Experiment on electron diffraction. 1
--- electron gun; 2 --- edge; 3 --- shadow; 4 ---
fluorescent screen.

An electron in an atom has both kinetic and potential energies. The kinetic energy is the energy of the motion. It is related to the momentum by the well-known formula: $E_k = m\,v^2\,/\,2 = p^2\,/\,2m$. The potential energy of the electron is the energy of interaction with the nucleus. In the *International System of Units (SI)* it is equal to $E_p = -\frac{1}{4\pi\varepsilon_0}\frac{e^2}{r}$ where e is the electron charge, r is the distance between the electron and the nucleus and the dimensional constant is called the permittivity of free space. It may be found from the equality $(4\pi\varepsilon_0)^{-1} = 9\cdot 10^9\,m\,/\,F = 9\cdot 10^9\,J\cdot m\,/\,C^2$.

The electron has a definite value of the full energy $E = E_k + E_p$ in every state. The state with minimal energy is called the *ground (nonexcited)* state. Let us estimate the atomic radius for the ground state.

Imagine that the electron is spread over a domain of dimension r_0. Attraction to the nucleus tends to make r_0 smaller and collapse the probability

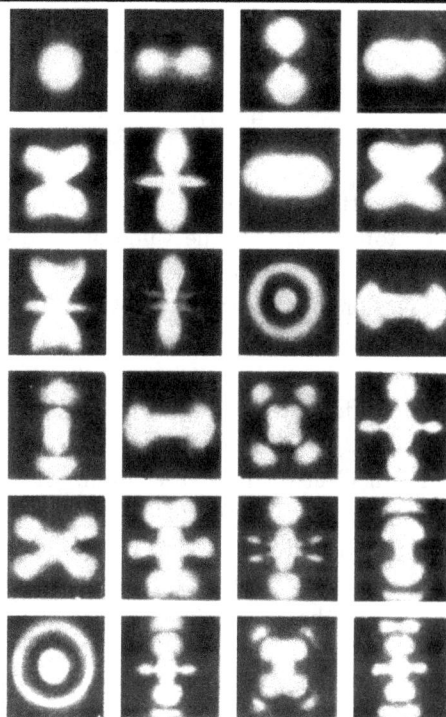

Fig. 28.3: The ''smearing'' of electrons in an atom. This is not a photograph but a result of calculations. Symmetries of these pictures strongly influence symmetries of molecules and crystals. One may even say that they are the keys to understanding the beauty of ordered life forms in nature.

cloud. This corresponds to lessening the potential energy which by an order of magnitude is $-\frac{1}{4\pi\varepsilon_0}\frac{e^2}{r_0}$ (the absolute value of the negative quantity grows as r_0 decreases). If that kinetic energy was not there the electron would fall into the nucleus. However, as you remember, localized particles always possess kinetic energy by virtue of the uncertainty principle. And this prevents electrons from falling inward! Decreasing r_0 enhances the minimal momentum of the particle $p_0 \sim \hbar / r_0$ and as a result the kinetic energy

$E_k \sim \hbar^2 / 2m\,r_0^2$ grows too. The full energy of an electron E is minimal when the derivative $dE\,/\,dr_0$ is zero, which results in

$$r_0 \sim \frac{4\pi\varepsilon_0\,\hbar^2}{m\,e^2}. \qquad (28.1)$$

This evaluation determines the typical dimension of the localization domain that is essentially the atomic radius. The value of r_0, obtained from Eq. (28.1), is 0.05 nm ($5 \cdot 10^{-11}$ m). As you know, within an order of magnitude this is the actual dimension of the atom. Obviously if the uncertainty principle makes it possible to correctly estimate the atomic radius it must belong to the most profound laws of the microworld.

Another principle that follows directly from the uncertainty relation pertains to complex atoms. *Ionization energy* E_i is defined as the work required for detaching an electron from the atom. It can be measured pretty accurately. Imagine that the product of $\sqrt{E_i}$ and the atomic size d is the same for absolutely different atoms up to 10–20 percent. You may have already guessed the reason: the momentum of the electron is $p \sim \sqrt{2mE}$ and according to the uncertainty relation the product $p \cdot d \sim \hbar$ must be constant.

28.3   Zero-point oscillations

Impressive results come out of applying the uncertainty principle to oscillations of atoms in the solid state. Atoms (or ions) oscillate about the nodes of a crystal lattice. Usually the oscillations are due to thermal motion and they increase as temperature rises. But what happens if the temperature is reduced? From a classical point of view the amplitude of oscillations will decrease and atoms will come to a halt at absolute zero. But is this possible from the point of view of quantum laws?

Shrinking the amplitude of oscillations means, in quantum language, compressing the probability cloud (or the localization domain) of a particle. We have seen, that because of the uncertainty relation, the price to pay will be the enhancement of the particle momentum. Attempts to arrest a quantum particle fail. It turns out that even at absolute zero temperature atoms in solids continue to oscillate. These *zero-point oscillations* give rise to a number of beautiful physical effects.

First of all let us try to estimate the energy of zero-point oscillations. In an oscillatory system a restoring force $F = -k\,x$ appears as the body is shifted to a small distance x away from the equilibrium. For a spring k is the elasticity coefficient while in a solid it is defined by forces of interactions between the atoms. The potential energy of the oscillator is

$$E_p = \frac{k\,x^2}{2} = \frac{m\,\omega^2\,x^2}{2},$$

where $\omega = \sqrt{k\,/\,m}$ is the angular frequency of oscillations.

This means that the energy of the oscillator may be expressed in terms of the amplitude of oscillations $x_{\max}$,

$$E = \frac{m\,\omega^2\,x_{\max}^2}{2}; \qquad x_{\max} = \sqrt{\frac{2E}{m\,\omega^2}}.$$

But in quantum language the amplitude of oscillations is the typical dimension of the localization domain that, because of the uncertainty relation, determines the minimal momentum of the particle. It comes out that, on the one hand, the smaller the energy of oscillations the smaller the amplitude must be. But, on the other hand, reducing the amplitude increases the momentum and, consequently, the kinetic energy of the particle. The minimal energy of a particle is given by the estimate,

$$E_0 \sim \frac{p_0^2}{2\,m} \sim \frac{\hbar^2}{m\,x_0^2} \sim \frac{\hbar^2}{m} \cdot \frac{m\,\omega^2}{E_0}.$$

Comparison of the last two expressions results in $E_0 \sim \hbar\,\omega$. The exact calculation gives half the value. The energy of zero-point oscillations is equal to $\hbar\,\omega\,/\,2$. It is maximal for light atoms that oscillate with higher frequencies.

Probably the brightest manifestation of zero-point oscillations is the existence of a liquid that does not freeze even at absolute zero. Obviously a liquid will not freeze if the kinetic energy of atomic oscillations is enough to destroy the lattice. It does not matter whether the kinetic energy appears due to the thermal motion or due to quantum oscillations. The most likely candidates for nonfreezing liquids are hydrogen and helium. The energy of zero-point oscillations is maximal in these lightest substances. Moreover, at normal temperature helium is an inert gas. The interaction between helium atoms is very weak and it is comparatively simple to melt the crystal lattice. It turns out that the zero-point oscillation energy is enough and helium does not freeze even at absolute zero. On the contrary the interaction of atoms in hydrogen is much stronger and it freezes despite the zero-point

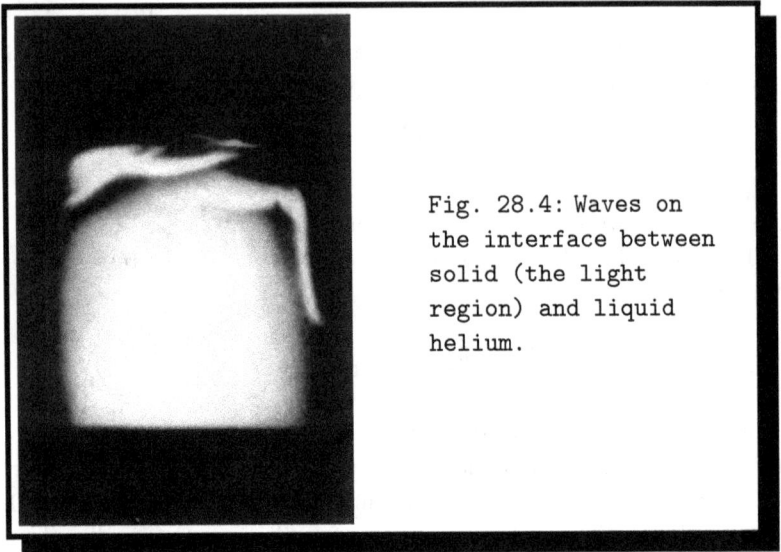

Fig. 28.4: Waves on
the interface between
solid (the light
region) and liquid
helium.

oscillation energy of atoms being greater in hydrogen than in helium. All other substances do freeze at absolute zero. So helium is the only one that always remains liquid at normal pressure. One may say that it is the uncertainty principle that prevents it from freezing. Physicists call helium a *quantum liquid*. Another exceptional property of helium is *superfluidity*, (see page 304), which is also a macroscopic quantum phenomenon.

Still, under a pressure of about 2.5 *MPa* liquid helium becomes solid. However, solid helium is not an ordinary crystal. For example the kinetic energy of atoms at the interface of solid and liquid helium is defined by the zero-point oscillations. This enables the crystal surface to perform gigantic oscillations as if it was an interface of two immiscible fluids, Figure 28.4. Physicists have graced solid helium with the title of *quantum crystal*.

Chapter 29

On snowballs, nuts, bubbles, and. . . liquid helium

I made myself a snowball
As perfect as could be.
I thought I'd keep it as a pet
And let it sleep with me.
I made it some pajamas
And a pillow for its head.
Then last night it ran away,
But first it wet the bed.

Shel Silverstein, *"Snowball"*

Helium, despite its second position in the periodic table of elements, has, since its discovery, been a source of a great many hassles for physicists due to its quite unorthodox properties. Yet these troubles and headaches were overwhelmingly outshone by the beauty and uniqueness of the physical phenomena occurring in liquid helium as well as by the opportunities it offers to researchers and engineers in the production of extremely low temperatures. Besides superfluidity, among the peculiarities of this quantum liquid is its specific mechanism of charge transfer which differs from other liquids. And here our story starts.

Physicists started tinkering with this question in the late 1950s. At that time, the most probable candidates for the role of charge carriers seemed to be electrons and positive ions produced by the ionization of helium atoms. A further assumption was that electric charge is transported not by the helium ions themselves (they are rather heavy and it would be too hard to accelerate them) but by the *"holes"*. To get an idea of what the are, you may imagine an electron sitting in a helium atom has leaped onto a positively charged helium ion which happened nearby.

When doing so, the electron must have left behind an empty place. However that "electron vacancy" may soon enough be occupied by another electron hopping from another atom. That new empty seat, in turn, will be taken by another electron continuing in this way. From aside, such an electron "leapfrog-game" looks as if a positively charged particle moved in the opposite direction. Yet, there is no real moving positive charge, just an absence of electron at its "dwelling place"; we can call this object a "hole". This charge transfer mechanism generally works fine in semiconductors, so it was considered quite plausible that it works in liquid helium.

As a rule, in order to measure the masses of charge carriers, both positive and negative, researchers study their trajectories in a uniform magnetic field. It is known that when a charged particle with some initial velocity enters a magnetic field it starts gyrating and its trajectory becomes a circle or a spiral. Knowing the initial velocity and the strength of the field one can easily find the mass of the particle simply by measuring the radius of gyration. However, the results of experiments turned out to be quite surprising: the masses of both the negative and positive charge carriers exceeded that of the free electron by tens of thousands of times!

Sure enough, in liquid, electrons and holes are surrounded and interact with atoms and, hence, their masses can differ from that of the free electron. Yet five orders of magnitude seemed way too much. Such a significant discrepancy of theoretical calculations and experimental results was considered unacceptable even for an extravagant element like helium. So there arose an urgent need to come up with a new model.

The American physicist Robert Atkinson soon proposed the correct explanation of the structure of the charge carriers in liquid helium. It is known that in order to transform a liquid into a solid one does not necessarily need to cool it –it is just as possible to solidify liquid by compressing it. The pressure at which the liquid becomes solid is called the *solidification pressure* (P_s). Naturally, P_s depends on temperature: the higher the latter is the more difficult it gets to solidify the liquid by compression, and, therefore, P_s goes up. It turns out that the whole "trick" with the structure of the positive charge carriers is explained by the rather low value of the solidification pressure of liquid helium: at low temperatures, $P_s = 25$ *atm*. And this causes the very unusual structure of the positive carriers.

We have mentioned earlier that positive ions, He^+, can commonly exist in liquid helium. When interacting with a neutral helium atom, a positive He^+ attracts the negatively charged electrons and repels the positively charged nucleus. The result is that the centers of positive and negative charges in the atom do not coincide anymore but get separated. Hence the presence of positive ions in liquid helium should lead to the *polarization* of its atoms. The polarized atoms are attracted by the positive ion and this, in turn, causes a rise of the local concentration of He atoms and the local density. As a result, the pressure around the ion increases. Graphically the dependence of the incremental pressure on distance from the positive ion is depicted in Figure 29.1.

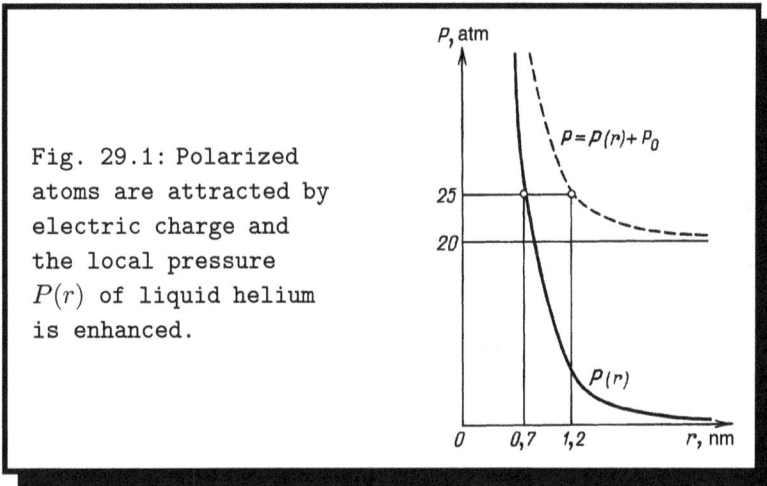

Fig. 29.1: Polarized atoms are attracted by electric charge and the local pressure $P(r)$ of liquid helium is enhanced.

Well, as we already know, at low temperatures liquid helium solidifies at 25 *atm*. As soon as the pressure in the vicinity of the positive helium ion reaches that critical value, a certain volume around it turns solid.[a] According to Figure 29.1 for low external pressures this solidification takes place within approximately $r_0 = 0.7\,nm$ $(0.7 \cdot 10^{-9}\,m)$ from the ion. Therefore it gets "frozen" in a "snowball" formed by solid helium. Now, if an electric field is applied, the "snowball" will start moving. But it is not going to move alone and will be accompanied by its new retinue, pulling along a whole "tail" of extra density.

Consequently, the total mass of the positive charge carrier will include the three major contributions. The first one is the mass of the "snowball" itself, which is the product of the density of solid helium times the "snowball" volume at normal ambient pressure. This gives $32\,m_0$ $(m_0 = 6.7 \cdot 10^{-27}\,Kg$, is the mass of a helium atom). The mass of the following retinue turns out to be just slightly less — the mass of the tail of extra density pulled by the ion amounts to $28\,m_0$.

---

[a]That, as you remember from Chapter 28, is rather flabby.

Besides these two, there must be one more mass added: when an object moves in a liquid, there always occurs some displacement of masses of the liquid around it. This, of course, takes some energy. That is, to accelerate a body in a liquid requires greater force than the same acceleration in a vacuum. So in liquid an object behaves as if its mass was somewhat greater. This additional mass due to the motion of liquid layers is called the *associated mass*.[b] For the "snowball" moving in the liquid helium at normal atmospheric pressure the associated mass turns out to be $15m_0$.

Finally, after summing up, the total mass of the positive charge moving in the liquid helium equals $75m_0$, a value which closely agrees with the experimentally measured one.

You see that the concepts of classical physics may deal successfully with the theory of positive charge carriers in liquid helium. But it is not so easy with negative ones. First of all, it happens that there are no negative ions in liquid helium at all (although a few of the negatively charged molecular ions He_2^- may be formed, they do not play any noticeable part in charge transfer). Hence, the electron is still the only remaining contestant for the role of the negative charge carrier. However, it catastrophically misses most of the mass required from the experimental data. And here is exactly the place for the idiosyncrasies of the quantum world to appear. Experiment shows that the electron, whom we have been so stubbornly intending for the negative charge carrier, cannot even freely penetrate into liquid helium.

To make sense of all this, we will have to digress and touch a little on the structure of atoms with several electrons. There is a paramount principle, unquestionably reigning in the microworld, determining the behavior of groups of identical particles. When applied to electrons, it is called *Pauli's exclusion principle*.[c] According to this rule, no two electrons can occupy the same quantum state at the same time. And we shall show that this explains the observed "aversion" of helium atoms towards free electrons and the troubles the latter run into when they try entering the liquid helium.

The energy of an electron in an atom, as we have noted already, can have only certain quantum values. And what is important is that for each

[b]We have encountered this concept already in Chapter 24 (see page 211).
[c]Wolfgang Pauli, (1900–1958), Austrian born, lived in Germany, Switzerland, and in the US; specialized in quantum mechanics, quantum field theory, relativity, and other fields of theoretical physics; 1945 Nobel Prize in physics.

such a value of energy there are several corresponding states available for electrons, varying by the character of their motion in an atom (for instance, by the electron orbit shape or, in quantum language, the shape of the probability cloud which defines the spread of electrons in space, see Figure 28.3). States of the same energy compose a so-called *shell*. According to the Pauli principle, when the number of electrons in an atom grows (as the atomic number goes up), they do not get "crammed" in the same states but fill new available shells one by one.

The first shell, corresponding to the lowest possible energy, must be occupied first. Located real close to the atomic nucleus, it can take in only two electrons. Thus, in helium, which is the second in the periodic table, the first shell is completely filled. There is no choice for the third electron other than to stay sufficiently far from the nucleus. When such an "undue" electron approaches a helium atom within a distance of the order of its radius, there appears a repulsion force precluding it from getting near.

Therefore, some "entrance work" is needed to ram an errant electron into the bulk of helium. Three Italian physicists, Carreri, Fasoli, and Gaeta, proposed the idea that as the electron entering helium cannot get too close to the atoms, it shoves them away and thereby forms a spherically symmetrical cavity, some sort of a bubble, Figure 29.2. And this bubble with the electron scurrying inside is indeed the negative charge carrier in liquid helium.

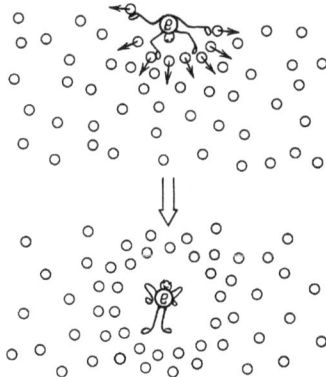

Fig. 29.2: Because of
quantum effects
electron cannot
approach the He atoms
and instead shoves
them away.

The size of the bubble can be rather easily estimated. The repulsion between electrons and helium atoms should decrease with distance. On the other hand, at large distances, electrons should act on helium atoms in exactly the same way the positive ions do, that is to polarize them. So, the interaction of electrons with helium atoms should be the same attraction as for the previous situation with "snowballs". Hence, when approaching a bubble with the electron captured inside, the incremental pressure in helium rises, following the law portrayed in Figure 29.1.

Yet, under normal conditions the pressure at the boundary of the bubble still remains far less than $25\,atm$ because of the comparatively large size of the bubble. Besides this pressure, resulting from the increased density of polarized helium, there is a force of surface tension. It acts at the bubble's border and is directed in the same way, to the center of the bubble. What would then balance these two external forces preventing our bubbles from collapse? It turns out that the required counteraction is created by that very same imprisoned electron.

Indeed, according to the uncertainty principle, which we have discussed in Chapter 23, the accuracy of measuring the momentum of the electron is directly related to the uncertainty of the electron's position in space, being $\Delta p \sim \hbar\,/\,\Delta x$. In our case the uncertainty in location of the electron is naturally defined by the size of the bubble, that is $\Delta x \sim 2R$. So, the movement of the captive electron should posses a momentum of the order of $\hbar\,/\,2R$ and, consequently, have a kinetic energy $E_k = p^2/2m_e \sim \hbar^2\,/\,8m_e\,R^2$. Resulting from the electron's collisions with the walls of the bubble, there should arise some outward pressure (remember the principal equation of the kinetic theory of gases, relating the pressure of the gas, P, to the average kinetic energy of its chaotically moving particles and their concentration: $P = (\frac{2}{3})nE_k)$. This pressure could very well balance the forces trying to squeeze our helium bubble. In other words, an electron confined in the bubble acts in exactly the same way as a gas isolated in a reservoir; however, this "electron gas" consists of a single particle! The concentration of such a gas is, obviously, $n = 1\,/\,V = 3\,/(4\pi R^3)$. After plugging that value and $E_k \approx \hbar^2\,/(8m_eR^2)$ into the expression for the pressure, we find that $P_e \approx \hbar^2/(16\pi m_eR^5)$. Precise quantum mechanical calculations lead to a similar answer:[d] $P_e \approx \pi^2\,\hbar^2\,/(4m_eR^5)$.

---

[d]The origin of the discrepancy is that the electron prefers to stay in the middle of the cavity rather than near the repelling walls. This effectively reduces the uncertainty of coordinate and enhances the pressure.

As long as the external pressure remains small, the dominating force trying to squeeze the bubble will be surface tension, $P_L = 2\sigma/R$, see Chapter 10. Hence, equating $P_e = P_L$, one can easily estimate the radius of a stable electron bubble in liquid helium:

$$R_0 = \left(\frac{\pi^2 \hbar^2}{8 \, m_e \, \sigma} \right)^{\frac{1}{4}} \approx 2 \, nm = 2 \cdot 10^{-9} \, m.$$

We see now that the negative electrical charge in liquid helium is carried by bubbles with electrons ensnared inside.

The total mass of such carriers can be calculated in the same manner as was carried out for the snowballs. Yet, now the bubble itself weighs almost nothing for the mass of the electron inside is negligibly small compared to the mass of the liquid dragged by the bubble (the retinue) plus the associated mass. So, the net mass of the carrier would be equal to the sum of the associated mass and the mass of the tail pulled by the drifting bubble. Because of the rather large size of the bubble, its resulting mass, $245 \, m_0$, turns out to be much greater than that of the snowball.

Now, let us consider how an increase in external pressure P_0, will effect the properties of the charge carriers. Figure 29.1 depicts the dependence of the total pressure (including the external one), $P = P(r) + P_0$, in the vicinity of an ion in liquid helium versus distance from the ion for $P_0 = 20 \, atm$. Such dependence for an arbitrary value of $P_0 < 25 \, atm$ can be graphed by simply shifting the curve for $P_0 = 0$, along the P-axis. As the picture shows, the higher the external pressure is, the further from the ion the total pressure attains $25 \, atm$. Hence, with the growth of the external pressure, the snowball behaves as if it was rolling down a snowy slope: it promptly lumps up on itself the "snow" – solid helium – and gets bigger. The relation between the size of the snowball and the value of external pressure $r(P_0)$, is presented in Figure 29.3. And what about the "bubble"? How does it behave as P_0 goes up? Well, for some while, like any bubble in liquid, it submissively shrinks while the surrounding pressure grows. Its radius $R(P_0)$ decreases as the upper curve in Figure 29.3 shows. And yet, further, at $P_0 = 20 \, atm$ the curves for $r(P_0)$ and $R(P_0)$ cross, meaning that at this point the sizes of the bubble and the snowball become the same and equal $1.2 \, nm$. We know already the future fate of the snowball: with the rise of P_0 it will rapidly grow at the expense of the helium solidifying on its surface. But what is the bubble to do — continue to shrink following the dashed line in Figure 29.3?

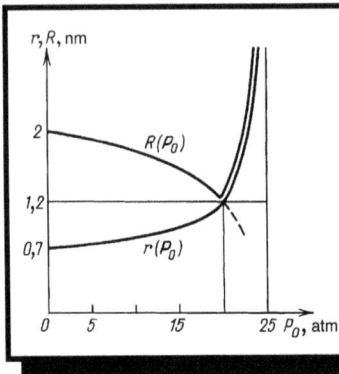

Fig. 29.3: Dependencies of the radii of the ''bubble'', R, and ''snowball'', r, on external pressure P_0.

No way! At this very moment, our bubble finally shows its real character. As the applied pressure continues to go up, the bubble begins to act as a snowball: it gets covered with an icy crust of the solid helium. Indeed, according to Figure 29.1 at $P_0 = 20\ atm$ the total pressure on the surface of the bubble becomes equal to 25 atm, reaching the solidification point for liquid helium. The internal radius of the bubble, protected now by its fancy icy attire, ceases to change and remains approximately the same regardless of the further increase of pressure, whereas its external radius equals that of the snowball at the corresponding pressure. Thus, at external pressures greater than $P_0 = 20\ atm$, our bubbles become coated with an ice shell and begin to resemble nuts. With that difference, though, that the kernel in the "nut" is of a quite peculiar nature – it is an electron rushing chaotically in its shell of solid helium.

One last thing worth mentioning here. As P_0 approaches 25 atm, the external radii of both the nuts and the snowballs continue to grow bigger (trying, in principle, to reach infinity). Finally all the helium in the reservoir becomes solid. The role of negative charge carriers in the solid helium is, therefore, played by the electron bubbles, frozen in the bulk of solid helium and having inherited their dimension of about 1.2 nm from the former nuts. The positive charge, on the other hand, must be transferred by helium ions, the remnants of the former snowballs. Of course, it is not so easy to carry anything in a rigid environment, the electric charges included. So the mobility of the carriers in solid helium will be by many orders of magnitude lower than that of the snowballs and the bubbles in the liquid phase.

Chapter 30

Superconductivity: a century of discoveries, dreams, and disappointments

This is why we're here; *unobtanium*, because this little gray rock sells for 20 million a kilo. That's the only reason. It's what pays for the whole party. It's what pays for your science.

Parker arguing with Dr. Augustine, *Avatar.*

Probably almost all our readers have heard of superconductivity. This phenomenon consists of the abrupt disappearance of electrical resistance of some pure metals and alloys at low temperatures. "Low temperatures" meaning the range of 10–$20\,K$, that is 10–20 degrees above absolute zero ($-273.15°\,C$). In order to cool to these low temperatures a sample is usually placed into liquid helium that at normal pressure boils at $4.2\,K$ and, as you already know, does not freeze down to absolute zero. Throughout the last century physicists and chemists in many laboratories all over the world were looking for compounds which become superconducting at high enough temperatures to be cooled, for instance, by comparatively cheap and widely available liquid nitrogen. So understand that the discovery of high-temperature superconductors, whose resistance becomes zero at temperatures above $100\,K$, was met as one of the greatest events in physics of recent years. Really, the practical significance of this discovery can be compared to that of magnetic induction at the beginning of the nineteenth century. It ranks with the discovery of uranium fission, the invention of the laser, and the discovery of the unusual properties of semiconductors in the twentieth century.

293

30.1   Starting from the end

The beginning of this exciting new stage in the development of super-conductivity was the work by K. A. Müller and J. G. Bednorz at IBM's lab in Switzerland. In the winter of 1985–86 they managed to synthesize a compound of barium, lanthanum, copper, and oxygen — the so-called metal oxide ceramic $La - Ba - Cu - O$, a compound which had superconducting properties at $35\,K$, the record at that time. The article, cautiously titled "The Possibility of High-Temperature Superconductivity in the $La - Ba - Cu - O$ System", was turned down by the leading American journal *Physical Review Letters*. In the previous 20 years the scientific community had gotten tired of receiving sensational reports about the discovery of high-temperature superconductors that turned out to be false, so it decided to avoid the trouble. Müller and Bednorz sent the article to the German journal *Zeitschrift fur Physik*. When the news about high-temperature superconductivity was finally heard and research began in hundreds of laboratories, every article devoted to the new phenomenon would start from a reference to this article. But in the spring of 1986 it passed practically unnoticed. Just one Japanese group checked the result and verified it. Soon the phenomenon of high-temperature superconductivity was corroborated by physicists in the United States, China, and the Soviet Union.

At the beginning of 1987 the whole world was in a fever, searching for new superconductors and investigating the properties of those already discovered. The critical temperature T_c increased quickly: it was $T = 45\,K$ for $La - Sr - Cu - O$ and it reached $52\,K$ for $La - Ba - Cu - O$ under pressure. Finally in February 1987 the American physicist Paul Chu got the idea to imitate the effect of external pressure by substituting La atoms by the smaller atoms of Y which is next in Mendeleev's table column. The critical temperature of the compound $Y - Ba - Cu - O$ (see Figure 30.1) broke the "nitrogen barrier", reaching $93\,K$. This was a long-awaited triumph but not the end of the story. In 1988 a five-component compound of the type $Ba - Ca - Sr - Cu - O$ with the critical temperature $110\,K$ was synthesized and a little later its mercury and thallium analogues with the critical temperature $125\,K$ appeared. The maximum critical temperature, under pressure of $30\,hPa$, of the mercury record-breaker looks impressive even on the Celsius scale $-108°\,C$!

The discovery of high-temperature superconductivity is unique in modern physics. First, it was discovered by just two scientists with very modest

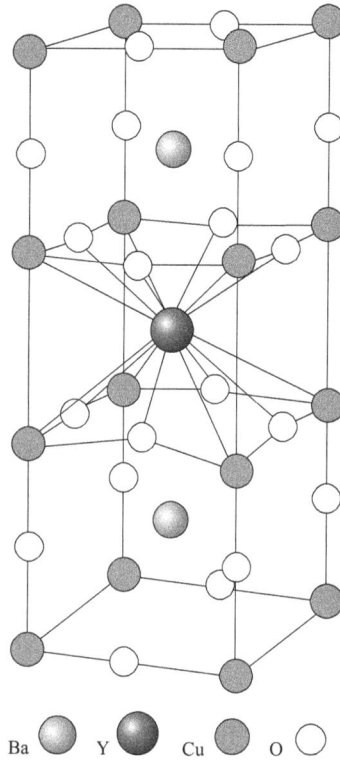

Fig. 30.1: Structure
of superconducting
$Y\,Ba_2\,Cu_3\,O_7$.

tools. Secondly, the compounds include easily accessible elements. As a matter of fact, these superconductors can be made in a high school chemistry lab in a day. What a contrast with discoveries in other areas such as high-energy particle physics which require massive collaborations and millions of dollars worth of equipment. There, a list of authors alone takes a whole page in a journal article. This discovery was a cause for optimism: the time of lone investigators in physics has not passed! And though the discovery had been anticipated for 75 years, it caught everyone by surprise. Theorists could only shrug their shoulders and continue to do so as the critical temperature went up.

So was the discovery by Bednorz and Müller a fluke or destiny? Could the discovered compound, with its unique properties, have been synthesized earlier? How difficult is it to answer these questions! We have long been accustomed to the fact that everything new is obtained on the edge of the impossible by using unique equipment, superstrong fields, ultralow temperatures, superhigh energies... There is nothing of the kind here. It is not too difficult to "bake" a high-temperature superconductor – a qualified alchemist of the Middle Ages could have managed it. It is worth recalling that about 10 years ago many laboratories of the world intensively investigated an unusual superconducting compound. This substance was called "alchemical gold" because of its yellow luster and high density, which made it resemble the noble metal. It was synthesized by medieval alchemists, passed off as true gold, and advertised as the result of successfully using the "philosopher's stone". Alchemic gold is a complex compound, and who knows, perhaps a high-temperature superconductor would have been baked in the Middle Ages if it had been blessed with a golden luster.

The dreams of the Middle Ages may take us too far. But you will be amused to hear that some of the present-day high-temperature superconductors were in fact stored on lab shelves since 1979! At that time they were synthesized in quite a different connection at the Moscow Institute of General and Inorganic Chemistry by I. S. Shaplygin and his collaborators. Unfortunately they did not measure the conductivity of the compositions at low temperature which would indicate the new phenomenon and the discovery did not happen.

30.2   Back to the beginning

On April 8, 1911, the Cryogenic Laboratory of Leiden University (Netherlands) supervised by Prof. Heike Kamerlingh Onnes had announced a discovery, which determined the course of the next century exploration for thousands of scientists all over the world. The measurements of resistance in a solid mercury wire showed a sharp drop, and completely vanished at temperatures close to liquid helium's boiling point.[a] Kamerlingh Onnes named this phenomenon "superconductivity". In order to understand why superconductivity was discovered in Kamerlingh Onnes' laboratory in the beginning of the 20th century, it is necessary to go back to the end of the previous century and immerse yourself into the atmosphere of that time.

At the end of the 19th century, due to the development of the world

---

[a]Under normal conditions this temperature equals $4.2\,K$ in the absolute scale of temperature, or $-268,95°\,C$ in the Celsius scale, well-known to reader.

telegraph system, the increase of metal conductivity became an important issue. Theoretical physics did not provide the answer to the question how to achieve this, because the mechanism of electrical conductivity had not yet been discovered. Soon, in 1897, after the discovery of the electron, the theory of electron conductivity started developing. It was started by a German physicist Paul Drude in 1900, who assumed that a metal consists of positively charged ions from which the "free electrons" are detached. As a result he succeeded in deriving the expression for conductivity:

$$\sigma = \frac{ne^2\tau}{m}, \tag{30.1}$$

where n is the concentration of electrons, e and m are the electron charge and mass, and τ is the time between two subsequent electron scatterings. It is noteworthy that even today Drude's theory is included in textbooks on physics as a classical theory of electrical conductivity in metals.

Fig. 30.2: Heike Kamerlingh Onnes (1853-1926) - Netherland physicist and chemist, Nobel Prize winner in Physics of 1913. Had a well deserved nickname ''The gentleman of absolute zero''.

According to the modern version of this theory, developed by Sommerfeld in 1928, after quantum mechanics was created, electrons were considered as a "degenerate quantum ideal gas" filling a crystal lattice in whose sites the ions of metal atoms and sometimes impurities are located. The definition of "quantum degeneration" applied to this gas means that, unlike a classical gas, its particles, electrons, remain in a state of chaotic motion with immense velocities, expressed as a percentage of the speed of light, even at a temperature equal to absolute zero. Applying an electric field results in the superposition of a certain ordered drift over this chaos: it resembles the wind blowing away a cloud of insects over the lake on a summer evening. It is surprising that being at zero temperature and moving around in a chaotic way, the electrons, due to the whims of quantum mechanics, manage to avoid scattering off the ions located at the lattice sites. In this case they are scattered only by impurities, whose positions in lattice are unpredictable. When the temperature is non-zero an additional scattering mechanism, caused by ions deviating from their equilibrium positions, occurs due to thermal fluctuations.

Obviously, in 1900 Drude was unaware of that — only in six years the "plum pudding" model was proposed, in thirteen years Bohr's postulates appeared, and more than a quarter of a century was remaining before the final formulation of the quantum theory. Namely the latter served as the basis for study of the degenerate electron gas behavior in the lattice's periodic potential. However, it appeared that even the model of a classical ideal electron gas being subjected to the effect of an external electrical field allowed one to obtain the excellent formula for conductivity, linking the current density to electrical field intensity. Following a well known at that time molecular theory, Drude suggested that in the course of their motion the electrons, like the molecules of an ideal gas, collide and randomly change their momentum direction during such collisions. He found that the metal conductivity is proportional to the time between such collisions. Their nature remained inessential for the validity of the obtained results — these could be electron-electron collisions, like between the molecules in the ideal gas, or the collisions between the electrons and other objects. Drude himself suggested that electrons scatter off the impenetrable ion cores.

It had been found experimentally that the electrical resistance of metals drops with decreasing temperature. However, it was unclear how it would behave at very low temperatures. There were different viewpoints on this issue. As it was mentioned before, the rigorous theory of electrical conductivity was only created a quarter of a century later, but scientists started to debate the dependence of conductivity on temperature soon after publication of the Drude theory.

The first theory of electrical resistance of metals, proposed by Lord Kelvin in 1902, maintained that with lowering temperature the electron motion slows down; they couple with the ions to form into atoms and are not able to move along the crystal, even under the effect of an electric field.

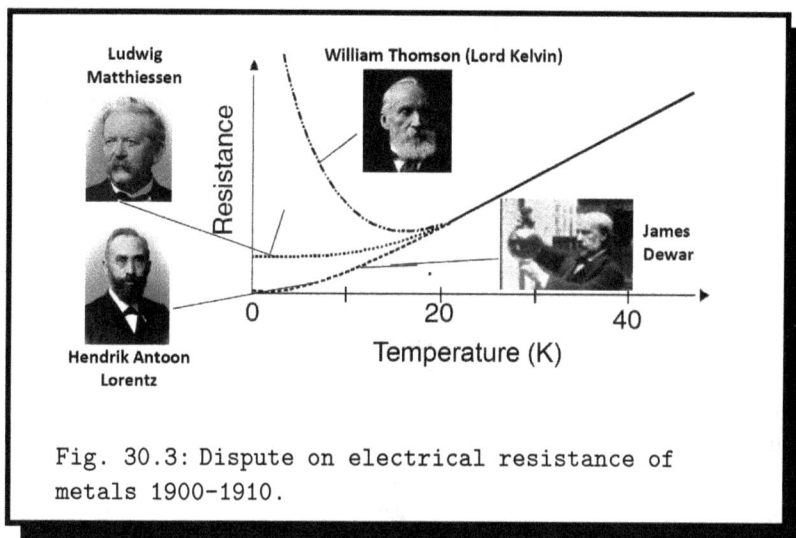

Fig. 30.3: Dispute on electrical resistance of
metals 1900-1910.

As a result the metal resistivity (the physical value inverse to conductivity) should tend to infinity at temperatures approaching absolute zero (see
Figure 30.3).

The alternative theory of the electrical resistance of metals (P. Drude
and H.A. Lorentz), was based on the Einstein's model of solids. In 1907,
Einstein, in order to explain the behavior of thermal capacity of solids at low
temperatures, proposed to consider the particles forming the lattice as the
oscillators experiencing the temperature vibrations. The Drude–Lorentz
theory, in contrast to Lord Kelvin's one, predicted that at extremely low
temperatures the resistance ($R \sim \sigma^{-1}$) of a clean metal should tend to zero.
Indeed, at zero temperature the classical oscillators do not vibrate (zero-
point oscillations were discovered much later), the electrons do not scatter
at all, thus the corresponding scattering time $\tau \to \infty$ and the conductivity
tends to infinity (see Eq. (30.1)).

In its turn Ludwig Matthiessen noticed that the presence of impurities
changes the situation since they present obstacles for electrons even at zero
temperature. As a result, the resistivity of an impure metal should remain
finite even in the absence of the thermal motion of the ions (see Figure
30.3).

Heike Kamerlingh Onnes takes part in this discussion, and he is among those who believe that the resistance of a pure metal tends to zero when the temperature approaches absolute zero. Unlike theoreticians, he is a "monopolist" in the field of the experimental study of ultra-low, for those times, temperatures. He had improved the cascade method of liquefaction of gases,[b] proposed in 1898 by the Scottish physicist James Dewar. Kamerlingh Onnes in 1906 had succeeded to liquefy helium, the last of elements from the Periodic Table which remained gaseous at normal pressure up to $4.2\,K$ $(-268,95°\,C)$. This value was determined, namely by Kamerlingh Onnes, as a result of thorough measurements.

Thus, having learned how to handle liquid helium and having at his disposal temperatures of only several degrees away from absolute zero, Kamerlingh Onnes advanced to studying pure metal behavior in this temperature range. He started with precious metals - gold, platinum, silver.[c] However, when studying the variation of resistance for gold, silver, and platinum samples versus temperature, Kamerlingh Onnes discovered that though all of them first linearly decrease with lowering temperature the values of resistance "arrive at saturation" (achieve certain fixed values) when the temperature approaches to absolute zero. He had not found the zero resistivity he was looking for.

To clarify this issue, it was necessary to experiment with highly purified metals; even the traces of foreign impurities would affect their electrical conductivity. Mercury was the easiest to purify. Purification was performed by the method of distillation: evaporation and condensation.

Already the first experiments in April of 1911 with a mercury conductor

---

[b]The cascade method of liquefying gases is based on a simple idea. In the process of pumping out of vapor phase of a substance which is in contact with its liquid phase, the fast molecules that escape from the liquid are carried away by the pump from the vessel. As a result, the average kinetic energy, and consequently, the temperature of the liquid phase decrease. Correctly selecting the set of gases with successively decreasing boiling temperatures and forcing them, one by one , to boil, it is possible to reach temperatures down to $1\,K$.

[c]Why gold? Our audiophile readers, who pay a lot of money for gilded wires, connecting amplifier with speakers, would understand. Gold is an excellent conductor of electric current. To get a sense of it, imagine yourself having tea at your Grandma's, where the tea is served with silver gilded family spoons and compare it to a cup of tea at the college buffet, where it is served with plastic spoons. Which of these spoons will be hot to touch? Of course, silver or golden. As far back as the middle of the XIXth century, Wiedemann and Franz discovered that thermal conductivity of metal is directly proportional to its electric conductivity - therefore, a good thermal conductor will also have a good electrical conductivity.

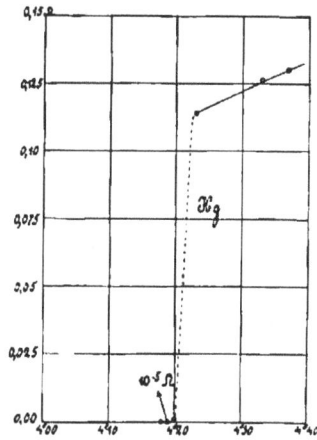

Fig. 30.4: Dependence
of the resistance of a
mercury wire on
temperature obtained
during Kamerlingh
Onnes' experiment.

showed that the electrical resistance at temperatures below 4.2 K would disappear. But it remained unclear which law was in play. Kamerlingh Onnes thought (and even derived an appropriate formula), that this decrease takes place gradually with the decrease of temperature. However, the next series of experiments, which were staged a month later, produced a surprising result: upon lowering the temperature to 4.15 K the electrical resistance of mercury dropped very abruptly. Kamerlingh Onnes' further studies showed that some other metals, e.g., tin and lead also produce an abrupt drop of electrical resistance (at 3.72 K and 7.19K respectively). This phenomenon, named superconductivity by Kamerlingh Onnes, did not fit into any of the models known at that time for electrical conduction in metals.

Seminal experiments began on April 8th, 1911. Gilles Holst, the assistant of Kamerlingh Onnes, started cooling the system at 4 a.m., and after lunch they proceeded to measure the resistivity. In the beginning, the resistivity of a thin mercury wire was going down linearly with temperature decrease, as it was with golden samples, however, at a temperature close to the liquid helium boiling point it abruptly dropped to zero!

Was it by accident or it was destined for Kamerlingh Onnes to discover the unanticipated phenomenon of superconductivity? The initial goal of his study was different. He wanted to make sure that pure metal resistivity would disappear at absolute zero temperature, according to the then naive classical understanding of the theory of metals. Today we can speak about

Fig. 30.5: A view of the cryostats a) used in
Kamerlingh Onnes' experiments (left) b) used in a
modern laboratory, one hundred years later (right,
courtesy of Dr. R. Natali)

his extraordinary good luck: the boiling point of liquid helium is $4.2\,K$, and mercury's transition into the superconducting state takes place at $T_{\rm c} = 4.15\,K$. Most of metals, if they enter the superconducting state at all, need much lower temperatures not available at Kamerlingh Onnes' time.

Yet, the most important role was played by systematic procedures for success. Kamerlingh Onnes had a multifaceted personality. First of all, he was a remarkable physicist, a researcher who was introspective about the essence of the phenomenon and was doing his best to understand its nature. Second, he was a brilliant experimentalist who managed to stage the experiment, which led him to a discovery.[d] Third, he was an excellent engineer, who, by improving the cascade method, managed to liquefy helium and thus paved his own way toward absolute zero. And finally, he was an outstanding organizer. The Leiden University laboratory, which he created, can set an example as to how a modern research facility should be organized and managed.

---

[d] In 1882, at the age of 29, Kamerlingh Onnes was appointed as professor of experimental physics at the University of Leiden. He became the head of the physical laboratory of that University. In his inaugural lecture he proclaimed the guidelines, which he himself followed during 42 years of his work in Leiden University: "Through measurement to knowledge". According to Kamerlingh Onnes, physical laboratories should take quantitative measurements and set up quality experiments; theoretical descriptions should be supported by accurate measurements conducted with astronomical precision.

Take a look at the system he designed (Fig. 30.5). All these tens of glass tubes had to be blown, connected, and be able to sustain 6 atmospheres pressure like a Champaign bottle. Of course, it was a highly skilled glass-blower who manufactured the tubes, the majority of the measurements of resistivity temperature dependence were performed by a talented assistant of Kamerlingh Onnes, but the ideas and planning of all the infrastructure of the experiment, as it is often said today, belonged to Kamerlingh Onnes. Consequently, the Nobel Prize received in 1913 "for his investigations on the properties of matter at low temperatures which led, inter alia, to the production of liquid helium" became an ultimate success and reward for talent, hard work, and curiosity.

30.3   From surprise to understanding

We would like to emphasize that the resistance of a sample in the supercon-ducting state is equal to zero not approximately but exactly. That is why electric current in a closed circuit can circulate as long as you like without damping. The maximum duration of a persistent superconducting current was recorded in England and lasted for about two years. (The current in the ring would have circulated up till now but for a transport worker strike which caused a break in the supply of liquid helium to the laboratory.) Even after two years, no damping of the current was detected.

Very soon superconductivity was discovered not only in mercury but in other metals as well. The prospects for practical applications of the dis-covered phenomenon seemed unlimited: power transmission lines without waste, superpowerful magnets, electric motors, new types of transformers, but there were two obstacles to such inventions.

The first were the extremely low temperatures at which superconduc-tivity was observed in all materials known at the time. To cool conductors to these temperatures, scarce helium is used (its stocks are limited, and even now producing a liter of liquid helium costs some dollars). This makes many projects using superconductivity simply unprofitable. The second obstacle discovered by Kamerlingh Onnes was that superconductivity had turned out to be rather sensitive to magnetic fields and to the value of current. In fact, it was destroyed by strong fields.

The next fundamental property of the superconducting state, discovered in 1933, was the Meißner-Ochsenfeld effect: the complete expulsion of a magnetic field from the volume of the superconductor. But again experimental investigations were complicated by the need to work with scarce liquid helium — before World War II it was produced in about 10 laboratories throughout the world (two of which were in the Soviet Union).

The fundamentals of superconductivity stayed absolutely out of reach of the classical theory of metals and the quantum one was still an embryo at that time. The so-called two-liquid model suggested a coexistence of two types of electrons in superconducting metals: normal electrons interact with the lattice but superconducting ones for some reason do not. This assumption let the London brothers[e] write down the equations of electrodynamics of superconductors which described the Meißner effect and some other features. Still the microscopic mechanism of superconductivity remained a mystery.

In 1938 P. L. Kapitza[f] discovered the phenomenon of superfluidity. It turned out that at temperatures below $2.18\,K$ liquid helium can flow through thin capillary tubes without any viscosity.[g] The theoretical explanation of this phenomenon was provided by L. D. Landau.[h]

It turns out that at such low temperatures the liquid helium is getting transformed into a new state: billions of its atoms begin a new life of a collective entity; their behavior is described by the unique wave function and is reminiscent of that one of an electron moving along Bohr's orbit inside the atom. They get accumulated at the lowest energy level (so called *Bose-condensation*). Landau showed that, as a result, a gap appears in the spectrum of excitations making possible the superfluid state. Discussing this macroscopic revelation of this entirely quantum effect, he called helium "a window to the quantum world".

Similarity of the phenomena of superconductivity and superfluidity gave rise to hopes that a theory of superconductivity was in the offing. Yet,

---

[e]H. London, (1907–1970), British physicist; F. London, (1900–1954), American physicist; specialists in low-temperature physics.

[f]P. L. Kapitza, (1894–1984), Russian physicist; 1978 Nobel Prize in physics.

[g]It is interesting to note that almost thirty years before, in the process of his seminal experiments on the superconductivity of mercury, Kamerlingh Onnes observed and described a preliminary manifestation of superfluidity: around $2\,K$ the effervescent surface of liquid helium suddenly subsided and flat calm settled down.

[h]L. D. Landau, (1908–1968), Russian physicist; 1962 Nobel Prize in physics.

the straightforward extension of these ideas to superconductivity failed. The reason was that electrons are particles with one-half spin (so-called *fermions*) and behave absolutely unlike helium atoms which possess a whole spin being *bosons*. In the quantum system of electrons, excitations with zero energy may appear even at zero temperature and the Landau criterion of superfluidity does not hold.

The natural desire to reduce the problem to that already solved inspired the idea to prepare from two fermions a composite boson with a whole total spin and after that to effect the Landau superfluidity scenario. However this was opposed by the Coulomb[i] repulsion of electrons which was too strong in spite of the screening that occurs in an electroneutral metal.

Ten years later, in 1950, the discovery of the *"isotopic effect"* first indicated the connection between superconductivity and the crystal lattice of a metal. Measurements of the critical temperature of lead proved that it depended on the mass number of the isotope under testing. Thus superconductivity ceased being a purely electronic phenomenon. A little later Frölich[j] and Bardeen[k] independently demonstrated that the interaction of electrons with lattice oscillations *(phonons)* may lead to an attraction. This could in principle overcome the electrostatic repulsion but one has to keep in mind the huge kinetic energies of electrons. At first glance those should overcome the just-mentioned weak coupling. Composite bosons did not work out.

Later in 1950, with the help of experimental data and the theoretical achievements of solid-state physics, based on quantum mechanics and statistical physics, Ginzburg[l] and Landau (USSR) developed a phenomenological theory of superconductivity, known as the Ginzburg–Landau theory. It proved so successful and predictive that even now it remains a powerful research tool despite the 50 elapsed years since its creation.

---

[i]C. A. de Coulomb, (1736–1806), French physicist and inventor; discovered the basic law of electrostatics in 1785.

[j]H. Frölich, (1905–1991), born in Germany, worked in Leningrad, 1933–1935 and since 1935 in Britain.

[k]J. Bardeen, (1908–1991), American physicist; 1956, 1972 Nobel Prize in physics(!)

[l]V. L. Ginzburg, (1916–2009), Russian physicist and astrophysicist; 2003 Nobel Prize in physics.

In 1957 the American scientists John Bardeen, Leon Cooper, and Robert Schrieffer[m] put together these ideas and hints and created a consistent microscopic theory of superconductivity. It was found that superconductivity is indeed linked with the appearance of a peculiar attraction of electrons in metals. This is an utterly quantum phenomenon.

We have already mentioned that the ground state of a fermionic system is characterized by big kinetic energies of electrons. Luckily those do not prevent the binding of low-energy excitations of the system that behave like *quasiparticles*. They have the same electric charge e as an electron and some effective mass but their energy may be arbitrary small. The attraction brings on a rearrangement of the quasiparticle spectrum and the long-awaited gap that was so crucial for the Landau superfluidity criterion opens at last.

The origin of the attraction may be understood with the help of an analogy of two balls lying on a rubber rug. If the balls are far from each other, each of them deforms the rug, making a little depression. But if we put a ball on the rug and place another one near the first, their holes will join, the balls will roll down to the bottom of the combined valley and lie together. In metals, the mechanism is realized by deformations of the crystal lattice. At low temperatures some quasiparticles (usually they are called, just the same, electrons) form bound pairs. These are called "Cooper pairs" after the man who discovered the binding. The size of the pairs on the atomic scale is really quite large, reaching hundreds and thousands of interatomic distances. According to the graphic comparison suggested by Schrieffer, they should be envisaged not as a double star composed of electrons but rather like a couple of friends in a discotheque who either come together or dance in different corners of the hall, separated by dozens of other dancers.

You see that it took almost half a century since the discovery of superconductivity to gain cardinal progress in understanding its nature and to develop a consistent theory. This period may be considered to be the first stage of superconductivity studies.

---

[m] L. Cooper (born 1930), and R. Schrieffer, (born 1931), American physicists; 1972 Nobel Prize in physics (together with J. Bardeen).

30.4 Chasing high critical parameters

The creation of the theory of superconductivity was a powerful impulse to investigate it in earnest. Without fear of overstatement, we can say that great progress has been achieved in producing new superconducting materials in the subsequent years. The Soviet scientist A. A. Abrikosov's[n] discovery of an unusual superconducting state in a magnetic field played a significant role in this development. Before then a magnetic field was thought to be incapable of penetrating the superconducting phase without destroying it (which is actually true for most pure metals).[o] Abrikosov theoretically proved that there was another possibility: under certain conditions a magnetic field could penetrate into a superconductor in the form of current vortices. The core of the vortex turned to the normal state but the periphery remained superconducting! Depending on their behavior in a magnetic field, superconductors were divided into two groups: superconductors of the first type (old) and those of the second type (discovered by Abrikosov). It is important that superconductors of the first type can be changed into one of the second type if we "spoil" them by adding impurities or other defects.

A real hunt for superconducting materials with high critical fields and temperatures started. The ingenuity of the pursuers was really boundless. Arc welding, instant cooling, and sputtering onto hot substrate were employed. The efforts resulted in the discoveries of, for example, the alloys $Nb_3 Se$ and $Nb_3 Al$ which have the critical temperature (that is the temperature of transition to the superconducting state) $T_c = 18\,K$ and the upper critical field of more than $20\,T$. Further progress was achieved lately with ternary compounds. Before the discovery of high-temperature superconductors the record value of the upper critical field ($60\,T$) belonged to the alloy $Pb\,Mo_6\,S_8$ with $T_c = 15\,K$.

Among superconductors of the second type, scientists managed to find compounds capable of carrying a high-density current and bearing gigantic magnetic fields. And although many problems had to be solved before they

---

[n]A. A. Abrikosov, *(1928–2017), American and Soviet physicist*, pupil of L. D. Landau, specialist in condensed matter physics; 2003 Nobel Prize in physics.

[o]Strictly speaking this is true only for cylindrical specimens placed in a magnetic field parallel to the axis of the cylinder. If either the specimen is not a cylinder or a strong enough field is oriented differently the so-called intermediate state may be realized. It is formed by alternating macroscopic layers of superconducting and normal phases.

could find practical application (the compounds were brittle, high currents were unstable), the fact remained: one of the two major obstacles to the widespread use of superconductors in technology was overcome.

But increasing the critical temperature was still problematic. If critical magnetic fields were increased thousands of times in comparison with Kamerlingh Onnes' first experiments, the changes in critical temperature were not too encouraging: at best it managed to reach $20\,K$. So for the normal operation of superconducting instruments the expensive liquid helium was still necessary. This was particularly vexing because a fundamentally new quantum effect, the "Josephson effect", had been discovered. This made it possible to use superconductors widely in microelectronics, medicine, instrumentation, and computers.

The problem of increasing the critical temperature was extremely acute. Theoretical evaluations of its peak value showed that in the context of normal phonon superconductivity (that is, superconductivity caused by electron attraction due to interaction with the crystal lattice) this temperature could not exceed $40\,K$. But the discovery of a superconductor with such a critical temperature would still be a great success, since that could be achieved with relatively cheap and available liquid hydrogen (which boils at $20\,K$). It would open the era of "mid-temperature superconductivity". This stimulated attempts to modify existing superconductors and create new ones by traditional methods of materials science. But the ultimate dream was to create a superconductor with a critical temperature of $100\,K$ (or, even better, above room temperature), which could be cooled by cheap and widely used liquid nitrogen.

The best result of the search was the alloy $Nb_3\,Ge$ with the critical temperature of $23.2\,K$. This record temperature was achieved in 1973 and stood for 13 years. Until 1986 the critical temperature could not be raised by even one degree. It seemed that the possibilities of the phonon mechanism of superconductivity had been exhausted. In view of this, in 1964 the American physicist Little and the Soviet scientist V. L. Ginzburg proposed the following idea: if the possibility of increasing the critical temperature is limited by the nature of the phonon mechanism of superconductivity then it should be replaced by something else – that is, electrons should form Cooper pairs by means of some other, nonphonon, attraction.

During the last 20 years many theories were proposed, and tens or hundreds of thousands of new substances were investigated in detail. In his work Little has drawn attention to quasi-one-dimensional compounds — long molecular conducting chains with side branches. According to theoretical evaluations, a noticeable increase in critical temperature could be expected there. Despite the attempts of many laboratories throughout the world, such superconductors were not synthesized. But on the way, physicists and chemists made many wonderful discoveries: they obtained organic metals, and in 1980 crystals of organic superconductors were synthesized (the current record for the critical temperature of an organic superconductor is over $10\,K$). They managed to obtain two-dimensional layered metal-semiconductor "sandwiches" and, at last, magnetic superconductors where the former enemies, superconductivity and magnetism, coexisted peacefully. But there were no new prospects for high-temperature superconductivity.

By this time, superconductors had extended the range of application, but the need to cool them with liquid helium remained the weak spot.

In the mid-1970s, strange ceramic compounds of the type $Pb - Ba - O$ appeared as candidates for high-temperature superconductivity. In their electrical properties, they were "poor metals" at room temperature but became superconducting not too far from absolute zero. "Not too far" means about 10 degrees below the record value of the time. But the new compound could hardly be called a metal. According to theory, the obtained value of the critical temperature was not by any means low, but surprisingly high for such substances.

This attracted attention to ceramics as would-be high-temperature superconductors. Since 1983, Müller and Bednorz worked like alchemists with hundreds of different oxides, varying their composition, quantity, and conditions of synthesis. According to Müller, they were led by some physical ideas that are now being validated by the most complicated experimental studies of the new materials. In this painstaking way, they stealthily approached the compound of barium, lanthanum, copper, and oxygen that showed superconducting properties at $35\,K$.

30.5 Quasi-two-dimensional superconductivity between antiferromagnetic and metallic states

A fair number of various chemical compounds exhibiting temperatures of superconducting transition higher than the record from 1973 have been synthesized by now. The chemical formulae of some that have critical

Temperature, K Reference temperature	Superconducting materials	Critical temperature T_c, K
220	H_3S	$T_c = 204$ K
200		(under the pressure of $90 \cdot 10^4$ atm)
180 — Record registered, at the "Vostok" station in Antarctica		≈ 20 K
160	$HgBa_2Ca_2Cu_3O_8$	$T_c = 164$ K (under the pressure of $30 \cdot 10^6$ Pa)
120 — Night temperature of lunar surface	$HgBa_2Ca_{n-1}Cu_nO_{2n+2}$	$n = 3$ $T_c = 132$ K
	$Tl_2Ba_2Ca_{n-1}Cu_nO_{2n+3}$	$n = 4$ $T_c = 122$ K
100	$Bi_2Sr_2Ca_{n-1}Cu_nO_{2n+4}$	$n = 3$ $T_c = 110$ K
— Oxygen boiling point		
80 — Nitrogen boiling point	$YBa_2Cu_3O_{6+x}$	$T_{c\,max} = 92$ K
60		
— Temperature of Pluto surface	$La_{2-x}Sr_xCuO_4$	$T_c = 40$ K $T_c = 23$ K
40	$Ba_{1-x}K_xBiO_3$	$T_c = 30$ K
— Hydrogen boiling point	Rb_3C_{60} Nb_3Ge	
20	Pb Nb	$T_c = 9$ K $T_c \approx 7$ K
— Helium boiling point	Hg	$T_c = 4.2$ K
	Al, Ga, Zn	

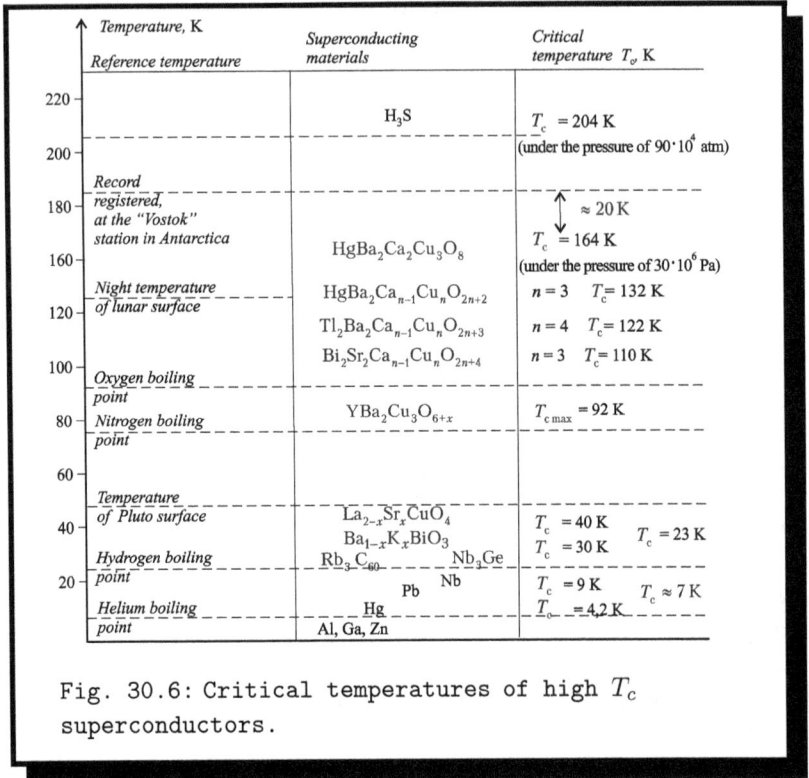

Fig. 30.6: Critical temperatures of high T_c superconductors.

temperatures above the boiling point for liquid nitrogen are summarized in Figure 30.6.

The feature shared by high-temperature superconductors is the layered structure. The best studied high-temperature superconductor is by now the compound $Y Ba_2 Cu_3 O_7$. Its crystal structure is illustrated in Figure 30.1 on page 295. It is easily seen that atoms of copper and oxygen are arranged in planes interspaced by others atoms. As a result the conducting layers are separated by insulating ones and the motion of charge carriers (those are usually not electrons but holes) is *quasi-two-dimensional*. Namely, holes migrate freely within $Cu O_2$ layers although hops between the layers are comparatively rare. Cooper pairs are also localized in the planes.

Apparently the quasi-two-dimensional nature of the electron spectrum in high-temperature superconductors is a key to the understanding of the microscopic mechanism of this wonderful phenomenon. This has yet to be realized. Nevertheless a brilliant phenomenological theory of the vortex state in high-temperature superconductors is already at hand. It proved so rich with diverse effects that now, in fact, it constitutes a new realm of physics, i.e., the physics of "vortex matter". A bedrock of that is the quasi-two-dimensionality of the electron liquid.

Indeed, once electrons and Cooper pairs are confined in two dimensions Abrikosov vortices consist of elementary vortices which are attached to the conducting planes. These elementary vortices are known among physicists as "pancakes". At low temperature the pancakes draw up in a line due to a weak attraction between them. Then the lines form a vortex lattice. As temperature rises thermal fluctuations make the vortex lines more twisted and at some point the vortex lattice melts almost as if it was an ordinary crystal. Thus in a high-temperature superconductor the ordered Abrikosov lattice gives way to a disordered "vortex liquid" phase formed by chaotically twisting tangled vortex lines. It is interesting that the further increase of temperature may break apart vortex lines and cause vortices to "evaporate" but at the same time preserve the superconductivity. Elementary vortices in the layers will become absolutely independent of each other and of vortex configurations in neighboring planes. Inhomogeneities of various kinds that are inevitably present in real crystals make the phase picture of the vortex matter even more complicated.

Despite significant progress in understanding of properties of high-temperature superconductors the mechanism behind the effect remains a secret. No less than 20 conflicting theories proclaim having explained high-temperature superconductivity but what we need is the one that is true.

Some physicists believe that Cooper pairs in these superconductors are formed because of a magnetic fluctuation interaction of some sort. An indication may be the fact that the critical temperature and the concentration of free electrons simultaneously drop down in crystals, $Y Ba_2 Cu_3 O_{6+x}$, impoverished in oxygen, that is at $x < 1$, Figure 30.7, (the right curve). At $x < 0.4$ one deals already with a dielectric, but at low enough temperatures a magnetic ordering of copper atoms takes place. The magnetic moments of neighboring atoms become anti-parallel and the total magnetization of the crystal stays zero. This type of ordering is well known in the physics of magnetism where it is called *antiferromagnetic ordering* (see the left curve in Figure 30.7; here T_N is the so-called Néel[P] temperature, that

---

[P]L. Néel, (1904–2000), French physicist; 1970 Nobel Prize in physics.

is the temperature of the transition to the antiferromagnetic state). One could believe that copper atoms retain the fluctuating magnetic moment in the superconducting phase and in the long run this gives origin to the superconducting attraction of electrons. This mechanism leans on specific properties of copper atoms which, depending on the valence, are either magnetic or not. The presence of $Cu - O$ layers in all high-temperature superconductors could be considered as an argument in favor of the theory. However quite recently the superconductivity of $W_3\,O\,Na_{0.05}$ at $90\,K$ was reported. The exact composition of the superconducting phase is not known yet; or for certain whether there are no "magic" copper atoms in it. Moreover, none of the elements in the formula of the new high-temperature superconductor show magnetic properties.

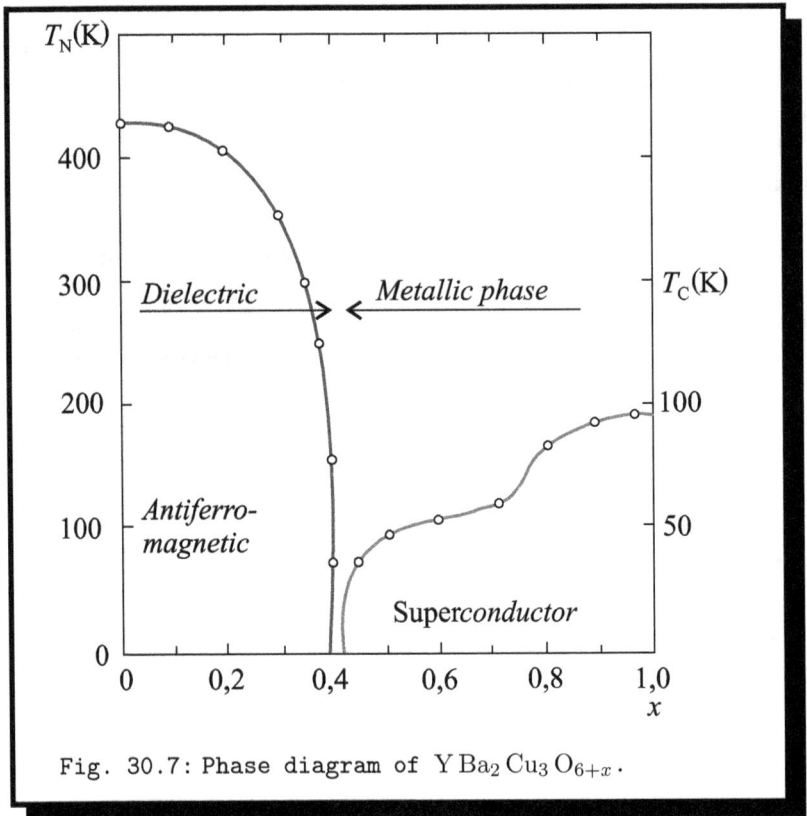

Fig. 30.7: Phase diagram of $Y\,Ba_2\,Cu_3\,O_{6+x}$.

In other theories physicists try to generalize in one way or another the classical theory of superconductivity, revise the very basics of the theory of the metallic state, "crossbreed" superconductivity and antiferromagnetism in spaces of higher dimension, separate the spin and charge of carriers, preform Cooper pairs in advance at temperatures higher than the critical one, and undertake other attempts to explain the unusual properties of the high-temperature superconductors in a universal manner.

The gauntlet thrown down by Nature itself remains unchallenged; theoretical physicists still cannot come to an agreement. In their quest for the truth they create beautiful "toy models", apply them to the experimental data and go away from them when they do not fit with the purpose to proceed further with their research. Sometimes, these abandoned "toys" turn into valid theories in quite unexpected areas of physics.

30.6   Superconductivity for penguins

Recently, after 30 years of research in the realm of high-temperature superconductors as well as the attempts to explain their superconductivity, a group of German scientists headed by M. Eremets took a look back into the past and came across a scientific research conducted by American physicist N. Ashcroft 50 years ago. He predicted the possible existence of superconductivity at room temperature in metallic... hydrogen. Yes, in metallic hydrogen! At normal conditions, we are used to having hydrogen in a gaseous state, after cooling at atmospheric pressure up to $-252.87°\,C$ in a liquid state, and at $-259.14°\,C$ in a solid state. But, it would still be a dielectric. It appeared that in 1935 Wigner and Huntington offered a hypothesis that at a super high pressure of about $2,500,000$ atm it is possible to expect hydrogen to transition into the metallic state. The role of ions in the lattice sites of such a metal are played by the protons. They possess a very small mass and, respectively, a high frequency of zero oscillations (see Chapter 27). This leads to an increase in propagation velocity and characteristic frequency of sound waves running along the lattice. According to the classical theory of superconductivity the quanta of the sound waves serve as a "glue" joining electrons into Cooper pairs. The specific phonons, corresponding to the sound waves in metallic hydrogen, turn out to be able to bind electrons stronger than in regular superconductors, and for this reason Ashcroft considered metallic hydrogen as the most likely candidate for a high temperature superconductor. Experiments with hydrogen metallization were going on for about half a century. Wigner and Huntington's estimate was too optimistic — today scientists hope to obtain metallic hydrogen at a pressure of the order of $4,000,000 - 5,000,000$ atm. It is way above the pressure in the center of the Earth, which is $3,500,000$ atm.

The authors of an article published in Nature magazine[q] instead of hydrogen embarked on studying sulphur hydride, which, according to their observation, was transformed into metal under pressure of $900,000$ atm. When this system was cooled down to $203\,K$ $(-70°\,C)$, the authors noticed clear signs of superconductivity: a sharp drop of resistance down to zero, and also a drop of the observed critical temperature under the effect of a magnetic field. On top of it all, when hydrogen in the compound was replaced with deuterium, the authors discovered a strong isotopic effect, which is indicative of an electron-phonon mechanism in the detected superconductivity.

This discovery creates hope that the much anticipated room temperature superconductivity will be achieved in the near future.

---

[q]A.P. Drozdov, M. I. Eremets, I. A. Troyan, V. Ksenofontov and S. I. Shylin, Nature, 525, 73-76 (2015).

Chapter 31

What is a SQUID?

In Montmartre, on the fourth floor of number 75 Rue
Orchampt, there once lived a fine fellow named Du-
tilleul who had the remarkable gift of being able to pass
through walls with perfect ease.

Marcel Aymé, *The Walker-Through-Walls.*

31.1 The quantization of magnetic flux

In the microworld, that is the world of atoms, molecules, and elementary
particles, many physical quantities may take only definite discrete values.
Physicists say that they are quantized (we have already mentioned that
according to the Bohr rule, energies of electrons in atoms are quantized.)
Macroscopic bodies consist of big collections of particles and random thermal motion leads to an averaging of physical quantities. This smears the
little steps and conceals quantum effects at a macroscopic level.

Now, what if the body is cooled to a very low temperature? Then arrays
of microparticles can move in accord and reveal quantization at macroscopic
scales. A good example is the fascinating phenomenon of quantization of
magnetic flux.

Everyone who has studied the laws of electromagnetic induction knows
the definition of the magnetic flux through a closed contour:

$$\Phi = B\,S,$$

where B is the value of magnetic field and S is the area encircled by the
contour (for simplicity let the field be normal to its plane). Nevertheless

it will be a discovery for many that magnetic flux produced by superconducting current, say, in a ring may assume only discrete values. Let us try
to understand this at least superficially. Presently it is sufficient to believe
that microparticles are moving along in quantum orbits. This simplified
image often substitutes probability clouds in blackboard discussions.

The motion of superconducting electrons in a ring, Figure 31.1, resembles that of electrons in atoms: it seems that electrons follow gigantic orbits
of the radius R without any collisions. Therefore a natural assumption is
that their motion obeys the same rule as in atoms. The Bohr postulate
states that only certain orbits of electrons are stationary and stable. They
are selected by the following quantization rule: products of the momentum
of an electron mv and the radius of the orbit R (this quantity is called the
angular momentum of the electron) form a discrete sequence:

$$mv\,R = n\,\hbar. \qquad (31.1)$$

Here n is a natural number and $\hbar$ is the minimal increment *(quantum)* of
the angular momentum, equal to the Planck constant $\hbar$. We have already
met it when talking about the uncertainty principle. It turns out that
the quantization of all physical quantities is determined by this universal
constant.

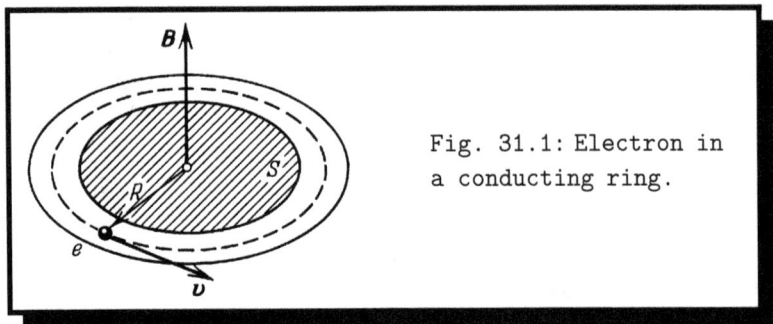

Fig. 31.1: Electron in
a conducting ring.

Let us find the value of the magnetic flux quantum. Consider a single
electron and let the magnetic flux through the ring gradually increase. As
you know the electromotive force of induction appears:

$$\mathcal{E}_i = -\frac{\Delta\Phi}{\Delta t},$$

and the strength of the electric field is

$$E = \frac{\mathcal{E}_i}{2\pi R} = -\frac{\Delta\Phi}{2\pi R \, \Delta t}.$$

By Newton's second law the acceleration of a charged particle is:

$$ma = m\frac{\Delta v}{\Delta t} = -\frac{e \, \Delta\Phi}{2\pi R \, \Delta t},$$

where e is the electric charge. After the obvious cancelation of Δt we obtain:

$$\Delta\Phi = -\frac{2\pi m \, \Delta v \, R}{e} = -\frac{2\pi}{e}\Delta(m\,v\,R).$$

You see that magnetic flux across the ring is proportional to the angular momentum of electrons. According to the Bohr quantization rule (31.1) angular momentum may take only discrete values. This means that magnetic flux through a ring with superconducting current must be quantized as well:

$$m\,v\,R = n\,\hbar \qquad \text{and} \qquad \Phi = (-)\frac{2\pi}{e}\,n\,\hbar. \tag{31.2}$$

The value of the quantum is extremely small ($\sim 10^{-15}$ Wb) but twentieth-century techniques make it possible to observe magnetic flux quantization. The studies carried out in 1961 by the Americans B. S. Deaver and W. M. Fairbank used a superconducting current circulated not in a ring but in a hollow superconducting cylinder.[a] The experiment confirmed that the magnetic flux through the cylinder changed stepwise but the measured value of the quantum was half of that obtained above. The modern theory of superconductivity gives the answer. Remember that in the superconducting state electrons join into Cooper pairs with the charge $2e$. Superconducting current is a motion of these pairs. Therefore the correct value of the magnetic flux quantum Φ_0 is obtained when substituting into the formula (31.2) the electric charge $2e$ of a pair:

$$\Phi_0 = \frac{2\pi\,\hbar}{2e} = 2.07 \times 10^{-15} \; Wb.$$

This is the way to recover the factor two. We were not the first to miss it. The English theoretician F. London had lost it as well. He predicted

---

[a]The quantization of magnetic flux was simultaneously observed by R. Doll and M. Näbauer. The two papers appeared in the same issue of *Physical Review Letters*.

the magnetic flux quantization as early as 1950, long before the nature of the superconducting state was understood.

It is worth saying that our derivation of the magnetic flux quantization certainly is too naive. It is rather surprising that we contrived to obtain the right meaning of the quantum this way. In fact superconductivity is a complicated quantum effect. Those who really want to comprehend it have a long and hard way ahead of them. It demands many years of resolute but rewarding work.

31.2 Josephson effect

Let us turn to another quantum superconducting phenomenon that lay a cornerstone for several unrivaled measuring methods. The Josephson effect was discovered in 1962 by a 22-year-old British graduate student and for this theoretical prediction he received the Nobel Prize 11 years later.[b]

Imagine a glass plate (that is called a substrate) supporting a super-conducting film. (Usually the superconducting material is sputtered in vacuum.) The surface of the film has been oxidized and the oxide forms a thin dielectric layer on it. Finally the superconductor is sputtered once again. The final outcome is a so-called superconducting sandwich interlaid by a thin insulating sheet. Sandwiches are widely used in observations of the Josephson effect. For convenience the two thin superconducting strips usually cross each other, see Figure 31.2.

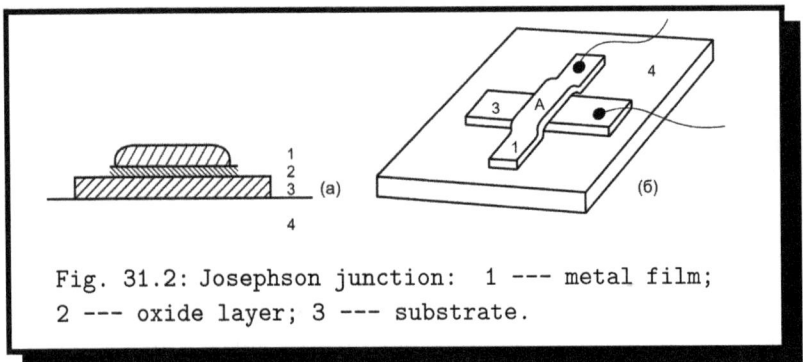

Fig. 31.2: Josephson junction: 1 --- metal film;
2 --- oxide layer; 3 --- substrate.

---

[b]B. D. Josephson, (born 1940), British physicist; 1973 Nobel Prize in physics.

We shall begin with a case when the metallic layers are in the normal, nonsuperconducting state. Is it possible for electrons to pass from one metallic film into another, Figure 31.3, *a*?

From the first sight, no, because of the dielectric in between. The dependence of the electron energy versus the *x*-coordinate (*X*-axis is perpendicular to the plane of the sandwich) is plotted in Figure 31.3, *b*. Electrons in metal move freely and their potential energy is zero. The potential energy of electrons in dielectric, W_u, surpasses their kinetic (and total) energy in metal W_e. The work to be done by electrons when exiting to the dielectric[c] is $W_u - W_e > 0$. Therefore one says that electrons in the two films are separated by the potential barrier of the height $W_u - W_e$. If the electrons

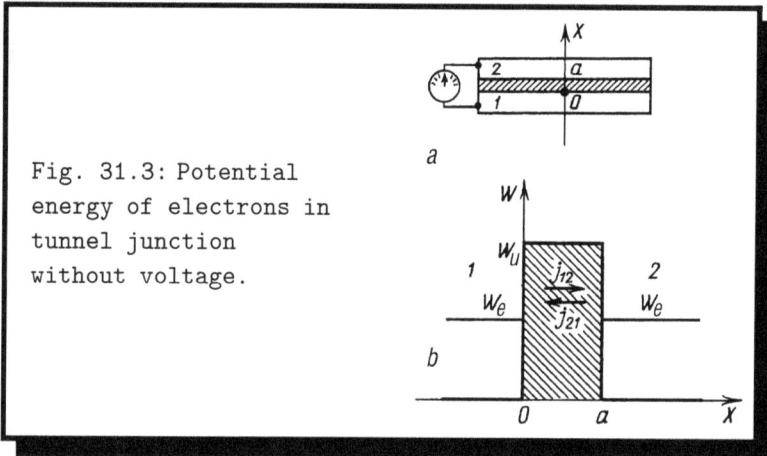

Fig. 31.3: Potential energy of electrons in tunnel junction without voltage.

obeyed the laws of classical mechanics the barrier would be insuperable. But electrons are microparticles and the specific laws of the microworld permit many things that would be ruled out for bigger bodies. For example neither man nor electron are able to mount a barrier higher than their energy. But an electron may simply penetrate through it! As if it tunneled under a mountain when the energy was not enough to climb it. This is called the tunnel effect or quantum tunneling. Of course you should not take this literally like really digging a hole. The true explanation comes from the wave properties of microparticles and their "spreading" in space,

---

[c]This resembles the heat of evaporation, that is the work done when extracting a molecule from liquid.

a deep understanding of which requires good command of quantum mechanics. But in a word electrons can with some probability pass through the dielectric from one metal film to another. The probability increases for smaller heights $W_u - W_e$ and widths a of the barrier.

Once the dielectric film is permeable for electrons we may ponder the electric current flowing through it. At the moment this so-called tunnel current is zero: the number of electrons coming to the upper electrode from below is equal to those going back.

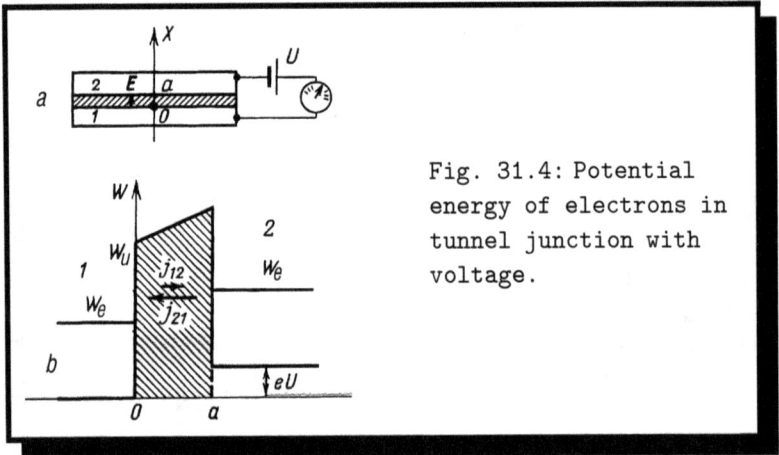

Fig. 31.4: Potential energy of electrons in tunnel junction with voltage.

What should we do to make the tunnel current nonzero? Simply break the symmetry. For example, let us connect the metal films to a battery with the voltage U, Figure 31.4, a. Then the films will act like two plates of a capacitor and the electric field of the strength $E = U/a$ will set up in the dielectric layer. The work done when moving a charge e a distance x along the field is $A = F\,x = eE\,x = eU\,x/a$ and the potential energy of electrons takes the form plotted in Figure 31.4, b. Evidently electrons from the upper film ($x > a$) penetrate the barrier more easily because those moving from below must jump to the higher level. Therefore even small voltages break the balance and give rise to a tunnel current.

Tunnel junctions of normal metals are used in electronic devices but do not forget that our aim was the practical applications of superconductivity. The next step is to assume that the metal strips separated by the insulating layer are superconducting. How does the superconducting tunnel junction behave? It turns out that superconductivity leads to quite unexpected results.

We said that electrons in the upper film possess the surplus energy eU with respect to the lower one. Upon coming down they must dump the energy and come to equilibrium with the others. This was not a problem in the normal state: several collisions with the crystal lattice would redistribute the extra energy and convert it to heat. But if the film is superconducting this way is not acceptable! It must emit the energy in the form of a quantum of electromagnetic radiation. The energy of the quantum is proportional to the applied voltage U:

$$\hbar \omega = 2e\, U.$$

You see that the electric charge in the right-hand side is twice that of the electron. This indicates that tunneling of superconducting pairs has taken place.

This was the dazzling prediction by Josephson. Applying constant voltage to a superconducting tunnel junction (sometimes called the Josephson junction) brings about the generation of electromagnetic radiation. The first experimental observation of this effect was performed in 1965 by I. M. Dmitrienko, V. M. Svistunov, and I. K. Yansons at the Kharkov Physical-Technical Institute of Low Temperatures.

The first use that comes to mind for the Josephson effect is the generation of electromagnetic waves. However it is rather difficult to extract the radiation from the narrow space between the superconducting films (this was a serious obstacle to experimental observation of the effect). Besides the emission is too weak. Now Josephson elements are used mainly as detectors of electromagnetic radiation because they are the most sensitive ones in certain frequency ranges.

This application exploits the resonance between the frequency of the external (registered) wave and the proper frequency of oscillations in the junction under a voltage. The idea of resonance is basic for most receivers: a set is "tuned in" when the proper frequency of the receiving contour is adjusted to that of the station. A Josephson junction makes a convenient receiving cell. The two advantages are: first, the frequency depends on the voltage and is easily varied; second, the resonance being very sharp results in high selectivity and precision. Josephson elements are employed in the most sensitive detectors for the observations of the electromagnetic radiation of the Universe.

31.3   The quantum magnetometer

The Josephson effect together with magnetic flux quantization provide a basis for a whole family of supersensitive measuring devices called *SQUIDs*. This abbreviation stands for *Superconducting Quantum Interference Devices*. We will now look at the quantum magnetometer, which measures weak magnetic fields.

The simplest quantum magnetometer consists of a superconducting ring with a Josephson junction, Figure 31.5. As you know, in order to create a current through a normal tunnel contact one must apply some voltage. But for a superconducting junction this is not necessary. Superconducting pairs may tunnel through the insulating layer and the superconducting current may circulate in the ring regardless of the Josephson junction. This is called the stationary Josephson effect. (In distinction to the nonstationary Josephson effect accompanied by emission that was described in the previous section.) However the current is limited by a maximum allowable value called the critical current of the junction, I_c. Currents exceeding I_c destroy the superconductivity of the junction and a voltage drop appears across it. The Josephson effect becomes nonstationary.

So, the insertion of a Josephson junction does not completely destroy superconductivity of the contour. Nevertheless a segment of imperfect

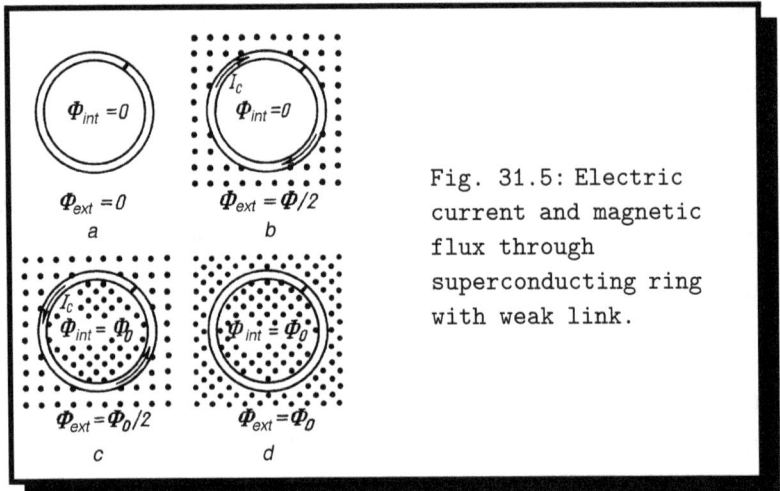

Fig. 31.5: Electric current and magnetic flux through superconducting ring with weak link.

superconductivity, the so-called weak link, appears. It plays a crucial role in the operation of a quantum magnetometer. Let us try to understand this.

If the entire contour was superconducting the magnetic flux through it, Φ_{int}, would be strictly constant. Indeed, by the law of electromagnetic induction any change of external magnetic field gives rise to the electromotive force of induction, $\mathcal{E}_i = -\Delta\Phi_{\text{ext}}/\Delta t$, that effects the electric current. The change of the current in its own turn generates the electromotive force of self induction, $\mathcal{E}_{si} = -L\,\Delta I/\Delta t$. The resistance of the superconducting contour and the voltage drop in it are zero:

$$\mathcal{E}_i + \mathcal{E}_{si} = 0,$$

and

$$\frac{\Delta\Phi_{\text{ext}}}{\Delta t} + L\frac{\Delta I}{\Delta t} = 0.$$

Remember that the magnetic flux through the contour that arises due to the current I is $\Phi_I = LI$. This means that $\Delta\Phi_{\text{int}} = \Delta\Phi_{\text{ext}} + \Delta\Phi_I = 0$ and the change of the superconducting current compensates for the change of the external field. The total magnetic flux through the contour remains constant, $\Phi_{\text{int}} = \Phi_{\text{ext}} + \Phi_I = const$. There is no way to change it without transferring the contour to the normal state. The magnetic flux is "frozen".

What happens if the contour contains a weak link? Then magnetic flux through a contour may change since the weak link allows magnetic quanta to penetrate inside the ring. (You remember that magnetic flux encircled by a superconducting current is quantized and is equal to a whole number of the quanta Φ_0.)

Let us watch the magnetic flux through a superconducting ring with a weak link and the electric current in it as the external magnetic field changes. Let the initial external field and current be zero, Figure 31.5, *a*. Then the magnetic flux through the ring is zero as well. If we enhance the external magnetic field a superconducting current will arise and the external flux will be completely compensated. This will happen until the electric current reaches the critical value I_c, Figure 31.5, *b*. To be definite we shall assume that this happens when the flux of the external field equals one half of the quantum: $\Phi_0/2$.[d]

---

[d]Critical current depends on many factors and in particular on the thickness of the dielectric. It is always possible to fit the latter so that the flux created by the critical current has the desired value: $LI_c = \Phi_0/2$. This simplifies the analysis but does not affect the essentials.

As soon as the value of the current reaches I_c the superconductivity of the weak link is destroyed and the quantum of magnetic flux Φ_0 enters the ring, Figure 31.5, *c*. The ratio Φ_{int}/Φ_0 stepwise increases by unity. (The superconducting contour passes to the next quantum state.) And what about the current? The value remains the same but the direction reverses. Judge for yourself, formerly the external flux was compensated by the field of the current: $\Phi_I + \Phi_{ext} = -L\,I_c + \Phi_0/2 = 0$. After the quantum has entered the ring the current and the external flux add up: $\Phi_{ext} + \Phi'_I = \Phi_0/2 + L\,I'_c = \Phi_0$. Thus, allowing the flux quantum in the system has instantaneously changed the direction of the current.

As the external field grows the current in the ring decreases and the superconductivity of the junction is restored. When the external flux is Φ_0 the current disappears, Figure 31.5, *d*. After that it changes direction again in order to screen the excess of magnetic flux. Finally, when the external flux comes to $3\Phi_0/2$ the current becomes I_c, the superconductivity of the junction is destroyed, and one more quantum of magnetic flux enters the ring, etc.

The dependencies of the magnetic flux across the ring, Φ_{int}, and the electric current I versus the external magnetic flux Φ_{ext} are shown in Figure 31.6. Both fluxes are measured in magnetic quanta Φ_0 which represents the natural unit. The stepwise shape of the dependence offers the possibility of "counting" individual flux quanta despite their extremely small $(2.06783383 \cdot 10^{-15}\ Wb)$ magnitude. The reason is quite clear. Even though the magnetic flux through the superconducting contour changes by a tiny amount $\Delta\Phi = \Phi_0$ this happens in a very short time, Δt, almost instantly. Therefore the rate $\Delta\Phi/\Delta t$ during this abrupt change may be really big. It can be measured, for example, by the electromotive force induced in a special measuring coil of the device. This is the principle of operation of a quantum magnetometer.

Construction of a real quantum magnetometer is much more complicated. Say, usually not one but several weak links are connected in parallel. This gives rise to interference of superconducting currents (or, to be exact, the interference of the quantum waves that determine locations of electrons). This helps to increase the precision of measurements. The collective name *SQUID* of such devices refers to the interference of quantum waves. The sensitive element of the device is inductively coupled with an oscillatory contour where the jumps in magnetic flow are converted into electric impulses to be amplified later. But these technical subtleties are far beyond the scope of the book.

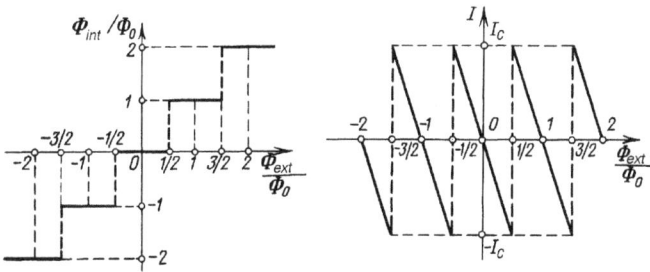

Fig. 31.6: Magnetic flux *(a)* through a superconducting ring with weak link and electric current *(b)* in it as a function of external magnetic flux.

The fact is that supersensitive magnetometers capable of measuring magnetic fields with 10^{-15} T accuracy are now widely applied in industrial production. Among other things they are used in medicine. It turns out that the working heart, brain, and muscles create weak magnetic fields. For example the magnetic induction due to activity of the heart is $B \approx 3 \cdot 10^{-11}$ T being a million times less than the field of the Earth. But still these fields, however weak they are, lie within the reach of SQUIDs. Records of rhythms of these fields are called magnetocardiograms, magnetoencephalograms, etc. Superconducting facilities offer new possibilities to register and study the most delicate signals of the human organism. This was a breakthrough in medical diagnostics for many diseases.

Experiments in the field started in the 1970s. In order to minimize the influence of the magnetic field of the Earth measurements were carried out in specially designed screened chambers. Their walls were made of three layers of metal with high magnetic permeability that presented efficient magnetic screening plus two layers of aluminum in between for electric screening. These precautions provided the means to reduce the magnetic field inside the chamber to several *nanoTesla* ($1\,nT = 10^{-9}\,T$) that is a thousand times less than that of the Earth. Clearly such chambers cost a fortune. Further development of this promising realm of SQUID application led to remarkable progress and greatly simplified the procedure. Modern superconducting techniques permit taking a distinct magnetocardiogram with no screens at all, Figure 31.7. The only imperative condition is to remove metal clips and the content of your shirt pocket.

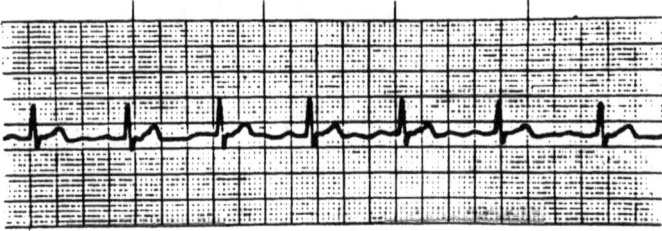

Fig. 31.7: A magnetocardiogram.

Chapter 32

Superconducting magnets

But the greatest curiosity, upon which the fate of La-
puta depends, is a loadstone of a prodigious size, in
shape resembling a weavers shuttle... By means of this
loadstone, the island is made to rise and fall, and move
from one place to another. For, with respect to that
part of the earth over which the monarch presides, the
stone is endued at one of its sides with an attractive
power, and at the other with a repulsive.

Jonathan Swift, *Gulliver's Travels.*

Strong magnetic fields can be obtained by passing strong electric currents through a coil. The greater the current the bigger is the field. If
the coil possesses electrical resistance, heat is released as the current flows.
Supporting the current requires enormous energy and, besides, one must
dissipate the heat which otherwise may damage the coil. In 1937 a magnetic field with the induction $10\,T$ was first realized. The field could be
supported only at night when few consumers were using power. The lib-
erated heat was removed by running water and 5 liters ($1.3\,gal$) of it were
brought to the boil every second. The heat release set the main limitation
to creating strong magnetic fields with ordinary coils.

As soon as superconductivity was discovered the idea appeared to exploit it in the production of strong magnetic fields. At first glance the
only thing to be done is to wind up a coil of superconducting wire, send
a sufficiently strong current, and short the circuit. Once the resistance of
the coil is zero no heat is released. The gains would justify the work done
when cooling the solenoid down to the temperature of liquid helium. But

unfortunately, the strong magnetic field destroyed the superconductivity of superconductors that were known at that time.

A way out was found. The help came from the laws of quantum mechanics. As you know, in superconductivity these laws may work on macroscopic scales.

32.1   The Meißner effect in detail

In Figure 32.1 you can see the scheme of the experiment that was performed by Kamerlingh Onnes in 1911 in Leiden. The Dutch scientist put a lead coil into liquid helium where it cooled down to helium's boiling point. The electrical resistance of the coil disappeared because it achieved the superconducting state. After that he reconnected the switch and closed the coil onto itself. The undamped current began to circulate in the coil.

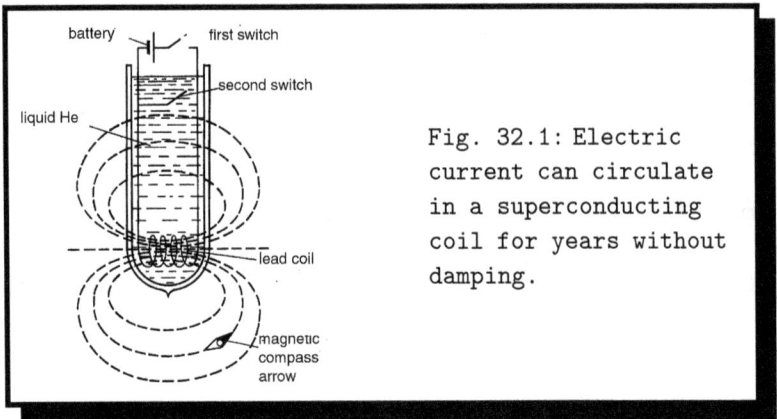

Fig. 32.1: Electric current can circulate in a superconducting coil for years without damping.

The current generates a magnetic field with the induction proportional to its strength. A naive assumption is that the larger the current in the coil the bigger the magnetic field it produces. But the results were discouraging: as the field reached several hundredths of a *Tesla* the solenoid passed to the normal state and electrical resistance appeared. Attempts were made to prepare coils of other superconductors but in those again superconductivity was destroyed at relatively weak fields. What was the rub?

The puzzle of this inconvenient behavior of superconductors was solved in 1933 in the laboratory of W. Meißner in Berlin. It was found that superconductors possess the property of expelling magnetic fields; the induction inside superconductors is zero. Imagine that a metal cylinder (a piece of wire) was cooled and became superconducting. Then one switched on a magnetic field with the induction $\vec{B}_{ext}$. By the law of electromagnetic induction this must cause circular currents at the surface of the cylinder, Figure 32.2. The magnetic field $\vec{B}_{cur}$ created by the currents inside the cylinder is equal to $\vec{B}_{ext}$ in magnitude but opposite in direction. The currents are superconducting and do not die out. Therefore the net induction in the superconductor is zero: $\vec{B} = \vec{B}_{ext} + \vec{B}_{cur} = 0$. Lines of magnetic induction do not penetrate superconductors.

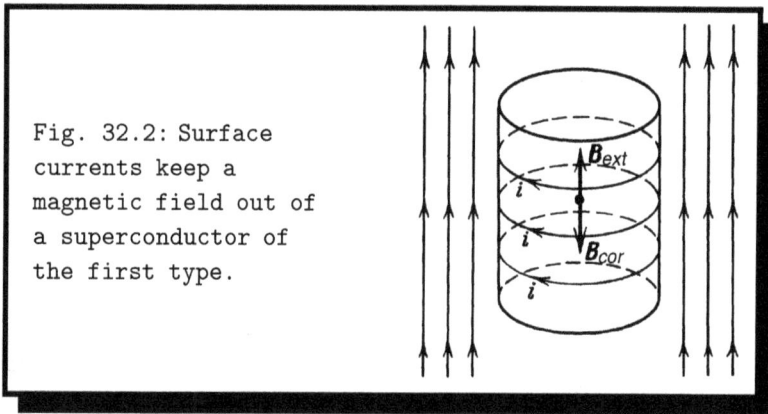

Fig. 32.2: Surface currents keep a magnetic field out of a superconductor of the first type.

But what if we change the order and apply the field before cooling the specimen to the superconducting state? It seems that the magnetic induction will not change and there would be no point in generating surface currents. This was the logic of Meißner when he checked calculations by Laue[a] concerning the first experimental procedure. But still he preferred to check. The result of the renewed experiment was stunning. It turned out that the magnetic field was just the same, forced out of the superconductor without penetrating it. This was called the Meißner effect.

Now it is clear why a magnetic field destroys superconductivity. Exciting surface currents takes energy. In this sense the superconducting state is less favorable than the normal one when the magnetic field enters the bulk and there are no surface currents. The higher the induction of the external

---

[a]M. von. Laue, (1879–1960), German physicist; 1914 Nobel Prize in physics.

field the stronger the screening current it demands. At some value of magnetic induction the superconductivity will inevitably be destroyed and the metal will return to normal. The value of the field when the destruction of superconductivity occurs is called the critical field of the superconductor. It is important that the presence of an external field is not a necessary condition of the destruction. Electric current in the superconductor produces a magnetic field of its own. At a certain intensity of current the induction of the field reaches a critical value and the superconductivity breaks down. The value of the critical field increases at low temperatures but even near absolute zero critical fields of pure superconductors are modest, see Figure 32.3. So it could seem a vain hope to obtain strong magnetic fields with the help of superconductors.

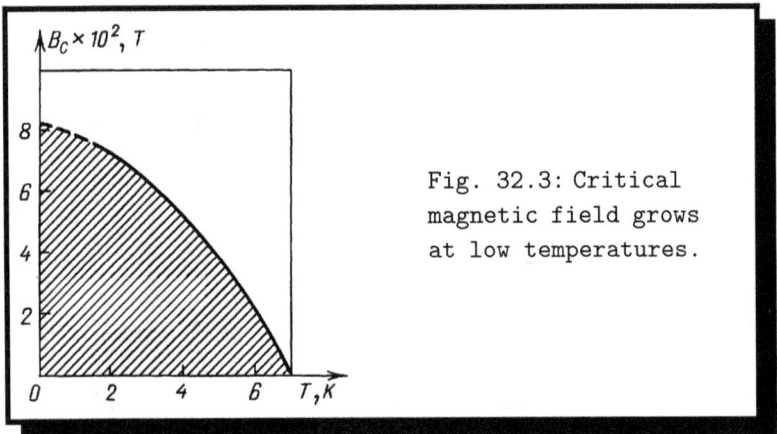

Fig. 32.3: Critical magnetic field grows at low temperatures.

But further investigations in the field proved that the situation is not desperate. It was found that there is a whole group of materials that stay superconducting even in very strong magnetic fields.

32.2   The Abrikosov vortices

As was already mentioned above, in 1957 the Russian theoretical physicist A. A. Abrikosov (see page 307) showed that a magnetic field does not destroy superconductivity of alloys so easily. Similar to the pure case, the magnetic field begins penetrating the superconductor at some critical value of induction. But in alloys the field does not occupy the entire volume of the superconductor at once. At first only detached bundles of magnetic lines are formed in the bulk, Figure 32.4. Every bundle carries an exactly fixed portion. It is equal to the quantum of magnetic flux, $\Phi_0 = 2 \cdot 10^{-15}$ Wb, that we have already met.

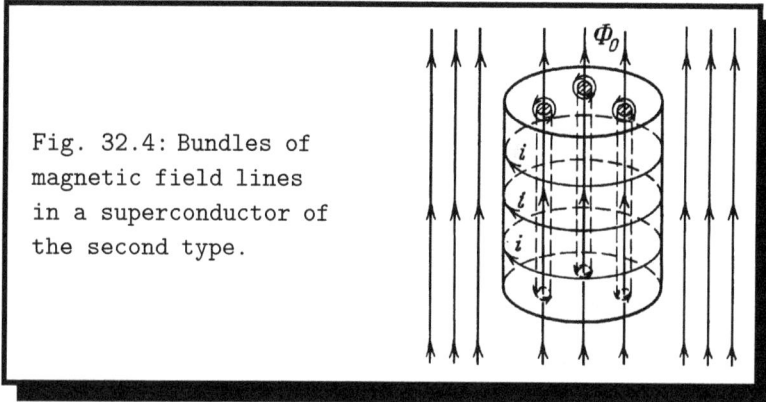

Fig. 32.4: Bundles of magnetic field lines in a superconductor of the second type.

The stronger the magnetic field the more bundles enter the superconductor. Each of them brings one magnetic quantum and the total flux changes stepwise. Again, as before, magnetic flux through a superconductor may take only discrete values. It is astonishing to see the laws of quantum mechanics working on macroscopic scales.

Each bundle of magnetic field lines piercing the superconductor is enveloped by undamped circular currents that resemble a vortex in a gas or liquid, Figure 32.4. For this reason the bundles of magnetic lines together with the superconducting currents around it are called Abrikosov vortices. Certainly in the core of the vortex, the superconductivity is broken. But in the space between the vortices it is conserved! Only in very strong fields when numerous vortices begin overlapping is the superconductivity destroyed completely.

This remarkable reaction of superconducting alloys to magnetic fields was first discovered at the tip of a pen. But modern experimental techniques make it possible to observe Abrikosov vortices directly. Fine magnetic powder is applied to the surface of a superconductor (for example, to the base of a cylinder). The particles gather at the places where the field enters the alloy. Electron microscope study of the surface reveals the dark spots.

Such a photograph of the structure of Abrikosov vortices is shown in Figure 32.5. We notice that the vortices are arranged periodically and form a pattern similar to a crystal lattice. The vortex lattice is triangular (this means that it is made up of periodically repeated triangles).

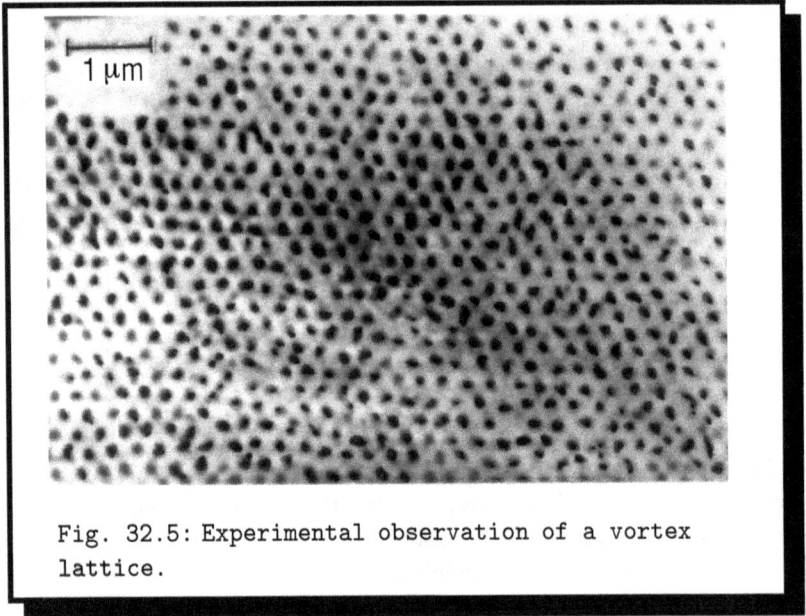

Fig. 32.5: Experimental observation of a vortex lattice.

So, in distinction to pure metals, these alloys possess not one but two critical fields: the lower critical field marks the moment when the first vortex enters the superconductor and the upper critical field corresponds to the complete destruction of superconductivity. Over the interval between these two the superconductor is pierced by vortex lines. This is called the *mixed state*. Superconductors exhibiting such properties are now called "type two". The first type refers to those where the magnetic destruction of superconductivity happens immediately.

It looked like the problem of producing superconducting magnets was solved. But nature had kept for researchers one more catch. The wire for superconducting solenoids must withstand not only strong magnetic fields but strong electric currents as well. As we shall see in the next section this happened to be different.

32.3 What is pinning?

It is well known that a force acts on an electric current in a magnetic field. But where is the counteraction force that must appear according to Newton's third law? When the field is caused by another current then, no doubt, it experiences an equal force in the opposite direction (the interaction of energized conductors obeys Ampère's[b] law). But our case is more difficult.

A current that flows in a mixed state superconductor interacts with the magnetic field in the cores of vortices. This affects the distribution of current but the domains where the magnetic field concentrate do not remain intact either. They start moving. Electric current compels Abrikosov vortices to move!

The force exerted onto a current by magnetic field is perpendicular to the magnetic induction and to the conductor. The force acting on Abrikosov vortices is also perpendicular to the induction of the field and to the direction of the current. Suppose that a current traverses the superconductor depicted in Figure 32.5 from left to right. Then the Abrikosov vortices will move either up- or downwards depending on the direction of the magnetic field. However, the transport of the Abrikosov vortex across the superconductor is the motion of the normal non-superconducting core. It undergoes friction which results in heat and dissipation. Current in the mixed state superconductor meets a resistance just the same. It could look like these materials are no good for solenoids.

What was the solution? It is to block the motion and hold the vortices in place. Fortunately this is possible. One has simply to build defects into the superconductor and worsen it. Usually defects appear by themselves as a result of mechanical or thermal treatment. Figure 32.6 demonstrates an electron-microscope photograph of niobium nitride film. The critical temperature of the film is $15\,K$. It was obtained by means of sputtering metal onto a glass plate. One clearly discerns the granular (or rather columnar) structure of the material. It is not so easy for a vortex to jump the boundary of a grain. Hence up to a certain current strength, the so-called critical current, vortices stay in place and the electrical resistance is zero.

This phenomenon is known as *pinning*, because the vortices are pinned by defects.

---

[b]A. M. Ampère, (1775–1836), French physicist, one of the founders of classical electrodynamics.

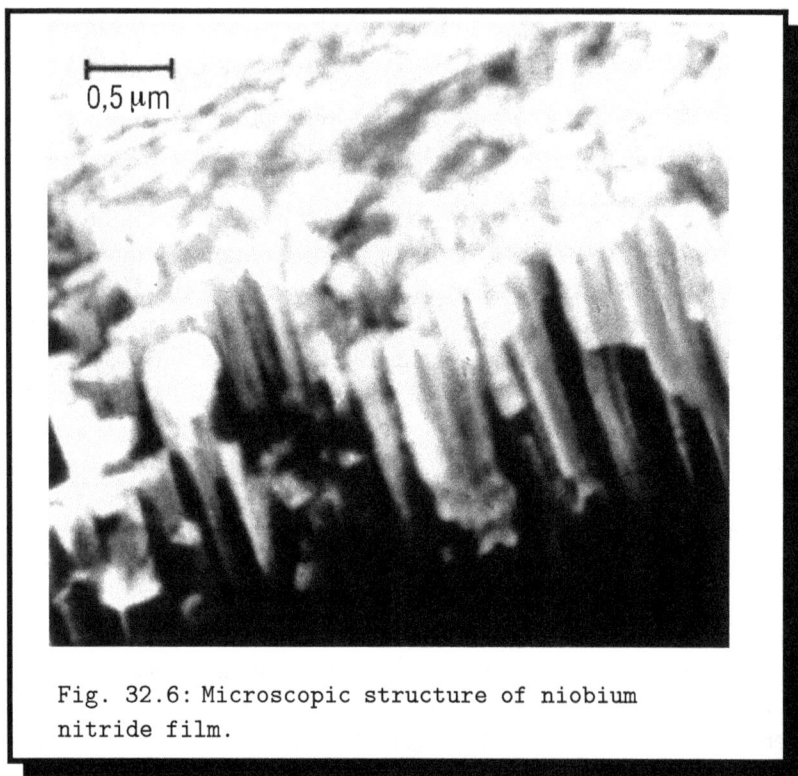

Fig. 32.6: Microscopic structure of niobium
nitride film.

Pinning offers the possibility of preparing superconducting materials exhibiting high critical values of both magnetic field and electric current. (It
would be more accurate to speak not of critical current but of critical current density, that is the current crossing a unit area of cross section.) The
critical field is determined by properties of the material. In the mean time
the critical current depends on methods used in preparation and treatment
of the conductor. Modern technology provides a means to obtain superconductors with high values of all critical parameters. For example, starting
from the tin-niobium alloy one can fabricate a cable with the density of the
critical current reaching several hundreds A/cm^2, with the upper critical
field equal to 25 T and the critical temperature being 18 K.

But this is not the end of the story. It is important that the mechanical
properties of the material allow it to make a coil. The tin-niobium alloy by
itself is too fragile and it would be impossible to bend such a wire. So the
following procedure was invented: a copper tube was stuffed with a mixture

of niobium and tin powders, then the tube was stretched into a wire and the coil was wound and heated making the powders fuse. This resulted into a solenoid of the Nb_3 Sn alloy.

Industry prefers more practical materials such as the more plastic niobium-titanium alloy Nb Ti. It is used as a base for so-called composite superconductors.

First one drills a number of parallel channels in a copper bar and inserts the superconducting rods. Then the bar is stretched into a long wire. The wire is cut and the pieces inserted into another drilled copper bar. That is once more stretched, cut into pieces, and so on... Finally one obtains a cable that contains up to a million superconducting lines, like those shown in Figure 32.7. This is used for winding coils.

The important advantage of such cables is that the superconducting current is distributed among all the lines. When compared to a superconductor copper behaves like an insulator. If copper and a superconductor are connected in parallel then the entire current will choose the path that has no resistance. There is a second advantage. Suppose that by accident superconductivity breaks down in one of the lines. This causes heat liberation and the danger that the whole cable will pass to a normal state. It is imperative to remove the heat. Copper is a good heat conductor and perfectly suits the purpose of thermal stabilization as well as having good mechanical properties for cable.

What are the specific issues involved in the use of high temperature superconductors (HTS) in superconducting wire manufacturing technology, and what improvements should be made in it as compared to existing traditional samples? Unfortunately, the brittleness of new materials constitutes a serious limitation. This circumstance is extremely important, because cable used in superconducting magnets is intended for the creation of very high fields and is subjected to the impact of enormous mechanical stresses.

Another limitation is a consequence of the weak pinning characteristics of HTS. The fact of the matter is that the benefits of using HTS are determined by their high critical temperatures. Unfortunately, the Abrikosov vortices in this region of temperatures appear to become systems of weakly connected disks instead of the elastic vortices due to the layered structure of perovskites. Their stiffness is loosened by thermal and quantum fluctuations, and the Abrikosov lattice is then melted and transformed into the vortex liquid (See Chapter 30).

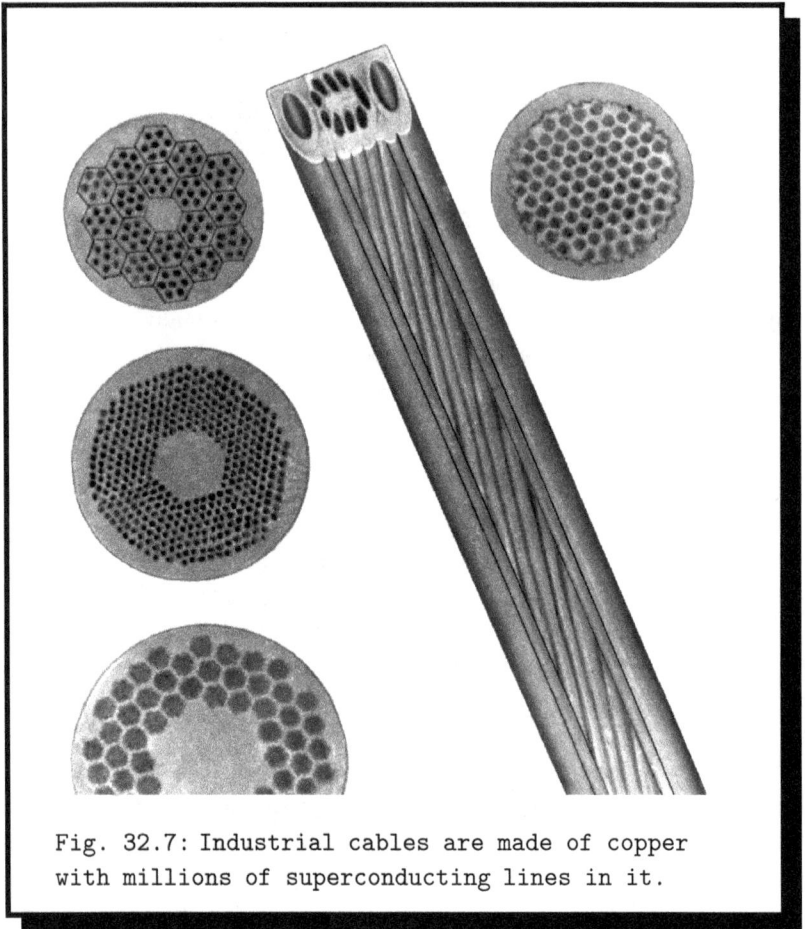

Fig. 32.7: Industrial cables are made of copper
with millions of superconducting lines in it.

Despite these fundamental limitations, the intensive research being carried out by the scientists in many countries during the last decades allowed to launch production of high-temperature superconducting cables, which found wide enough applications today. Thus, composite superconducting cables of up to a kilometer in length were created based on $Bi_2Sr_2Ca_2Cu_3 O_{8+x}$ superconductor mixed with silver.

Commercial cables with HTS are becoming more cost-effective than traditional ones when the transfered power is above 1 gigawatt. In Long Island (USA), a long high-temperature superconducting cable transferring 574 megawatts has been in operation for several years and provides electrical energy for around 300,000 homes. Cables used within the cities are, as a rule, layed underground; they can carry 3-5 times more energy than copper ones of the same diameter. In the Large Hadron Collider CERN (Switzerland) the enormous currents which feed the magnets are transferred through superconducting cables. Cables based on HTS are used to improve the parameters of devices used for MRI-tomography, in engines, generators, and other devices.

Superconducting tapes with high current carrying capacity are being created based on $YBa_2Cu_3O_{7-x}$ superconductor. In telecommunications, cellular telephony and in many other applications the superconducting tapes, whose critical current density reaches up to $10^7\,A/cm^2$, are being used. The kilometer length superconducting tapes based on middle temperature superconductor MgB_2 (critical temperature $T_c = 40\,K$) are manufactured in Italy. Magnetic fields up to $60\,T$ can be generated with their help.

32.4 *Postscriptum* for taxpayers

Let us turn to the prospects. These are really fantastic. Many global projects of the past are put back on the agenda because of the advent of high-temperature superconductivity. For example, at present 20–30 percent of all produced electrical energy is wasted in power transmission lines. Using high-temperature superconductors in energy transmission could eliminate these losses.

All projects involving thermonuclear synthesis need giant superconducting magnets that keep high-temperature plasma away from the walls of the chamber. Streams, if not rivers, of liquid helium are necessary to maintain the superconducting state. The helium would be replaced by nitrogen at a tremendous saving in cost.

Gigantic superconducting coils would serve as accumulators of electrical power, which would share the load during peak periods.

Supersensitive equipment for making magnetocardiograms and magnetoencephalograms, based on the use of superconducting Josephson elements, would come to every hospital.

Magnetic cushions created by superconducting coils would support intercity express trains commuting at speeds of $600 - 700\,Km/h$.

A new generation of supercomputers based on superconducting elements and cooled by liquid nitrogen would be constructed.

Do not think we have lost our heads over high-temperature superconductivity. Since its discovery, the ardor of many investigators has notably cooled down. The same happens when an Olympic record stays out of reach for years. But as soon as the record has been set it serves a benchmark. The possibility of producing materials with unique characteristics has been confirmed. Certainly economic considerations will effect realization of projects and it is unlikely that we will surpass the records tomorrow and make them a routine. But today we know for sure that the impossible has become accessible. And this has irreversibly changed the reference point in our attitude toward superconductivity.

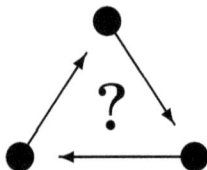

Why do superconducting transmission lines not require expensive high-voltage equipment?

Chapter 33

What is magnetic resonance imaging?

> Come, come, and sit you down; you shall not budge;
> You go not till I set you up a glass
> Where you may see the inmost part of you.

> William Shakespeare, "*Hamlet.*"

Today, there is nothing unusual about patients being given a prescription for Magnetic Resonance Imaging, or magnetic resonance tomography instead of an X-ray exam or Electrocardiography(ECG). To find out what stands behind these terms, we should start from the basics: understanding the magnetic nature of the nucleus. But, before this, we need to introduce some important concepts missing from the high school physics curriculum.

33.1 Magnetic moment

The magnetic properties of a small flat current loop placed into a magnetic field are determined by the magnetic moment of the current, which is equal to

$$\vec{\mu} = IS\vec{n},$$

where I is the current, S is the loop area, and $\vec{n}$ is normal to a loop vector, chosen by the rule-of-thumb (see Figure 33.1). In particular, the energy of a loop in a magnetic field with induction $\vec{B}$ equals

$$W = -\left(\vec{\mu} \cdot \vec{B}\right) = -\mu_z B \qquad (33.1)$$

(z-axis is directed along $\vec{B}$). To rotate the loop to give a change of vector $\vec{\mu}$ projection from μ_z to $-\mu_z$, work $A = 2\mu_z B$ has to be performed.

The electrons orbiting around the atomic nucleus may be considered equivalent to a circular current and a magnetic moment can be assigned to it. The fact that electrons may have such an "orbital" magnetic moment is made apparent in the atomic energy change when the latter is placed into a magnetic field (see equation (33.1)).

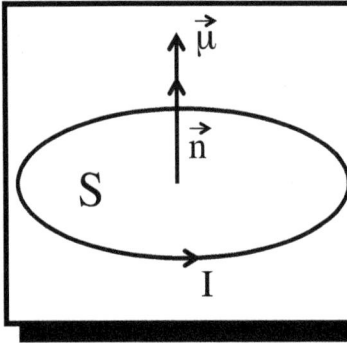

Fig. 33.1: Magnetic moment of the current loop.

In the course of careful experimental data analysis, it appeared that the atomic properties of the external magnetic field were determined not only by the electron's circulation around the nucleus, but also by the presence of a hidden "internal rotation" in the electron, which was named "spin". All elementary particles possess spin (in some of them it is equal to 0). The intensity of such "rotation" is described by the spin number s, which can be only an integer or half-integer. For electrons, protons, and neutrons, $s = 1/2$. "Internal rotation", similar to orbital rotation, results in the possession of a spin magnetic moment. The projection of a spin magnetic moment to the z-axis (magnetic field direction) gives values of

$$\mu_z = \gamma m_s \hbar,$$

where $\hbar$ is the Planck constant, γ is called the gyromagnetic factor, while the quantum number m_s can take $(2s+1)$ integer values: $-s, -s+1, ..., s-1, s$. The modulus of the vector $\vec{\mu}$ is larger than its maximum projection: $\mu = \gamma\sqrt{s(s+1)}\hbar$, i.e., $\vec{\mu}$ in all stationary states is positioned at some angle with respect to the z-axis and it is rapidly spinning around this axis (see Figure 33.2). The projection of a spin magnetic moment for electrons, protons, and neutrons may take only two values: $m_s = \pm 1/2$. The gyromagnetic factor is for electrons $\gamma_e = -\frac{e}{m_e c}$, for protons $\gamma_p = 2.79\frac{e}{m_p c}$, where m_e and m_p are the electron and proton masses, and c is the speed of light. Even neutrons possess a spin magnetic moment, despite the fact that they are electrically neutral in general.[a] The gyromagnetic factor for a neutron is $\gamma_n = -1.91 \frac{e}{m_n c}$ (the neutron mass m_n is slightly larger ($\approx 3\%$) than for protons). Evidently, proton and neutron magnetic moments are three orders ($\sim 10^3$) smaller than the electron magnetic moment (their masses are approximately 2,000 times bigger). Within an order of magnitude, almost all other atomic nuclei consisting of protons and neutrons should have approximately the same magnetic moment.

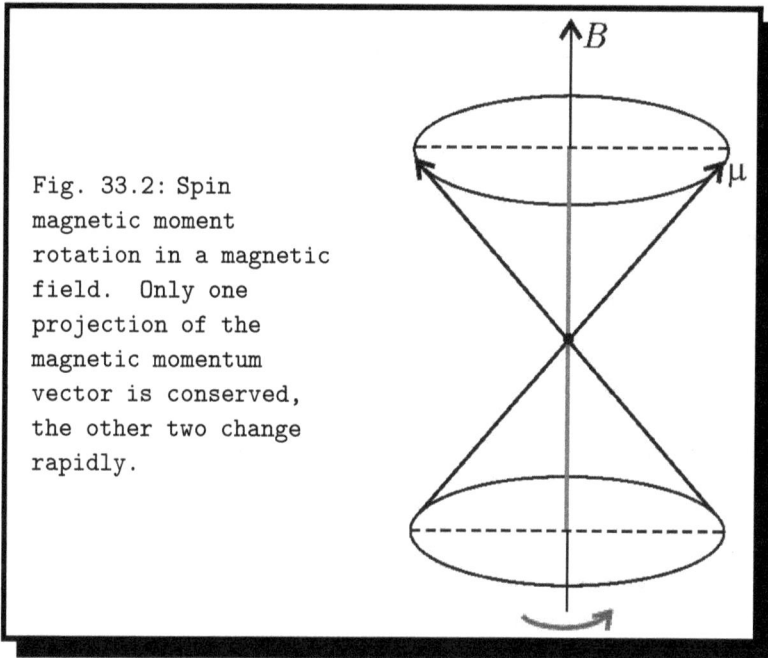

Fig. 33.2: Spin magnetic moment rotation in a magnetic field. Only one projection of the magnetic momentum vector is conserved, the other two change rapidly.

Magnetic moments of nuclei were measured with great precision. It is the presence in nuclei of these tiny (as compared to atomic) magnetic moments, whose values vary for different nuclei, which is fundamental for nuclear magnetic resonance (NMR) phenomenon as well as for magnetic resonance tomography. We will be primarily discussing hydrogen nuclei — the protons which most frequently occur in nature. Deuterium, which is a hydrogen isotope with mass number 2, also possesses a nuclear magnetic moment.

---

[a]This testifies to the fact that neutrons should have an internal structure. Just like protons, they consist of charged quarks.

33.2   What is nuclear magnetic resonance?

Let us consider a nucleus of a hydrogen atom (proton) placed in an external magnetic field, $\vec{B}$. The proton can exist only in two stationary quantum states: in one of them, the projection of the magnetic moment upon the magnetic field direction is positive:

$$\mu_z = \frac{2.79e\hbar}{2m_pc}$$

and in the other it is negative, with the same magnitude. In the first state, the nucleus's energy in the magnetic field equals $-\mu_z B$, in the second, it equals $+\mu_z B$. Initially, all nuclei are in the first state;[b] to transfer to the second state requires an amount of energy

$$\Delta E = 2\mu_z B$$

per each nucleus. It is easy to see, then, that one can force a nucleus to change its magnetic moment projection just placing it in the electromagnetic field of frequency ω, corresponding to the transition between these states:

$$\hbar\omega = 2\mu_z B. \tag{33.2}$$

Substituting here the value of the proton's magnetic moment, one obtains

$$2\pi\hbar\nu = \frac{2.79e\hbar}{m_pc}B,$$

from where for $B = 1\,T$ one finds for the electromagnetic field frequency $\nu \approx 4 \cdot 10^7\,Hz$ and corresponding wavelength: $\lambda = c/\nu \approx 7m$ (these are the typical frequency and wavelength of broadcast band). The absorption by the nuclei of electromagnetic field quanta (photons) of this very wavelength results in magnetic moment flips towards the field direction. The result of these flips is an increase in the energy of each nucleus by such a quantized amount (see Eq. (33.2)).

First of all it should be noted that in NMR experiments, i.e., for typical frequencies of medium range, the electromagnetic waves are not used in the form we are accustomed to when discussing light propagation or absorption and atomic light radiation. In the simplest case we are dealing with a coil,

---

[b]At zero temperature, $T = 0$.

through which flows an alternating current created by the radio frequency generator. A sample containing the examined nuclei we want to expose to electromagnetic field effects is located along the coil axis. The coil axis is directed perpendicularly to a static magnetic field B_0 (the latter is created by means of an electromagnet or superconducting solenoid). When alternating current is running across the coil, an alternate magnetic field B_1 is induced along its axis. The amplitude of B_1 is chosen much smaller than B_0 (usually 10,000 times less). This field is oscillating at the same frequency as the current, i.e., at the frequency of the generator.

If the generator frequency is close to the one calculated above, intensive absorption of the radiation quanta by the hydrogen nuclei with a transition of the latter into the state with negative projection μ_z takes place. If the generator frequency is different from the calculated frequency, the absorption of quanta does not occur. It is due to the sharp dependence of intensity versus the frequency of an alternating magnetic field, in the process of energy transmission to the nuclei, that this phenomenon is called nuclear magnetic resonance (NMR).

How is it possible to notice these nuclear moment flips? It is easy, if you are equipped with state-of-the-art NMR technology: when you turn off the radio frequency generator creating the field B_1, you should simultaneously turn on the receiver using the same coil as an antenna. In this case, it will be registering radio waves radiated by the nuclei as soon as they come back to the initial orientation along the field B_0. This signal is induced in the same coil which was used earlier to rotate the magnetic moments. Its time dependence is processed by computer and represented in the form of a corresponding spectral distribution.

From this description you can see that NMR-spectrometers differ essentially from regular spectrometers, which perform measurements in the range of visible light.

Heretofore, we were examining a simplified picture: the behavior of an isolated nucleus in a magnetic field at zero temperature. At the same time, it is clear that the nuclei are not isolated either in solid matter or in liquids. They can interact with each other, as well as with the other excitations whose energy distribution is determined by the temperature and statistical properties of the system. Interactions between different types of excitations, their origin, and their dynamics are a subject of study of modern condensed matter physics.

33.3 How NMR was discovered

The first signals associated with nuclear magnetic resonance were obtained in the middle of the XXth century by the research groups of Felix Bloch[c] at Stanford and Edward Purcell[d] at Harvard University. At that time, the experimental problems were overwhelming. All the equipment was made by researchers in the laboratories on site. The exterior of the devices made in those times bears no resemblance to the modern MRI devices (with powerful superconducting solenoids) you see at hospitals and clinics today. Suffice it to say, the magnet used in Purcell's experimental studies was manufactured from scrap metal found in the back yard of the Boston Streetcar Depot. In addition, it was calibrated so poorly that the magnetic field, in fact, had a greater magnitude than necessary for a nuclear moment flip under the radio frequency radiation $\nu = 30\,MHz$.

Purcell and his young colleagues sought the proof that the phenomenon of NMR had indeed taken place in their experiments in vain. After many days of fruitless efforts, feeling sad and disappointed, Purcell came to the conclusion that the predicted NMR phenomenon could not be observed and gave an order to shut down the power supply. While the magnetic field was decreasing, the disenchanted researchers continued to look at the oscilloscope screen, hoping to see the signals they were waiting for. At some point, the magnetic field achieved the magnitude necessary to create the resonance effect, and the NMR signal showed up on the screen.

From then on, NMR technology began moving at a quick pace. It has been widely used in scientific research in the fields of condensed matter physics, chemistry, biology, meteorology, and medicine. But its most notable application is the imaging of the internal organs of the human body obtained with the help of NMR.

33.4 How the imaging of internal organs is performed by NMR

Until now, we were implying without saying that if the influence of the weak currents in the coils is neglected, the magnetic field where the nuclei are positioned will be homogenous, i.e., it will have the same value at every point. In 1973 Paul Lauterbur[e] suggested that NMR experiments be

---

[c]Felix Bloch (1905–1983), American physicist; 1952 Nobel Prize in physics.

[d]Edward Purcell (1916–1997), American physicist; 1952 Nobel Prize in physics.

[e]Paul Lauterbur (1929–2007), American chemist; 2003 Nobel Prize in medicine.

performed by placing a sample into an inhomogeneous magnetic field. It is obvious that in this case, the resonant frequency for the studied nuclei will change point-to-point, which makes it possible to estimate the spatial positioning of small parts of the sample. Considering that the intensity of a signal emitted from a certain spatial domain is directly proportional to the number of hydrogen atoms in this domain, one can obtain information about the matter density spatial distribution. Basically, this is what lies at the root of magnetic resonance imaging (MRI) methodology. As the reader may see, the underlying principle is simple, although, to obtain real images of internal organs, powerful computers controlling pulsed radio frequency radiation were necessary, as well as plenty of time to improve the methodology for the creation of magnetic field profiles and for processing the NMR signals received from the coils.

Let us imagine that small, water-filled spheres are positioned along the x-axis (Figure 33.3). If the magnetic field does not depend on x, a single signal occurs (see Figure 33.3, a). Then, let us assume that with the extra coils (with respect to the coil creating the main magnetic field oriented along the z-axis) we create an additional magnetic field which changes along the x-axis, with its magnitude growing from left to right. In this case, it is clear that for the spheres with different coordinates, the NMR signal now corresponds to different frequencies, and the measured spectrum contains five characteristic peaks (see Figure 33.3, b). The height of the peaks is proportional to the number of spheres (i.e., water mass), which have a corresponding coordinate, and thus, in the previous case the peak intensities relate as 3:1:3:1:1. If the magnetic field gradient is known (i.e., the rate of its change along the x-axis), one can present the measurable frequency spectrum in the form of the hydrogen atomic density dependence on coordinate x. In this case, it is possible to say that wherever the peaks are higher, the density of hydrogen atoms will be greater: in our example the number of hydrogen atoms corresponding to the positions of the spheres indeed relates as 3:1:3:1:1.

In a constant magnetic field, let us position a somewhat more complex configuration of small water-filled spheres and apply an additional magnetic field changing along all three axes. Measuring NMR radio frequency spectra and keeping in mind the values of the magnetic field gradients along the coordinates, we can create a three-dimensional map of the sphere distribution (and hence, hydrogen density) in the studied configuration. This will be much more difficult to do than in the previous one-dimensional case, but

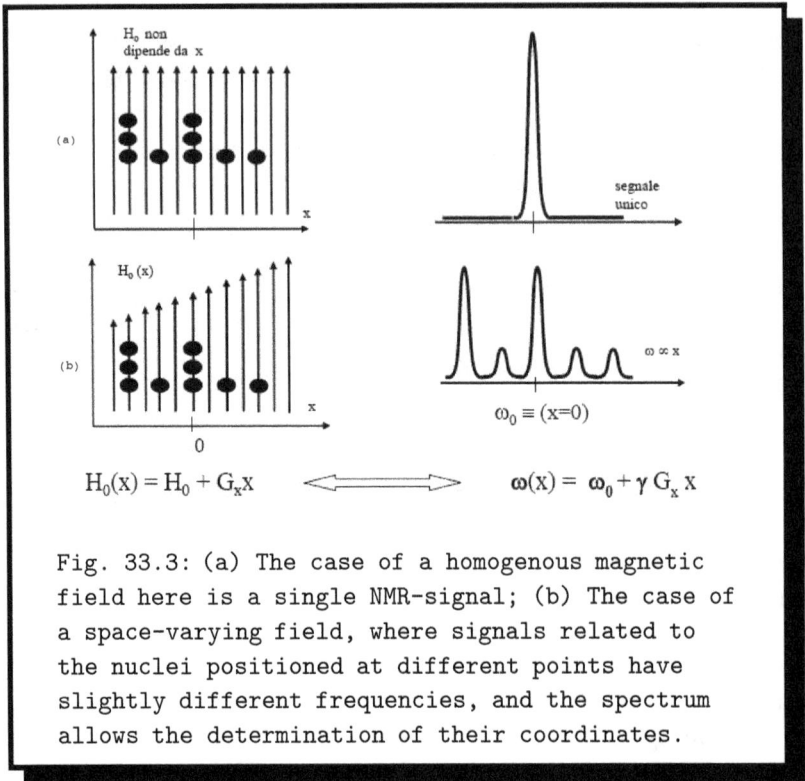

Fig. 33.3: (a) The case of a homogenous magnetic field here is a single NMR-signal; (b) The case of a space-varying field, where signals related to the nuclei positioned at different points have slightly different frequencies, and the spectrum allows the determination of their coordinates.

the nature of the process can be understood intuitively.

An image-recovery procedure similar to the one we have described is performed during MRI tomography. When the data collection is finished, the computer begins signal "processing" with very fast algorithms and establishes a connection between the intensity of measured signals at certain frequencies and the resonating atom density at given points on the body. When this procedure is completed, a two- (or even three-dimensional) "image" of a certain organ or a body part appears on the computer screen.

33.5 Striking "images"

In order to fully appreciate the results of an MRI exam of human internal organs (e.g., different sections of the brain, which today a medical physicist can obtain without even touching a cranium!) it should be understood that we are talking here about a computer recreation of "images", as opposed to real "shadows" appearing on photosensitive films, as that due to X-ray absorption during an X-ray exam.

The human eye is a sensitive meter of electromagnetic radiation in the visible range. Fortunately or not, we cannot see radiation emitted by internal organs – we can only see the human exterior. At the same time, as we have just discussed, under certain conditions the nuclei of atoms in human internal organs can emit electromagnetic waves in the radio frequency range (i.e., much smaller frequencies than necessary for visible light) and this frequency can be slightly variable depending on the radiation point. Such radiation is not visible to a human eye, therefore it is registered by sophisticated devices and then special computer processing turns it into a consolidated image. Nevertheless, we are talking about seeing the internal part of an object or of a human body in real life.

This extraordinary progress made by humanity became possible due to a number of breakthrough achievements in fundamental science: among them are quantum mechanics and its angular momentum theory, a theory of interaction between radiation and matter, digital electronics, mathematical algorithms of signal transformation, and computer technology.

The benefits offered by MRI tomography as compared to other methods of diagnostics are numerous and significant. A technician can easily choose which sectors of the patient's body to scan; he can also examine several views of a selected organ simultaneously. Among other things, proper selection of magnetic field gradients may help obtain images of vertical views of the skulls interior (see Figure 33.4). It can obtain a central view or the view can be shifted to the right or to the left. Such examination is practically impossible to perform with X-ray tomography. Technicians can also "narrow" the observation area, visualizing the NMR signal emitted by one selected organ only, or by a certain part of it only, thus increasing image resolution. The possibility of taking direct measurements of local viscosity as well as flow direction of blood, lymph, and other body fluids is another important advantage of this exam. Selecting a required correlation between parameters, e.g., impulse time and frequency, for every condition, a technician can achieve optimum image quality, e.g., to improve the sharpness of the image (see Figure 33.4).

Summing up the above, one can say that for every imaging point (pixel)

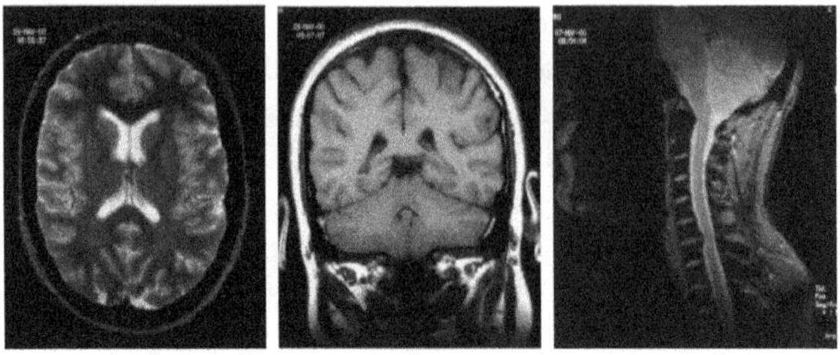

Fig. 33.4: The image of a skull and vertebral
column which, depending on the contrast dye, shows
white or grey brain tissue, vertebrae, and
cerebrospinal fluid with anatomic precision.

corresponding to a tiny volume of the object under study, it is possible to extract a variety of useful information, in some cases even the distribution of certain chemical element concentrations in the human body. To improve measurement sensitivity, i.e., to increase the ratio between signal intensity and noise, a huge number of signals should be accumulated and integrated. In this case we will obtain a quality image which adequately shows a real object. This is the reason why an MRI session takes a long time — the patient has to stay in the solenoid producing the field, motionless, for several scores of minutes.

In 1977, Peter Mansfield[f] invented the combination of magnetic field gradients which, although not providing high-quality imaging, nevertheless allowed one to receive the image very quickly: one signal is enough for appropriate plotting (in real time it takes 50 milliseconds). This technique – today it is called eco-planar imaging – allows us to trace heartbeats in real time: in this "movie", heart contractions and expansions are alternating on the screen.

Could anybody imagine in the early days of quantum mechanics that 100 years later a scientific breakthrough would make all these miracles possible?

---

[f]Peter Mansfield (1933), British physicist; 2003 Nobel Prize in medicine.

Chapter 34

Towards a quantum computer

Small is Beautiful.

E. F. Schumacher. ""

In Figure 34.1 you can see the famous Rosetta Stone, which was found in 1799 in Egypt, close to the town Rosetta. "In the reign of the young one — who has received the royalty from his father — lord of crowns, glorious, who has established Egypt..." was cut in a dark granite slab with dimensions of $114 \times 72\ cm^2$ in honor of King Ptolemy V on the occasion of the anniversary of his coronation. Fortunately, the Rosetta Stone bears three inscriptions: the top register in Ancient Egyptian hieroglyphs; the second one, in the Egyptian demotic script; and the third one, in ancient Greek. Since ancient Greek is well known, the Rosetta Stone gave Jean-François Champollion the key to explaining the secret of Egyptian hieroglyphs in 1822.

In 1970, engineers at the Nippon Electric Company (NEC) created a dynamic memory element, which had 1,024 memory cells (small rectangles) arrayed in four grids of thirty two columns with an equal number of rows (see Figure 34.2). Its size was $0.28 \times 0.35\ cm^2$ and there was enough room for all the information inscribed on the Rosetta Stone. It took about 300 microseconds to upload the Rosetta Stone text on this chip.

34.1 Milestones of the computer era

The story of a radically new approach to storing and managing information began in 1936, after publication of the article "On Computable Numbers

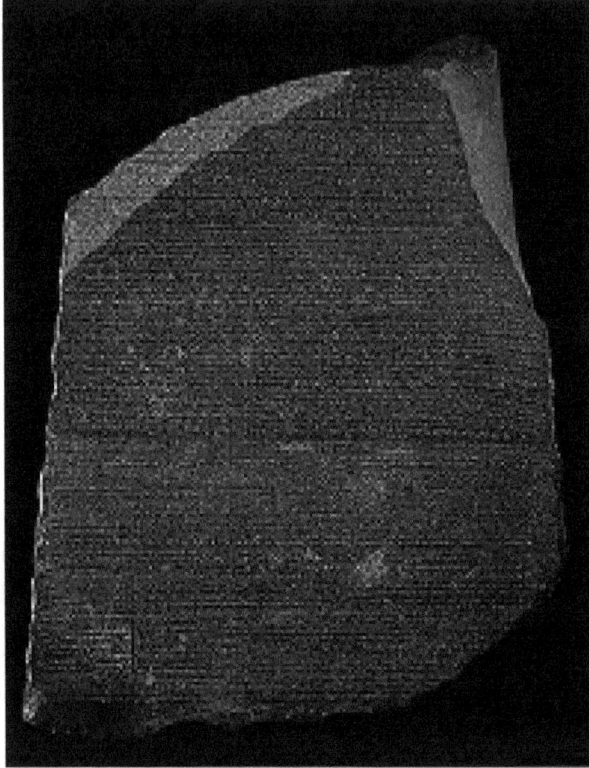

Fig. 34.1: Rosetta stone.

with an Application to the Entscheidungsproblem" by Alan Turing. The paper contained the main concepts of the logical design of a universal computer, which is now called the Turing machine.

The first generation of practically realized computers based on vacuum tubes (such as had been used in old TV sets) appeared ten years later. These computers formed the prehistory of the computer era. They were

Fig. 34.2: The first dynamic memory element.

mostly used to verify this or that theoretical concept. In essence, the first computers served as experimental facilities. The weight and the size of these "dinosaurs" was enormous; often special buildings were built for them to be placed in.

At the same time, considerable progress was achieved in electronics. In 1947, Walter Brattain,[a] William Shockley,[b] and John Bardeen[c] invented the first semiconductor transistor. It was realized by means of a point contact at the semiconductor substrate and had dimensions of the order of centimeters. Later, transistors were widely used in electronics as amplifiers of electromagnetic signals or controllable switches operated by electric signals. They replaced vacuum tubes worldwide.

In 1958, Jack Kilby created the first semiconductor integrated circuit (chip), which contained only two transistors. Soon after that the microcircuits integrating tens and hundreds of transistors on the same semiconductor substrate appeared.

---

[a]Walter Brattain (1902–1987), American physicist; 1956 Nobel Prize in physics.

[b]William Shockley (1910–1989), American physicist; 1956 Nobel Prize in physics.

[c]John Bardeen (1908–1991) American physicist; he was awarded two Nobel Prizes in physics: for the invention of the transistor in 1956 and for the creation of the theory of superconductivity in 1972 (see Chapter 25).

These discoveries initiated the second period of the computer era (1955–1964). At the same time, magnetic cores and magnetic drums (the ancestors of modern hard discs) came into use as computer data storage (often called the *storage* or *memory*).

Side by side with the rapid development of computer hardware, the ideology and architecture of logical facilities leaped forward. In June 1954, John von Neumann[d] presented the work entitled, "The first draft of a report on the Electronic Discrete Variable Automatic Computer (EDVAC)", which fully described the operation of the digital electronic computer. Moreover, in this paper von Neumann discussed in detail the fundamentals of the logic used in computer operation, and justified the use of the binary numeral system.[e] Since then computers have been recognized as a subject of scientific interest. This is why even today some scientists call computers "von Neumann machines".

All these fundamental discoveries and inventions outlined the beginning of the second period of the computer era. Computers of the third generation (1965–1974) were created on the basis of integrated circuits. More miniature and capacious semiconductor memory devices replaced bulky magnetic ones. They are still used in personal computers as the random access memory (RAM). A breakthrough in theoretical cybernetics, physics, and technology achieved in the last third of the twentieth century facilitated rapid development of computer technology allowing a tremendous decrease in computer sizes and prices making them affordable for the public. That is why since the mid-1970s computers have become an important element of everyday life. Incredible progress in technology allowed tens of millions of transistors to be placed inside an integrated device within an area of 3 cm^2 (the same area was used by J. Kilby). Modern memory devices allow a single laptop computer to store the contents of the entire US Library of Congress (150 million books and documents). The well-known Moore's law (an empirical observation made by Gordon E. Moore in 1965) in one of its forms has established exponential growth of the memory capacity and the

---

[d] John von Neumann (1903–1957) Hungarian-born American mathematician and physicist who made major contributions to a vast range of fields, including set theory, functional analysis, quantum mechanics, ergodic theory, continuous geometry, economics and game theory, computer science, numerical analysis, hydrodynamics (of explosions), and statistics, as well as many other mathematical fields. He is generally regarded as one of the greatest mathematicians in modern history.

[e] The binary numeral system, or base-2 number system, represents numeric values using two symbols: 0 and 1.

number of transistors in a processor. One can see in Figure 34.3 how it behaved over a 40-year period.

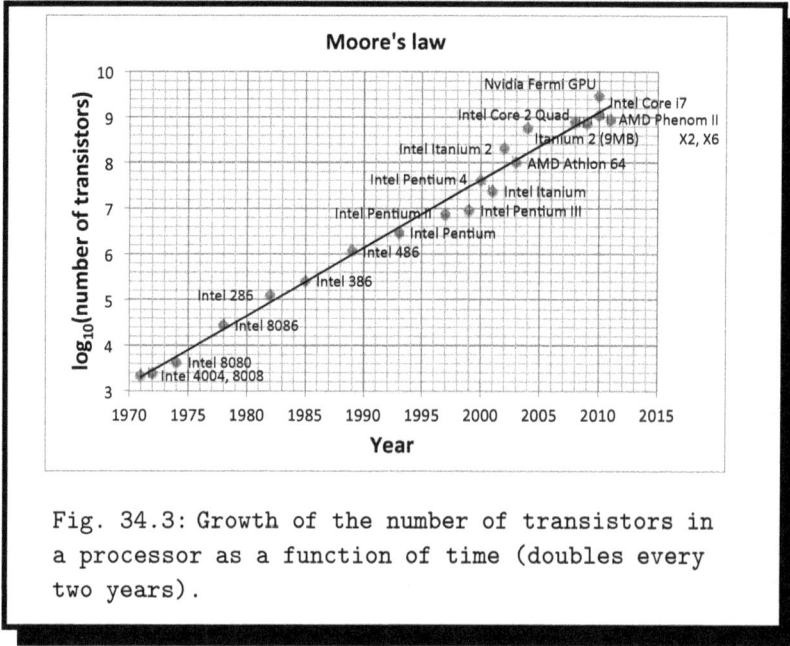

Fig. 34.3: Growth of the number of transistors in a processor as a function of time (doubles every two years).

Fitting more transistors inside the same area and higher clock frequency (characterizing computer speed) became an obsession of computer manufacturers. In 2004, Intel's Prescott processor designed with 90 nm technology (this means that several million transistors were placed on $1 cm^2$ of its area) contained 150 million transistors and operated with the clock frequency at 3.4 GHz. In 2007–2008 computer manufacturers switched to 45 nm technology, and today the newest Apple chips are created using 20 nm process technology, Samsung is on the verge of 14 nm.

Nevertheless, this straightforward method of computational power increase is almost exhausted. For example, one of the obvious restrictions is caused by the finiteness of the electromagnetic signal propagation speed in circuits: its value depends on the technical realization of the processor, but certainly cannot exceed the speed of light $c = 3 \cdot 10^5 \ Km/s$. This value seems to be enormous, but let us make a simple estimate. The clock frequency of $\nu = 3.4 \ GHz$ means that the interval between two operations is of the order of $\delta t = \nu^{-1} \approx 3 \cdot 10^{-10}$ s, i.e., the space separation between the units involved in two sequential operations cannot exceed $L = c\delta t \approx 10$ cm. Note that we obtained this result for the ideal case of signal propagation in vacuum at the speed of light! Hence, the inevitable condition of the further increase of the clock frequency (which means a corresponding increase in the number of elementary transistors in the processor) is the further miniaturization of the device. Engineers chose this path a long time ago.

Yet, we are approaching the day when further miniaturization will come into conflict with the physical properties of the basic elements of modern processors. The major scourge caused by miniaturization is the processor overheating due to an increase of released Joule heat — even when currents passing through hundreds of millions of transistors are extremely small. Another essential restriction on miniaturization is caused by the growth of the electric field in the oxide layer[f] between electrodes of the transistor: when its thickness decreases, dielectric breakdown may occur. The third limitation is posed by the typical scale of fluctuations in the impurities' distribution density. These fluctuations should be able to self-average over the size of the smallest elements of the processor. Hence the tinier the processor the weaker the currents that should be used. However, it is impossible to reduce the voltage to zero: beyond a certain limit the useful currents (signals) become so small that they simply get lost against the background of various noise signals always present in electronic circuits. This is the reason why researchers are not interested in a purposeful voltage decrease; however, they have to do it for the sake of further miniaturization. Therefore, the concept of a direct clock frequency growth has been diminished.

34.2   Twenty-first century: looking for a new paradigm

Due to the problems outlined above facing the direct clock frequency growth in traditional processors, two major manufacturers (Intel and AMD) decided in 2005 to switch to a radically new architecture of the "computer brain" — to a multicore processor. In the beginning there were two-core processors, and now we are witnessing the advent of processors with twenty four or more even cores (recently Intel has produced a 48-core processor for research in cloud computing). Such devices unify two or more independent

---

[f]The insulation barrier separates the electron systems of two electrodes.

processors while the scratch-pad memory and its controller remain common. This architecture allows a substantial increase in processing speed without an increase in the clock frequency, and thus helps to avoid the problems mentioned above. Nevertheless, one should not think that the two-core processor is twice as fast than the one-core one. The situation is more complex: the productivity of the multicore processor essentially depends on the type of problem under consideration, and on the use of a brand new software, based on the advantages of parallelism. In some cases, the computation speed increases almost twice; in other cases, this is not so.

Although the development of the multicore technology is an effective engineering solution, it still remains a palliative measure to a clear new mainstream concept for computer technologies. It should be recognized that we have practically reached the limits of the macroworld, where the laws of classical physics still govern the behavior of electrons in the processor circuits.

This is why during the last decade scientists have been trying to work out the concepts of fundamentally new computing devices, based on new logics (different from those by Turing and von Neumann) and new basic components. Typical examples of these approaches are the development of new, *quantum*, computing algorithms (quantum computing) and of nanometer-sized devices (nanoscience, nanotechnology). To date, the idea of the realization of a "quantum computer" has resulted in the development a new interesting field of math, which, however, has still failed to provide significant practical implementations. This is why we will now review several relatively novel devices that follow the "classical" computer logic, but exploit essentially quantum phenomena.

34.3 Where does the border between the macro- and microworld lie?

As was just mentioned above, the transfer to 14 nm design rules will become the next breakthrough in computer technology. Is this scale small or large with respect to the quantum world? The characteristic scales relevant to nanoscience are presented in Figure 34.4.

The first microscopic (but still classical) length scale which we have to bear in mind while decreasing sizes of the processor's elementary blocks is the *electron mean free path*, l_e, in metal layers. Let us recall that Ohm's

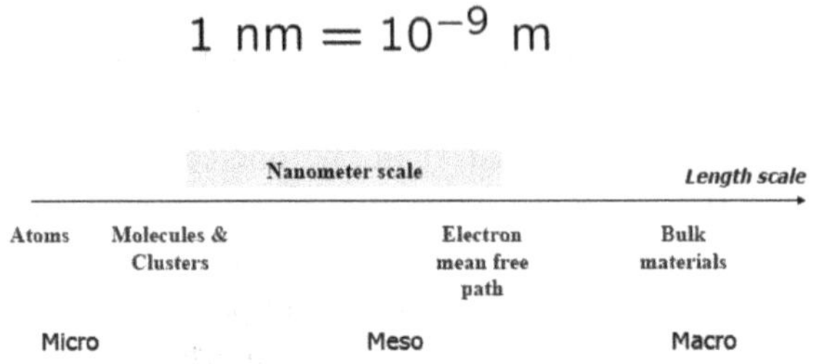

$$1 \text{ nm} = 10^{-9} \text{ m}$$

Fig. 34.4: The characteristic scales of
nanophysics.

law is based on the *diffusive* character of electron motion when the charge
transport is dominated by multiple electron scattering at impurities or other
defects. Since in the process of deriving Ohm's law, the electron motion
is averaged over positions of defects, the specifics of their locations is not
felt in the final result. This approximation remains valid until the number
of defects in a sample becomes large. Hence, the connecting wires between
the circuit transistors can be considered as classical Ohmic resistors, when
their size L is much larger than the electron mean free path l_e: $L \gg l_e$
(in sputtered metal films l_e is of the order of several nanometers). At the
opposite limit, $L \ll l_e$, the electron motion is *ballistic* — electrons are
accelerated by the electric field over the whole device rather than diffuse
across it. One can expect that the properties of such wires would deviate
considerably from the classical Ohmic behavior.

The next important scale characterizing the limits of the quantum world
is the electron's *de Broglie wavelength*, $\lambda_F = h/p_e$. We already know that
quantization and interference are characteristic features of the quantum
world. When even the smallest dimension of a resistor becomes comparable
with λ_F, electrons obey quantum mechanics. In particular, their motion
in the corresponding direction is quantized, and the whole resistor can be
described as a quantum well. The value of λ_F depends strongly on the
electron concentration and, for a normal metal, turns out to be of the
atomic scale. However, in semiconductors λ_F can be much larger, and
the quantum confinement in corresponding nano-objects becomes crucially

important. In particular, quantum confinement along one direction allows fabrication of a new object – two-dimensional electron gas, which serves as a building block for many modern electronic devices.

Another purely quantum scale absent in classical physics is the so-called *phase coherence length*, l_ϕ. Let us note that an important characteristic of a quantum particle is the phase of its wave function, which is closely related to its energy. When an electron is scattered by static impurities, only the *direction* of its momentum, $\mathbf{p}_e$, changes but it does not change its absolute value, $|\mathbf{p}_e|$ or its energy. Therefore, such scattering is called *elastic*, the corresponding length scale being l_e. Even after many elastic scattering events, the electron phase remains fully determined. Hence, it can take part in quantum interference processes. However, sometimes due to a scattering event the electron spin flips, or the electron scatters by lattice vibrations with a finite energy transfer. In these cases, the coherence of the final state with the initial one is broken. The characteristic length at which the electron phase changes by 2π due to random scattering events (that means complete loss of the phase memory) is called the phase coherence length, l_ϕ. It determines a new quantum scale: when the geometrical dimensions of the device become shorter than the phase coherence length ($L < l_\phi$), quantum interference may occur in the device.

Finally, recall that the value of the electron charge in classical physics is considered infinitesimally small, so the charge flow through the capacitor plates in the process of charging or recharging is assumed to be continuous. When the device capacitance, C, is so small that the one-electron electrostatic energy, $e^2/2C$, becomes comparable to other energy scales of the system (the thermal energy $k_B T$ or the electron energy in the gate potential, eV_g) new phenomena relevant to single-electron tunneling can take place.

Interplay between the different scales discussed above leads to a variety of regimes of quantum transport through nanosystems. The specific properties of different regimes can be used for various applications. Obviously, there is not enough room here for a detailed discussion of the physical properties of the new contenders for the roles of the basic elements for new computers, and we will briefly discuss only some of them.

Table 34.1: Relevant Length Scales in Mesoscopic Devices.

Conventional device	Mesoscopic device
diffusive, $L \gg l_e$	ballistic, $L \ll l_e$
incoherent, $L \gg l_\phi$	phase coherent, $L \ll l_\phi$
no size quantization, $L \gg \lambda_F$	size quantization, $L \ll \lambda_F$
no single electron charging, $\frac{e^2}{2C} \ll k_B T$	single electron charging, $\frac{e^2}{2C} \gg k_B T$

34.4   Quantum wires and quantum point contacts

Let us consider how electric charge flows through the so-called quantum wire, i.e., a very thin channel which contains charge carriers but is free of impurities or other defects of the crystal lattice (see Figure 34.5).

As has already been mentioned, under the conditions when the transverse size w of a conductor becomes comparable to the electron wavelength λ_F, the electron is confined in the quantum well and its motion in the appropriate direction is quantized. This means that the part of the energy corresponding to transverse motion can take only certain discrete values, E_n, while the electron motion along the wire remains free. Since the full energy E of the electron is conserved, the larger the value of E_n, the smaller the amount of available energy left for longitudinal motion. The smaller the energy, the longer the corresponding wavelength. Hence, each energy level E_n corresponds to the specific plane wave with its characteristic wavelength λ_n (wave mode). Thus, the electron propagating in a quantum wire resembles an aggregate of wave modes in a waveguide (see Figure 34.6,a) rather than a particle diffusing in a scattering media[g] (see Figure 34.6,b).

Each mode contributes to the process of charge transfer, with the to-

---

[g]Let us recall that in Chapter 3, we discussed sound wave propagation in the underwater waveguide without taking quantization into account. The fact of the matter is that we were not even considering such a possibility, whereas it does in fact exist. Indeed, when the sound wave wavelength is comparable to the sea depth, the underwater waveguide becomes too narrow for such waves, and only certain definite sound modes can propagate through. So the waveguide will not work well, like the Bell's water mic (Chapter 12). The characteristic wavelength must be of the order of 1 km, i.e., sound frequency $f = c/\lambda \approx 1 \ Hz$. As frequency grows, the number of nodes in the standing transverse wave increases and we return to the continuous picture of Chapter 3.

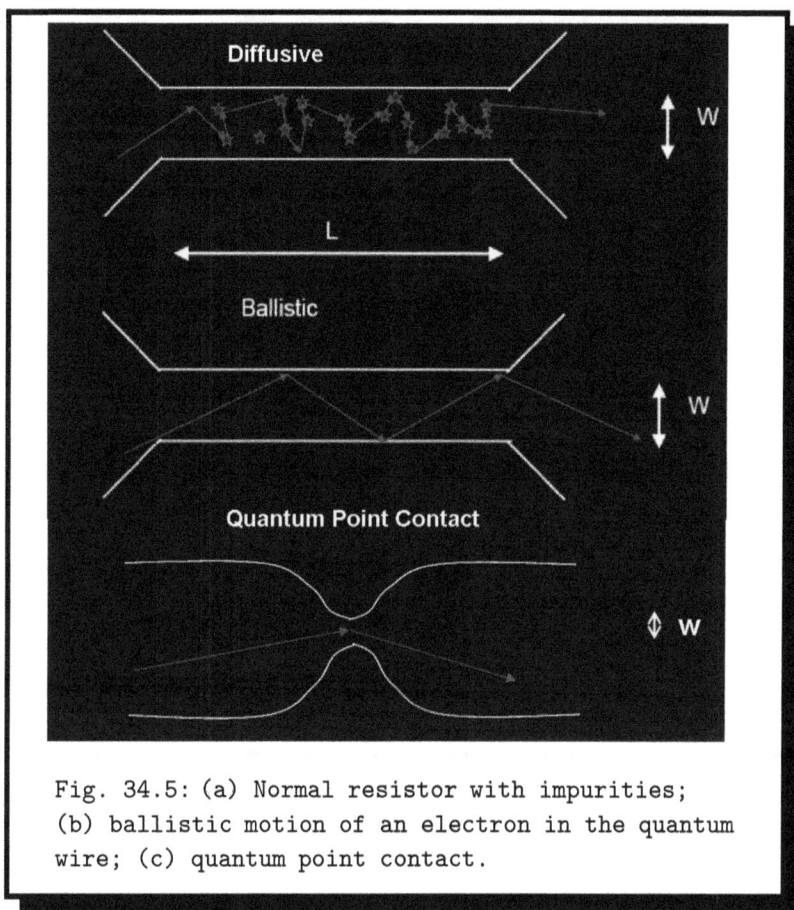

Fig. 34.5: (a) Normal resistor with impurities;
(b) ballistic motion of an electron in the quantum
wire; (c) quantum point contact.

tal conductivity being determined by the sum of all their contributions. The contribution of a single quantum channel to the charge transfer can be evaluated, as always, based on dimensional analysis. In an ideal case, it does not depend on the wire properties and is purely a combination of the universal constants. The only available combination with the dimensionality of resistance is h/e^2, which could not appear in the classical theory. It turns out that e^2/h is the maximum conductance value, which can be realized by a single mode propagating along a quantum wire.

In Figure 34.7, one can see a real experimental realization of the quantum point contact. Its conductance at low temperatures changes discretely

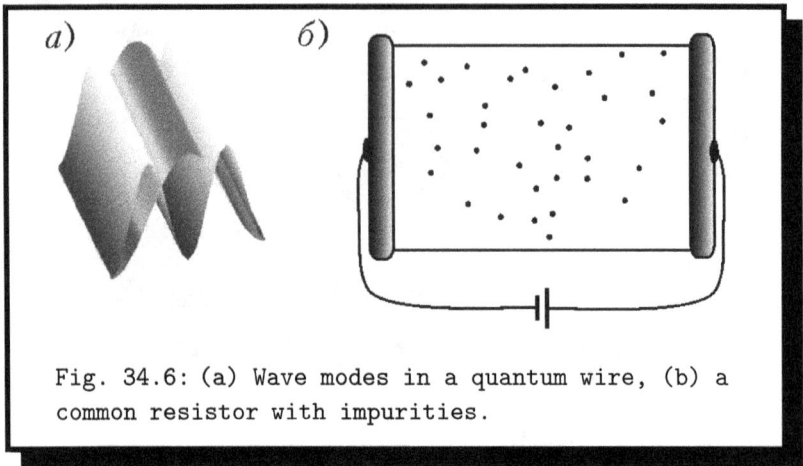

Fig. 34.6: (a) Wave modes in a quantum wire, (b) a common resistor with impurities.

versus gate voltage V_g (which controls the effective width of the conducting channel) instead of being constant (in accordance with classical behavior). The height of each step corresponds to the value $2e^2/h$, in accordance with our speculation based on dimensional analysis.

The conductance quantization in point contacts usually disappears at $L > 2\ \mu m$; it also smears out with an increase of the temperature.

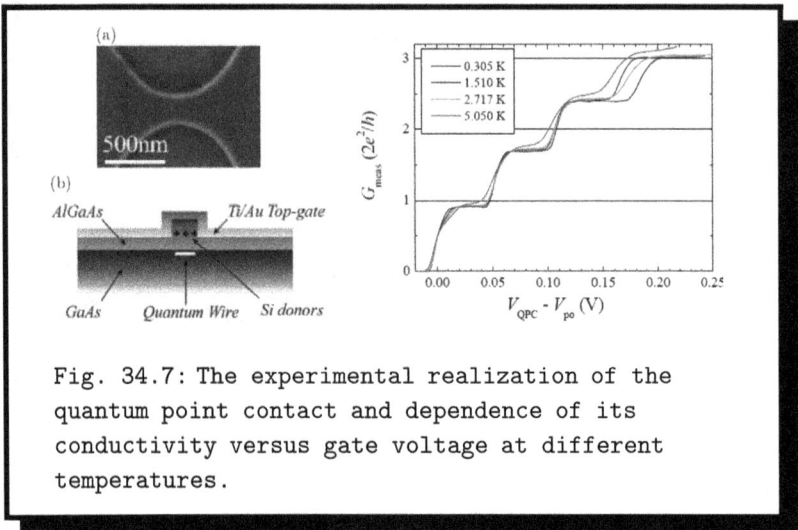

Fig. 34.7: The experimental realization of the quantum point contact and dependence of its conductivity versus gate voltage at different temperatures.

34.5 "Coulomb blockade" and single electron transistors

Let us consider properties of the so-called *"quantum dot"*: a small metallic island with size $\ll 1\,\mu m$ placed on an insulating substrate. Metal electrodes perform different functions. The first one, the "gate", changes the electrostatic potential of the dot. The second pair, the *"source"* and the *"drain"*, supply and remove electrons to and from the dot. They are arranged as shown in Figure 34.8.

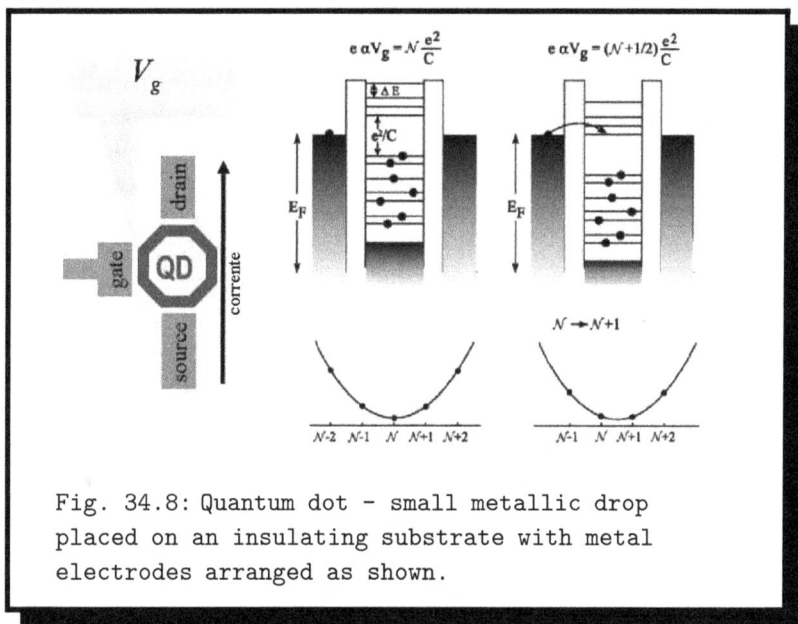

Fig. 34.8: Quantum dot - small metallic drop placed on an insulating substrate with metal electrodes arranged as shown.

Let us assume that the dot contains N excess electrons, hence its total charge is $Q = Ne$. The corresponding electrostatic energy of the dot in the absence of the external potential is $Q^2/2C = N^2 e^2/2C$, where C is capacitance of the dot (usually very small). This energy is due to repulsion of the excess charges distributed over the dot surface.

When the potential V_g is applied to the gate, the total electrostatic energy of the quantum dot consists of two terms: the energy of the electrostatic repulsion of the excess electrons found above and the work of the external field on the charge transferred to the dot from infinity

$$E(N,V) = -V_g Ne + \frac{N^2 e^2}{2C}.$$

This is a quadratic function of the number of electrons, N. Its minimum is formally reached at $N = CV_g/e$, where the derivative $dE/dN = 0$.

Now let us recall that N must be an integer number. Consequently, depending on the gate voltage V_g, different situations are possible (see Figure 34.8). In the first case, when $V_g = Ne/C$, the minimum of the parabola indeed corresponds to a real state with the integer number of electrons N. However, when $V_g = (N + 1/2)e/C$, the minimum is formally realized at a semi-integer number of electrons, which is impossible. The nearest possible state to this formal energy minimum are those with the integer numbers N and $N + 1$. It is important that both states have exactly the same energy:

$$E(N = \frac{CV_g}{e} - \frac{1}{2}, V_g) = E(N = \frac{CV_g}{e} + \frac{1}{2}, V_g) = \frac{CV_g^2}{2} + \frac{e^2}{8C}.$$

We conclude that when the gate voltage is $V_g = eN/C$, the states with N and $N + 1$ electron at the dot are separated by the energy $e^2/2C$, and the energy conservation law "blocks" the charge transport through the dot (between the source and the drain). On the contrary, at $V_g = e(N+1/2)/C$ electron transfer without any energy cost is possible, and the quantum dot is "open".

We see here that a small dot can operate as an efficient transistor. At low temperatures, $k_B T \ll e^2/2C$, the current flowing through such a "single electron transistor" almost vanishes at any gate voltage except specific values, $V_g(N) = e(N + 1/2)/C$. In the vicinities of these points the conductance shows sharp peaks. The experimental dependence of the quantum dot conductance (dI/dV_g) as a function of the gate voltage is presented in Figure 34.9.

One can hope that in the future, single electron transistors will allow development of logical circuits operating with the smallest possible currents and having extremely low dissipation.

34.6 What are physicists working at today in nanoscience?

We presented here a very brief, superficial, and far from complete review of the fascinating new research area of nanoscience, which has embraced a

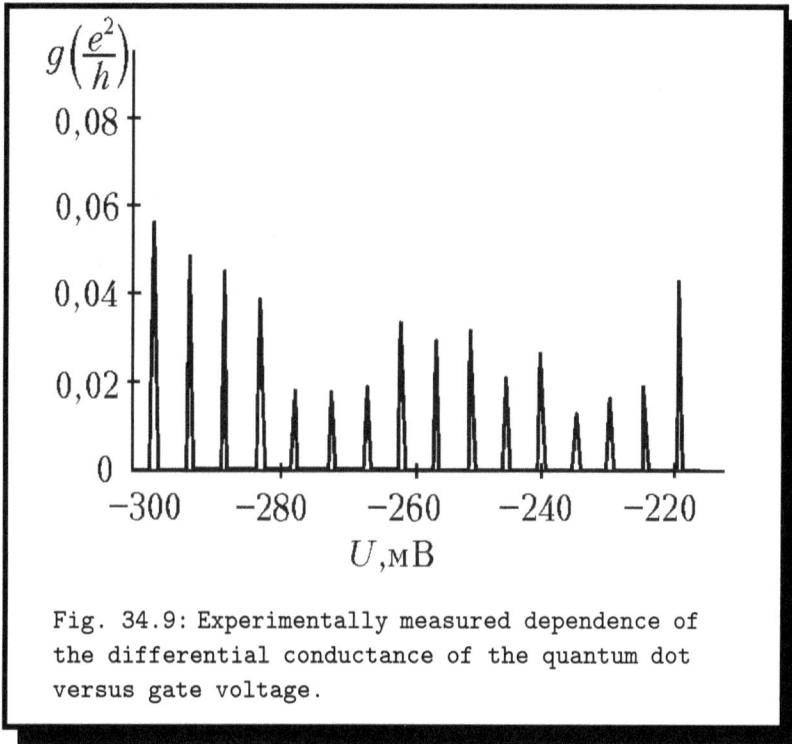

Fig. 34.9: Experimentally measured dependence of the differential conductance of the quantum dot versus gate voltage.

lot of findings from physics and chemistry. Let us list other possible candidates for the building blocks of nanodevices, which belong to the front end of the present research. These are: graphene[h] — a single layer of graphite, where the electrons form a two-dimensional electron gas with a relativistic spectrum; so-called Dirac materials – more complex graphene-like systems

---

[h]Graphene is a thin flake of ordinary carbon, just one atom thick. A. Geim and K. Novoselov have shown that carbon in such a flat form has exceptional properties that originate from the remarkable world of quantum physics. As a material, graphene is completely new – not only the thinnest ever but also the strongest. As a conductor of electricity, it performs as well as copper. As a conductor of heat, it outperforms all other known materials. It is almost completely transparent, yet so dense that not even helium, the smallest gas atom, can pass through it.

Andre Geim (1958), Dutch physicist; director of the Manchester Centre for Mesoscience and Nanotechnology at the University of Manchester.

Konstantin Novoselov (1974), British–Russian physicist; researcher at the same centre.

Both were awarded the 2010 Nobel Prize in physics "for groundbreaking experiments regarding the two-dimensional material graphene".

with specially designed electron spectra; molecular devices such as carbon and non-carbon nanotubes and nanoelectromechanical systems; devices for quantum computation based on various principles; spintronic devices based on manipulation of individual spins; superconductivity and magnetism at nanoscale, topological insulators — materials with non-trivial topological order which behave as insulators in their interior but as conductors along the surface, etc. Nanoscale systems require an understanding of the peculiarities of quantum transport and the interplay between electron–electron interaction and disorder, as well as the role of contacts and the electromagnetic environment of nanodevices. All these issues are far from being fully understood yet.

Chapter 35

A great and terrible nuclear energy

The release of atomic power has changed everything except our way of thinking... The solution to this problem lies in the heart of mankind. If only I had known, I should have become a watchmaker.

Albert Einstein.

Once, Ernest Rutherford, a prominent New Zealand physicist, the founder of nuclear physics, was asked what practical use could his discoveries in radioactivity be applied to? The scientist said: none. At that time, it was impossible even to imagine how intranuclear processes and phenomena could be applied. Nowadays, everyone is aware of the impact nuclear energy has on humanity.

The most important part in the discovery of possible ways of nuclear energy utilization belongs to a division capacity of certain heavy nuclei. In 1938 it was discovered that uranium nuclei, when bombarded by neutrons, form nuclei fragments of lighter elements. This process is always accompanied by emission of several new neutrons.

Why are nuclei fissible? The first theory of nuclear fission was developed in 1939 by Danish theoretical physicist Niels Bohr and American physicist John Wheeler, and also, independently from them, it was discovered by Soviet theoretical physicist Yakov Frenkel. The explanation they presented was based on a nuclear drop model. According to this model, the behavior of a nucleus, which is a cluster of nucleons, is similar to a drop of electrically charged liquid. Let us review this in every detail, and begin with clarifying the question what nuclear-binding energy depends on and how this dependence works. Nuclear-binding energy is the energy necessary to split the nucleus into its nucleon components.

Nuclear forces attracting nucleons to one another occur at very small distances (their working radius is about $10^{-15}\,m$), therefore every nucleon is interacting with its immediate neighbors only, instead of being involved with all nucleons found within this nucleus. The same phenomenon takes place in a drop of tap water. Due to the fact that intramolecular gravitational forces occur at distances greater than the distances between molecules, we have to take into account the interactions between the immediate neighbors only. The number of neighbors can be constant for every nucleon. Thus, the contribution to the binding energy, determined by nuclear forces, is proportional to the number of nucleons found in the nucleus, i.e., mass number A: $E_n \sim A$.

However, a "nuclear drop" surface of nucleons contains less neighbors than inside the nucleus interior, therefore they make a somewhat lesser contribution to binding energy than we ascribed to them. This can be accounted for if we subtract from E_n the surface energy E_{srf}, which is proportional to the number of nucleons on the surface of nucleus, and thus, to the surface of its area:

$$E_{srf} \sim S_{srf} \sim R_n^2 \sim A^{2/3},$$

(here we used the previous experimental data that nucleus radii are exactly proportional to the cube root of mass number). As we can see, here is a perfect analogy to a drop of tap water - molecules that are on the surface eager to leave deep down what creates the forces of surface tension.

Finally, to obtain the resultant expression of binding nuclear energy we have to consider the fact that some nucleons (protons) are charged. This means that in addition to nuclear attraction to immediate neighbors, they will be effected by the electrostatic repulsion forces in accordance with Coulomb's law. Mutual repulsion of protons tends to break the nucleus, and thus to reduce its binding energy. Apart from nuclear forces, Coulomb's forces are long-range ones — every proton is interacting with all protons of this nucleus. The energy of interaction between two protons is e^2/R_n. If there are Z protons in the nucleus, each of them is interacting with $(Z-1)$ of the rest, and the number of interacting couples is equal to $Z(Z-1)/2$ (for big Z this number can be considered as proportional to Z^2). Thus, we have that the energy of proton-proton Coulomb repulsion within the

nucleus is

$$E_{\text{Coul}} \sim Z^2 \cdot e^2/R_{\text{n}} \sim Z^2 \cdot e^2/A^{1/3}.$$

Consequently, we can conclude the following for the nucleus binding energy:

$$E_{\text{b}} = E_{\text{n}} - E_{\text{srf}} - E_{\text{Coul}} = \alpha A - \beta A^{2/3} - \gamma Z^2/A^{1/3},$$

where α, β and γ are certain constants. This formula provides a valid explanation of the experimental curve of specific binding energy dependence on mass number, although we left out from consideration several other less important contributions. In the beginning, the binding energy is also increasing with the increase of mass number. At some point in the middle of the Periodic Table of Elements, E_{b} will reach its maximum, and then will start decreasing (see Fig. 35.1.)

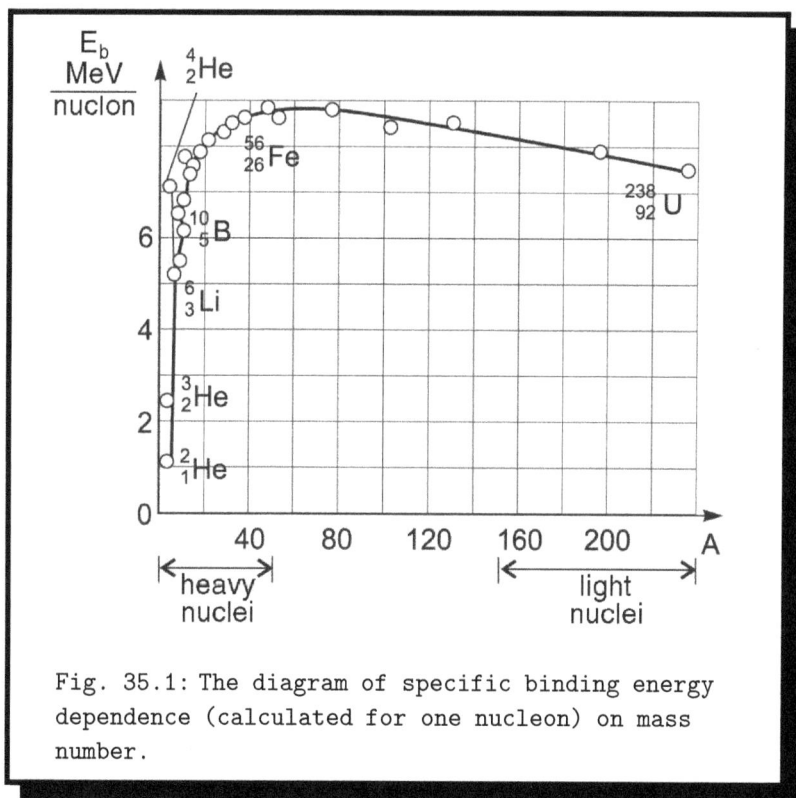

Fig. 35.1: The diagram of specific binding energy dependence (calculated for one nucleon) on mass number.

Let us review a nuclear division process. After adsorbing an outside neutron, a nuclear "drop" starts oscillating and, at some point, stretching. A spherical drop has a minimal surface area at a given volume. In a stretched drop the surface energy is increasing, and the nuclear binding energy is decreasing respectively. On the other hand, when the nucleus is stretched, the average distance between the nucleons is increasing, and their energy of electrostatic repulsion is decreasing, which will cause an increase of binding energy. If electrostatic interaction will win this battle, the nucleus will burst; if the surface binding will be the winner, the nucleus will return to its original state. It is clear from the equation received for

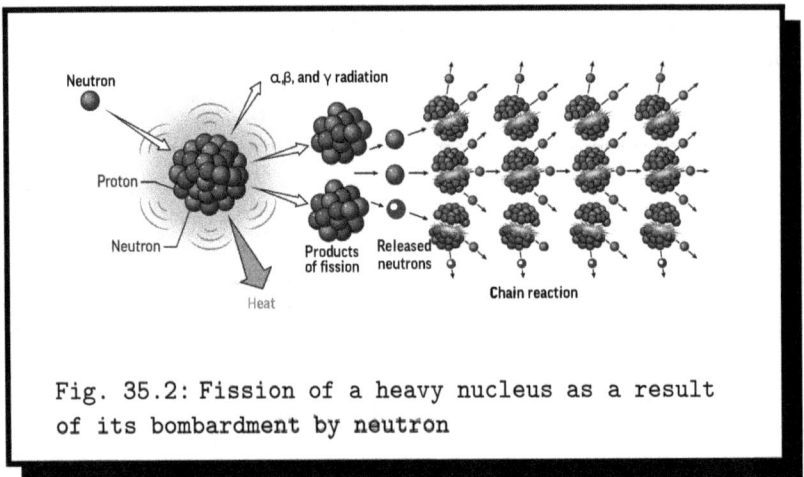

Fig. 35.2: Fission of a heavy nucleus as a result of its bombardment by neutron

E_b that the destiny of a nucleus depends to a great extent on the number of protons (Z) and total number of nucleons (A). With the growth of the element's order number the energy of electrostatic repulsion will be growing faster than that of the surface energy, therefore only heavy nuclei may be fissile.

Fission of heavy nuclei can be both spontaneous and resultant from neutron bombardment (see Fig. 35.2). Spontaneous fissions caused the transuranian elements to vanish during the early age of the Earth; now they can only be created artificially. As it has already been mentioned, in the course of nuclear fission several neutrons were always emitted (usually 2-3 for uranium). These very neutrons, in their turn, can trigger further nuclear fissions, releasing more free neutrons, and this process can go on and on: a growing chain reaction will take place. It was the practical implementation of a nuclear fission chain reaction, caused by bombarding uranium with neutrons, which released huge amounts of energy that refuted Rutherford's pessimistic prediction.

However, this self-sustaining process will take place only if the sample mass will exceed a certain "critical" amount. Indeed, if a chosen sample mass is small, then the neutrons irradiated during fission, before colliding with another nucleus and triggering a new division, will have a high probability of getting out of its scope. If the mass of a fissile sample exceeds a critical value, then the chain reaction will "accelerate": more and more nuclear fissions will take place within the time unit, and all the more heat will be released. If there is no control over heat removal an explosion will be inevitable.

The estimate of uranium-235 critical mass necessary for chain reaction became the task of utmost importance for the physicists on both sides of the frontline during World War II. German scientists, including a prominent physicist Werner Heisenberg, seemed to have overestimated this value — the future bomb appeared to be too heavy to be delivered by plane. This was probably the reason why their nuclear bomb project was only one of several lines of research towards the creation of "wunder waffe" (miracle weapon).

On the other side of the English Channel, a German born physicist Rudolf Peierls, who fled from Germany to Great Britain after Adolf Hitler came to power in 1933, came to much more optimistic conclusions in his estimate of critical mass. He even published his calculations, without mentioning the possibility of a military application of his results. Austrian physicist Otto Frisch, who also emigrated to Great Britain, came up with the plan how to use a chain reaction to create a super weapon. In 1940 Frisch and Peierls wrote a highly confidential memorandum and presented it to British authorities. In this document, the authors described the main stages of the process of nuclear bomb construction and gave the estimate of its destructive power.

Soon the Allies got down to business in a serious way. A vast and ambitious program of nuclear research named Manhattan Project was developed. It started in the USA in 1942 and was realized within three years as a result of joint effort of a big group of scientists, among whom, in addition to Frisch and Peierls, were such prominent minds as Enrico Fermi, Robert Oppenheimer, Richard Feynman and many others. The explosion of two nuclear bombs — in Hiroshima and Nagasaki in August of 1945 brought to an end World War II and opened a nuclear era in the history of mankind.

In our times, nuclear fission has found its peaceful implementation in the energy policy of many countries. The transformation of nuclear energy into easy to use electric power is going on, as a rule, the following way. Inside the reactor, nuclear fuel, usually it is uranium, when subjected to bombardment by neutrons, is getting split into the nuclei of lighter elements of the Periodic Table of Elements. Every such act of fission produces a significant amount of energy, and this is the bottom line of the process. If we compare the volume of power produced by uranium nuclear fission with the volume of power received from oil fuel burning, one gram of uranium will replace more than one ton of crude oil.

In a nuclear reactor the heat produced as a result of uranium nuclear fission is distributed through the liquid, the so called coolant, flowing inside the pipes stretched around the reactor. In its turn, the coolant transfers energy into water, which then converts into a steam. Steam will then rotate the turbine, the same way as in a steam engine. Finally, the mechanical energy of the turbine will be transformed into electrical energy with the help of a dynamo. The resultant energy is transferred through the high voltage power lines directly to the users hundreds of miles away from the nuclear power plant.

Not every isotope of uranium can undergo fission inside the reactor. Only U^{235} has this property. Its volume in the uranium ore excavated from the uranium mines makes 0.71% only. U^{238} is the most frequently encountered isotope on the Earth, but it would not be responsive to the process of fission inside the nuclear reactor. Therefore, to be used in nuclear power industry natural uranium has to be "enriched": the required isotope U^{235} is then extracted out of it after centrifuging at specialized plants. For modern reactors it is enough to use uranium with 2.5 % enrichment.

Safety is the major issue when nuclear reactors are being designed and running their operation cycle. As was mentioned above, in a reactor heat is emitted when U^{235} nuclei are split into two lighter elements. Such fission takes place when nuclear fuel is being bombarded by neutrons, which is followed by emission of several additional neutrons, and in response, the further fission takes place: the chain reaction occurs. However, such a self-sustained process will take place only if the fuel mass exceeds the "critical" value (it will take tens of kilograms of pure U^{235} to achieve critical mass). It is crucially important to prevent the acceleration of the chain reaction - even if the smallest error occurs, the reactor will be transformed into... a nuclear bomb!

To provide control over nuclear fission in a reactor, control rods made of efficient neutron absorbing materials (they could be cadmium, boron carbide, silver and indium based materials) are set up inside the reactor. Some of these rods are being suspended above the fuel, to be ready to fall into the respective cylindrical holes and stop the accelerating chain reaction. It was this out of control chain reaction that caused the disaster in April of 1986 due to a series of high-profile mistakes made at Chernobyl Nuclear Power Plant. As a result, the reactor was damaged and radioactive substances were spread over the vast area. In 2011 a similar disaster, caused by a tsunami, took place in Fukushima Nuclear Power Plant (Japan). But there, the chain reaction was timely stopped by protective devices. It caused a lot of damage, but nothing compared to Chernobyl and the "nuclear mushrooms" in Hiroshima and Nagasaki.

Afterword

Little by little our tale about physics comes to the end. We told you how physics helps to explain so many things all around us. Remember meandering rivers and the blue sky, think of coalescing droplets and hissing tea-kettles, do not forget the singing violin and the chime of goblets. Still the magic of physics is not solely the power to explain what happens, but the ability to foresee what will happen even if it never has before. This has put physics in the forefront of scientific and technical progress.

Modern physics has opened the way to the amazing quantum world. There, prisoners of potential wells flee away from their dungeons like the Count of Monte Cristo; magnetic fields make vortices to pierce superconductors; and a volatile amalgam of wave and particle entities of light quanta bring to mind mythical centaurs. The wonders of the quantum world are beyond imagination. But using its mathematical arsenal, theoretical physics succeeds in describing behavior of quanta so accurately that the results of experiments exactly coincide with theoretical predictions. This capability to correctly represent phenomena which escape even mental visualization is, in the opinion of the world-renowned physicist L. D. Landau, the greatest triumph of theoretical physics of the twentieth century.

Index

About the Authors

Lev Aslamazov (1944 – 1986)

Lev Aslamazov was born January 6, 1944 in Batumi (Georgia). In 1967 he was awarded Masters degree *cum laude* at Moscow Institute for Physics and Technology. In 1970, under the supervision of Anatoly Larkin, he received his Ph.D. in Condensed Matter Physics at Landau Institute for Theoretical Physics. From 1970 onwards he worked in the same institute as a researcher, and then senior researcher, before in 1976 he became an associate professor (1976 - 1980) of the Moscow Institute for Steel and Alloys (Technological University). In 1980 he was awarded the degree of Doctor of Sciences (Habilitatus) in Condensed Matter Physics and became full professor of the Department of Theoretical Physics of the same university. His main fields of scientific interests were: superconductivity, theory of metals, and theory of phase transitions. He authored classic works on superconductivity and three books. For many years he executed the duties of vice-Editor in Chief of the popular science magazine of the Russian Academy of Sciences "Kvant".

Andrey Varlamov

Andrey Varlamov was born April 25, 1954 in Kiev (Ukraine). He achieved his Masters degree *cum laude* at the Moscow Institute for Physics and Technology. In 1980, under the supervision of Alex Abrikosov, he was awarded Ph.D. in Condensed Matter Physics. He worked as a researcher, associate professor, and full professor at the Department of Theoretical Physics of the Moscow Institute for Steel and Alloys (Technological University), then as a fellow at the Condensed Matter Theory Group of Argonne National Laboratory (USA). Since 1999 he has been the principal investigator of the Institute of Superconductivity, Innovative Materials, and Devices of the Italian National Research Council (SPIN-CNR). His main fields of scientific interests are: superconductivity, theory of metals, theory of phase transitions, and nanophysics. He is the author of fifteen books, five monographic review articles, and 160 scientific papers.

Andrey Varlamov was awarded the USSR State Prize in Physics for young scientists (1986), the degree of Doctor of Sciences (Habilitatus) in Condensed Matter Physics (1988), degrees of *Doctor Philosophiae Honoris Causa* of Bogolyubov Institute for Theoretical Physics (Ukraine, 2011), Mediterranean Institute of Fundamental Physics (Italy, 2014), and the title of *Honorable Scientist of the Russian Federation* (2016). He is also a correspondent member of Istituto Lombardo Accademia di Scienze e Lettere (Milano, 2009). His book "Le kaleidoscope de la Physique" was awarded *Prix Grand Public Roberval* (France, 2015).

Andrey Varlamov is a devoted popularizer of science. Since 1984 he has been a member of the Editorial Board of the Russian Academy of Sciences popular science magazine Kvant (where for many years he executed the duties of the vice-Editor in Chief).